Biosensors and Molecular Imprinting

Special Issue Editors

Bo Mattiasson

Gizem Ertürk

MDPI • Basel • Beijing • Wuhan • Barcelona • Belgrade

MDPI

Special Issue Editors
Bo Mattiasson
Lund University
Sweden

Gizem Ertürk
Lund University
Sweden

Editorial Office
MDPI AG
St. Alban-Anlage 66
Basel, Switzerland

This edition is a reprint of the Special Issue published online in the open access journal *Sensors* (ISSN 1424-8220) in 2017 (available at: http://www.mdpi.com/si/sensors/BMI).

For citation purposes, cite each article independently as indicated on the article page online and as indicated below:

Author 1; Author 2. Article title. *Journal Name* **Year**, *Article number*, page range.

First Edition 2017

ISBN 978-3-03842-562-5 (Pbk)
ISBN 978-3-03842-563-2 (PDF)

Table of Contents

About the Special Issue Editors

Bo Mattiasson Professor Emeritus at Department of Biotechnology, Lund University, Lund, Sweden. PhD 1974 and Professor of Biotechnology 1985 when the department of biotechnology was started. Research areas involved enzyme technology, bioseparation, environmental biotechnology, extremophilic microorganisms, biosensors. During the years > 75 PhDs have finished their thesis work with BM as main supervisor. A large number of postdocs have also been trained. Collaboration with Swedish aid organization led to training of many students from developing countries from Africa, Latin- and Central America as well as South and Southeast Asia. Editor in chief of Biotechnology Reports, and member of 7 well established international biotech-journals. Has edited 10 books. Published more than 750 peer reviewed scientific papers. Collaboration with industry and also founding of 7 biotech companies during the years.

Gizem Ertürk received a master's degree in biology, specializing in biochemistry from Hacettepe University, Turkey in 2010. She completed her PhD in 2015, at the same university specializing on development of diagnostic kits by using biosensors for early detection of prostate cancer. She continued her postdoctoral studies in Lund University, Department of Biotechnology between 2015 and 2016 and at the same time in Malmö Högskola, Biofilms Research Center for Biointerfaces between 2015–2017 specializing on development of diagnostic tools for predicting oral infections based on artificial receptors and electrochemical sensing concept. Her research interests are biosensor based detection systems, affinity chromatography and molecular imprinting technology. She is now working as a postdoctoral researcher in Lund University, Department of Clinical Sciences, Division of Infection Medicine.

Preface to "Biosensors and Molecular Imprinting"

Sensitive and robust assays are important in many sectors of society. When the technology that was developed for clinical laboratories started to be implemented in other sectors of society such as environmental monitoring, food quality control, process monitoring and control, then some of the shortfalls became apparent. This initiated intense development of alternative routes to efficient analyses. Key challenges were storage- as well as operational stability besides selectivity of the biorecognition elements.

With the introduction of biosensors a new era in biosensing started. It became possible to carry out continuous assays in order to follow a process, more or less in real time. The first immune-based assays were ELISA implemented into a sensor format, but soon also direct binding assays were developed.

All through the rich development, there was a problem with stability and reusability of the affinity binder. Almost simultaneously as the biosensors were developed, one also saw an intense development in synthesis of artificial affinity binders. These came to be called molecularly imprinted polymers (MIPs).

In recent years one has seen descriptions of biosensors/chemical sensors based on exploiting the affinity of MIPs. The present volume presents the concept of synthesis of MIPs, their applications in constructing biosensors and also applications of such sensors in many different areas.

The development in the area of MIPbased biosensors has just began and one can forecast an exciting future for this niche of chemical analyses.

This volume is covering the development up to now and also indicates future development.

Bo Mattiasson and Gizem Ertürk

Special Issue Editors

sensors

MDPI

Editorial

Why Using Molecularly Imprinted Polymers in Connection to Biosensors?

Bo Mattiasson [1,2,*] **and Gizem Ertürk** [1,2]

1   CapSenze Biosystems AB, Scheelevägen 22, 22363 Lund, Sweden; ge@capsenze.se
2   Department of Biotechnology, Lund University, Box 117, 221 00 Lund, Sweden
*   Correspondence: bo.mattiasson@biotek.lu.se; Tel.: +46-070-605-98-30

Academic Editor: Nicole Jaffrezic-Renault
Received: 14 December 2016; Accepted: 19 January 2017; Published: 27 January 2017

The area of biosensor-oriented research has grown rapidly during recent years. From a start with enzyme-based sensors, it soon moved over to affinity-based sensors, e.g., immuno-based sensors. As long as the analytical device was used for a single measurement, i.e., as a disposable, stability problems were connected to storage stability, but not to operational stability. However, when repeated assays were to be used, or even sensors for on-line monitoring, then operational stability also came into play [1].

To secure high operational stability, one often operates in enzyme technology studies with a surplus of enzymes. That means that the process is diffusion-controlled, and when enzyme molecules denature and lose activity, then earlier resting enzymes come into action [2].

For immuno-based assays, one can approach the stability problems in two ways. The initial studies were made with competitive binding assays using labeled antigens mixed with free antigens. Then, the number of antibody binding sites was limiting, and denaturation then resulted in decreased sensitivity of the assay. One can compensate for these negative effects by measuring the binding of a standardized samples of labeled antigen [3]. If this is taken as 100% signal, then assay of the native antigen in the competitive assay will stay constant, as seen in Figure 1. For direct binding assays, an excess of binding sites is needed, and denaturation of a fraction thus does not influence sensitivity since earlier resting antibodies come into play instead of those that denatured.

Most binding assays involving MIPs are at present direct binding assays; thus, one shall use a surplus of the binding sites.

There are many reasons why a sensor with an immobilized biomolecule (enzyme, antibody, etc.) is losing activity. Denaturation due to temperature effects is often regarded as a causative; therefore, a refrigerator is recommended for storage of the bioactive components. The low temperature tolerance restricts the use of enzyme-based sensors in, e.g., fermentation control. It has so far, in essence, been impossible to operate with enzyme-based biosensors directly in fermentation broth. Off-line analysis has been the alternative choice. Thermostable enzymes from extremophilic microorganisms may help to some extent, but not enough for standing heat sterilization.

In areas where an unbroken cold-chain cannot be guaranteed, then biosensors based on selectivity of biological macromolecules constitute a problem. That may be a problem in developing countries, esp. when analysis is planned in remote areas.

However, there are other problems that might appear, e.g., the presence of proteases in the medium to be analyzed (that was what happened in the experiment illustrated in Figure 1). Such enzymes may destroy the bioactive component on the sensor surface, thereby reducing the sensitivity and definitely destroying the reproducibility. There are examples when infections of microorganisms have led to the presence of proteases that quickly destroyed the sensor device (Figure 1).

One more factor that is causing problems is the presence of heavy metal ions. It has been documented that even when using *pro. analysis* quality of buffer chemicals used in enzyme technology processes when the enzymes are exposed to buffer for extended periods of time, the enzymes lose the activity as a result of reaction with heavy metal ions present in the buffer used.

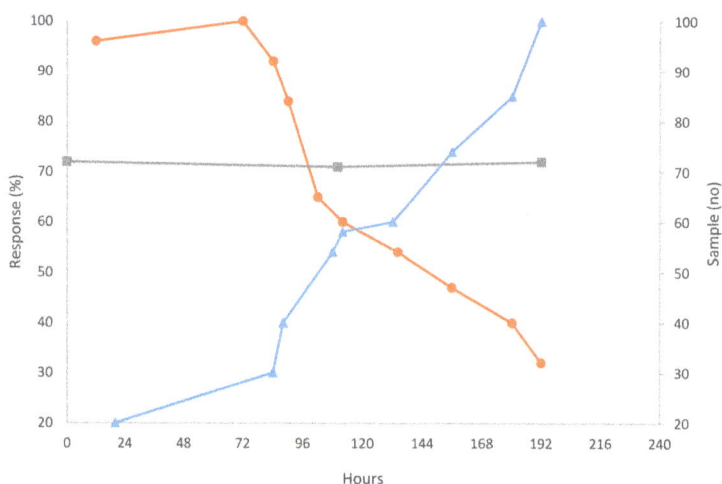

Figure 1. Response stability of antibody-Sepharose CL 4-B preparation. Orange line (circle) shows the peak height obtained when assaying a 100% sample. Blue line (triangle) shows number of assay cycles run. Grey line (square) shows the peak height in percent of a preceding pure aggregate pulse obtained for reference samples containing 40 µg HSA/mL sample (Modified from [3]).

Means to avoid the problems of bio-macromolecules have involved the use of small molecules with selective binding properties for certain analytes. Thus, amino-phenylboronate, which selectively binds vicinal hydroxyl groups, has been used for designing sensors for, e.g., glucose. It has even been proven that one can heat-sterilize a sensor surface with immobilized amino-phenylboronate and maintain the response pattern to the analyte, as seen in Figure 2 [4]. There are of course also some limitations with regard to selectivity. Synthesis of such low-molecular-weight compounds with selectivity is cumbersome and therefore expensive.

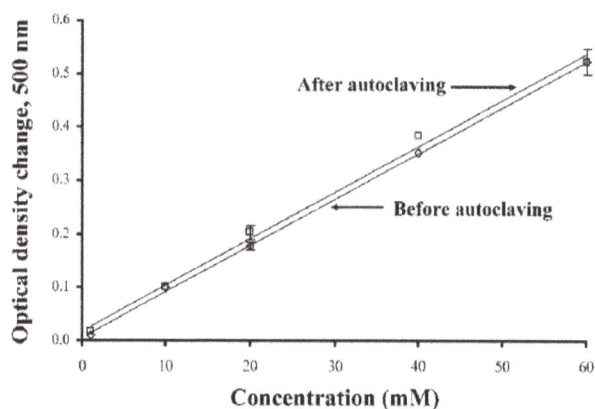

Figure 2. Sensitivity of N-acryloyl-m-aminophenylboronic acid–acrylamide copolymer (NAA PBA-Am copolymer) gel to glucose in 50 mM sodium phosphate buffer pH 7.3 before (◊) and after (□) autoclaving. Gel thickness is 350 nm. (Reproduced from [4] with permission).

Yet another alternative was presented by Nieland and Enfors [5]. They used the approach to add the enzymes to the space between the sensing part of an oxygen electrode and the protecting membrane covering the electrode surface. By that arrangement, the enzyme was excluded from being exposed to high temperatures, as seen in Figure 3.

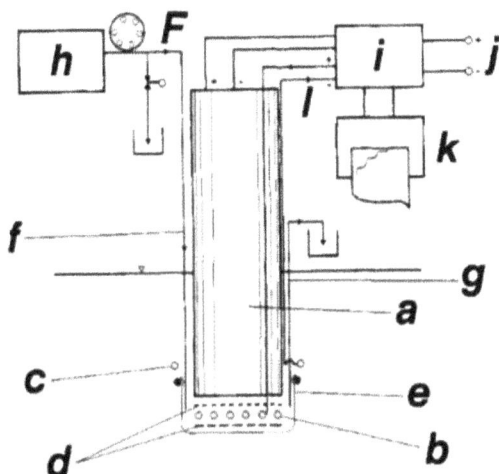

Figure 3. Main parts of the externally buffered enzyme electrode: a. oxygen electrode; b. Pt gauze with immobilized enzymes; c. Pt cathode; d. nylon nets; e. dialysis membrane; f. in-going buffer stream; g. buffer effluent; h. buffer reservoir; i. PID controller; j. reference potential; k. recorder; l. electrolysis current; F. buffer flow. (Reproduced from [5] with permission).

Molecularly Imprinted Polymers (MIPs)

Molecularly imprinted polymers are relatively easy to produce—at least compared to the efforts one needs to put in for producing enzymes or antibodies. Most enzymes used are of microbial origin, and very often much screening work is needed before a suitable enzyme is recognized. Cloning, expression, and production are most often utilized in order to produce large enough amounts of enzymes. With regard to antibodies, there are two possibilities: one is the immunization of animals and subsequent purification of the polyclonal antibodies. The other alternative is the production of monoclonal antibodies—a time consuming and usually expensive process.

When producing MIPs, a few criteria are of special interest to meet. Selectivity, binding strength (which is reflected in the sensitivity of the assay), the ability to regenerate the MIP after usage, and stability, both operational and storage.

In the by-now classical work on "plastibodies," it was found that MIPs were fully comparable to regular antibodies with regard to sensitivity [6,7]. Similar results have been reported for microcontact imprinting of proteins where sensitivity and selectivity was even better than what was measured for the commercial antibodies [8,9].

Molecularly imprinted polymers are said to be stable and can stand treatment of extremes of pH, organic base, and autoclave treatment [10]. An interesting aspect of MIPs is that one can modify the polymer properties by changing monomers to build the polymer. When introducing monomers that are often used for making stimuli responsive polymers, it is possible to facilitate release of bound material by simply changing one environmental parameter, often temperature. Thus, when N-isopropylacrylamide (NIPAm) is one of the co-monomers building up the MIP, then temperature sensitive conformation is obtained. However, when temperature returns to the starting point, then the conformation also returns to what was relevant before binding took place.

When MIPs were first described, block polymerization was used with a subsequent crushing and fractionation [11]. The smaller the particles were, the larger the surface area that was exposed and that could be used in the binding assays. Later on, spherical microparticles were produced [12], and more recent development has been described for the production of imprinted nanoparticles, with sizes of, e.g., those of antibodies [7]. Moreover, surface imprinting offers an option that can be used in connection to biosensors [13–15]. All of these alternatives can be used when designing biosensors, and papers in this issue will illustrate several of these.

MIPs can be applied in sensor systems in different ways. Forming the MIPs directly on the electrode surface is a common method; alternatively, small MIP particles can be added and fixed to the electrode, either by forming a composite via polymerization or by covalent coupling [16]. A recent development is the design of nano-MIPs—structures of sizes comparable to that of antibodies [7]. When small MIPs are designed, the number of crosslinks may become important for the stability of the MIP [17].

Still another area of applications of MIPs in analysis is the use as solid phase extraction medium. After a proper period of capture of the target molecule, an enriched fraction can be washed off the MIP and subsequently analyzed.

Experiences from affinity separation using MIPs will certainly stimulate to broaden the analytical applications of MIP.

The present volume is focused around the combination of biosensors and MIPs. A range of different targets is analyzed, and several sensor configurations are presented. From the data presented, it looks as if MIPs will become important when manufacturing selective sensors in the future. MIP-based sensors are often classified as "biosensors," even if the MIP is not actually a bio part.

Conflicts of Interest: The authors declare no conflict of interest.

References

1. Mattiasson, B.; Nilsson, M.; Berdén, P.; Håkanson, H. Flow-ELISA: Binding assays for process control. *TrAC Trends Anal. Chem.* **1990**, *9*, 317–321. [CrossRef]
2. Guilbault, G.G. Enzyme electrodes and solid surface fluorescence methods. *Methods Enzymol.* **1976**, *44*, 579–633. [PubMed]
3. Borrebaeck, C.; Börjesson, J.; Mattiasson, B. Thermometric enzyme linked immunosorbent assay in continuous flow system: Optimization and evaluation using human serum albumin as a model system. *Clin. Chim. Acta* **1978**, *86*, 267–278. [CrossRef]
4. Ivanov, A.E.; Thammakhet, C.; Kuzimenkova, M.V.; Thavarungkul, P.; Kanatharana, P.; Mikhalovska, L.I.; Mikhalovsky, S.V.; Galaev, I.Y.; Mattiasson, B. Thin semitransparent gels containing phenylboronic acid: Porosity, optical response and permeability for sugars. *J. Mol. Recognit.* **2008**, *21*, 89–95. [CrossRef] [PubMed]
5. Cleland, N.; Enfors, S.O. Externally buffered enzyme electrode for determination of glucose. *Anal. Chem.* **1984**, *56*, 1880–1884. [CrossRef]
6. Vlatakis, G.; Andersson, L.I.; Müller, R.; Mosbach, K. Drug assay using antibody mimics made by molecular imprinting. *Nature* **1993**, *361*, 645–647. [CrossRef] [PubMed]
7. Poma, A.; Guerreiro, A.; Whitcombe, M.J.; Piletska, E.V.; Turner, A.P.; Piletsky, S.A. Solid-Phase Synthesis of Molecularly Imprinted Polymer Nanoparticles with a Reusable Template—"Plastic Antibodies". *Adv. Funct. Mater.* **2013**, *23*, 2821–2827. [CrossRef] [PubMed]
8. Ertürk, G.; Berillo, D.; Hedström, M.; Mattiasson, B. Microcontact-BSA imprinted capacitive biosensor for real-time, sensitive and selective detection of BSA. *Biotechnol. Rep.* **2014**, *3*, 65–72. [CrossRef]
9. Ertürk, G.; Hedström, M.; Tümer, M.A.; Denizli, A.; Mattiasson, B. Real-time prostate-specific antigen detection with prostate-specific antigen imprinted capacitive biosensors. *Anal. Chim. Acta* **2015**, *891*, 120–129. [CrossRef] [PubMed]
10. Svenson, J.; Nicholls, I.A. On the thermal and chemical stability of molecularly imprinted polymers. *Anal. Chim. Acta* **2001**, *435*, 19–24. [CrossRef]

11. Kriz, D.; Ramström, O.; Mosbach, K. Peer Reviewed: Molecular Imprinting: New Possibilities for Sensor Technology. *Anal. Chem.* **1997**, *69*, 345A–349A. [CrossRef]
12. Yilmaz, E.; Ramström, O.; Möller, P.; Sanchez, D.; Mosbach, K. A facile method for preparing molecularly imprinted polymer spheres using spherical silica templates. *J. Mater. Chem.* **2002**, *12*, 1577–1581. [CrossRef]
13. Kempe, M.; Glad, M.; Mosbach, K. An approach towards surface imprinting using the enzyme ribonuclease A. *J. Mol. Recognit.* **1995**, *8*, 35–39. [CrossRef] [PubMed]
14. Hayden, O.; Dickert, F.L. Selective microorganism detection with cell surface imprinted polymers. *Adv. Mater.* **2001**, *13*, 1480–1483. [CrossRef]
15. Wang, Y.; Zhang, Z.; Jain, V.; Yi, J.; Mueller, S.; Sokolov, J.; Liu, Z.; Levon, K.; Rigas, B.; Rafailovich, M.H. Potentiometric sensors based on surface molecular imprinting: Detection of cancer biomarkers and viruses. *Sens. Actuators B Chem.* **2010**, *146*, 381–387. [CrossRef]
16. Sener, G.; Ozgur, E.; Yılmaz, E.; Uzun, L.; Say, R.; Denizli, A. Quartz crystal microbalance based nanosensor for lysozyme detection with lysozyme imprinted nanoparticles. *Biosens. Bioelectron.* **2010**, *26*, 815–821. [CrossRef] [PubMed]
17. Beyazit, S.; Ambrosini, S.; Marchyk, N.; Palo, E.; Kale, V.; Soukka, T.; Bui, B.T.S.; Haupt, K. Versatile synthetic strategy for coating upconverting nanoparticles with polymer shells through localized photopolymerization by using the particles as internal light sources. *Angew. Chem. Int. Ed.* **2014**, *53*, 8919–8923. [CrossRef] [PubMed]

sensors

MDPI

Review

Molecular Imprinting Techniques Used for the Preparation of Biosensors

Gizem Ertürk [1,*] and Bo Mattiasson [1,2]

[1] CapSenze Biosystems AB, Lund SE-22363, Sweden; bo.mattiasson@biotek.lu.se
[2] Department of Biotechnology, Lund University, Lund SE-22369, Sweden
[*] Correspondence: ge@capsenze.se; Tel.: +46-720-254-921

Academic Editor: Flavio Maran
Received: 5 December 2016; Accepted: 28 January 2017; Published: 4 February 2017

Abstract: Molecular imprinting is the technology of creating artificial recognition sites in polymeric matrices which are complementary to the template in their size, shape and spatial arrangement of the functional groups. Molecularly imprinted polymers (MIPs) and their incorporation with various transducer platforms are among the most promising approaches for detection of several analytes. There are a variety of molecular imprinting techniques used for the preparation of biomimetic sensors including bulk imprinting, surface imprinting (soft lithography, template immobilization, grafting, emulsion polymerization) and epitope imprinting. This chapter presents an overview of all of these techniques with examples from particular publications.

Keywords: molecular imprinting; biomimetic sensors; bulk imprinting; surface imprinting; epitope imprinting

1. Introduction

Molecular imprinting is the technology of creating artificial recognition sites in polymeric matrices which are complementary to the template in their size, shape and spatial arrangement of the functional groups. Recently, molecular imprinting technology has been used for the creation of biorecognition surfaces on biosensors. There are different types of molecular imprinting techniques, including bulk imprinting, surface imprinting and epitope imprinting.

The template molecule is imprinted as a whole in the polymer matrix in bulk imprinting, whereas only a small part or a fragment of a macromolecule is imprinted as a representative of the whole molecule in epitope imprinting. Surface imprinting is a popular and convenient technique, especially for biomacromolecules including proteins. There are different types of surface imprinting like soft lithography, template immobilization including microcontact imprinting, grafting and emulsion polymerization which have been used successfully in previous reports. All of these techniques have some advantages and at the same time limitations when it comes into applications, however, in general; molecular imprinting technology is very promising to create highly specific, sensitive and long-term stable biorecognition cavities on biosensor surfaces to be used in many applications.

2. Molecular Imprinting Techniques

Molecularly imprinted polymers (MIPs) are prepared by co-polymerization of cross-linking monomer and a complex which is pre-formed between the template molecule and functional monomers using covalent, non-covalent or semi-covalent interactions. When the template molecule is removed from the imprinted material after polymerization, it leaves behind specific cavities that are complementary to the template in size, shape and chemical functionality [1]. MIPs have been used successfully in many applications, including purification [2], isolation [3], chiral separation [4], catalysis [5] and in biosensors [6,7]. For the imprinting of molecules on biosensor surfaces, there are different types of molecular imprinting techniques that can be classified in three main categories; bulk imprinting, surface imprinting and epitope imprinting.

2.1. Bulk Imprinting

In bulk imprinting, a template molecule is imprinted as a whole in the polymer matrix and after polymerization it needs to be wholly removed from the molecularly imprinted material. In the next step, to form small particles from these bulk polymers, the bulk polymer is crushed mechanically and the formed particles are fractionated. By this way, readily accessible, template-specific 3D interaction sites are obtained within the selective, imprinted polymeric material [8–10]. Bulk imprinting is generally preferred for the imprinting of small molecules since, at least in theory, the adsorption and release of the template molecule are faster and reversible which is an advantage for the re-usability of the imprinted support for several rounds of analyses [9].

The use of whole proteins as a template presents some advantages over other methods especially in sensor applications [11,12]. The template structure will most accurately be the same with the target structure since the template protein (the target, at the same time) is imprinted wholly. However, in case of larger structures including proteins, living cells and microorganisms, the method has some drawbacks. Maintaining the conformational stability of a protein during the polymerization process is a big challenge [13]. Moreover, when large imprinted sites are formed due to the size of the template, these sites might also be attractive for smaller polypeptides which results in cross-reactivity and reduced selectivity [13]. Due to the thick morphology of bulk imprints, big template molecules are embedded in the matrices too deeply that lead to restriction or no access for the target molecules to bind to the sites. The fact of low accessibility leads to much longer response times, drift problems and also poor regeneration [9,14]. A more extreme position is that the target molecule is problematic to remove from the MIP. This will lead to hampered binding and in extreme cases to no binding at all. To overcome these limitations, alternative imprinting techniques including surface and epitope imprinting have been developed.

2.2. Surface Imprinting

High affinity recognition sites are formed at the surface of a substrate in surface imprinting [15]. Due to this the recognition sites are more easily accessible with favorable binding kinetics. In other words, template/polymer interactions are not diffusion limited to the same extent as was a general problem encountered in bulk imprinting [16–18]. Therefore, the technique is popular and most applicable especially for imprinting of biomolecules including proteins. In surface imprinting, less template molecules are used as compared to what is used in conventional imprinting techniques because template is only used in the surface coating step in the technique [14]. The main drawback of the method in sensor design is the possibility of lower sensitivity as compared to bulk imprinting owing to the reduced number of imprinted sites [9]. Surface imprinted polymers have been widely used for different types of analytes, including proteins [19,20], microorganisms and cells [10,21].

2.2.1. Soft Lithography

Soft lithography or stamping is one of the techniques that has been used for the preparation of (biomimetic) sensors by surface imprinting [22,23]. Micro and nano-scaled patterns can be prepared by this technique without any need for expensive and specialized equipment [9]. The technique was first introduced by Bain and Whitesides in 1989 [24]. In the general procedure of the method, micro- or nano-scaled patterns ranging from 30 nm to 100 μm are formed on the solid substrates by using a soft polymeric stamp [23]. A pre-polymerized layer is coated on a transducer surface and the template stamp is pressed over this surface for a certain period of time. Self-assembling of template structures including proteins, microorganisms onto a smooth support are used to produce the template stamps which are complementary to the template in their structural, geometrical and also chemical features.

At the end of the procedure, highly selective and specific surface imprinted films are obtained via this technique.

For the preparation of stamp, one of the important points that has to be considered is the amount of template on the stamp which has a strong influence in sensor response. Dickert et al. [25] reported that multi-layer stamps showed better sensor response than mono-layer covered polymers. For the selection of solid support, a glass slide can be preferred owing to its rigid structure which means mechanical stability [26]. A relatively flexible material poly (dimethyl siloxane) is also used for more fragile structured templates [27]. Furthermore, polymer's composition, polarity and methodology are also important parameters to obtain the optimal recognition between the template and the imprinted cavities [9,28]. After template removal process, it is important for the stability of the sensor surface to retain the shape of the imprinted cavities, which means the stability of the sensor surface is important for the binding efficiency.

A solvent-assisted soft lithography and UV-initiated polymerization were used to develop striped poly(methacrylic acid-ethylene glycol dimethacrylate) patterns for detection of traces of atrazine in aqueous solutions by using gravimetric QCM (Figure 1). This patterned MIP matrix provided more nanocavities owing to the increased surface area and a thin residual layer among patterns compared to planar MIP system. This increased surface area contributed to the rapid transport of atrazine in MIP films and this leads to faster sensing responses. To evaluate the selectivity, herbicides analogous in structure to atrazine were used and the sensing behaviors were found almost identical with that in each herbicide analogues because of the limited non-specific chemisorption regardless of concentration. Once adding trace amounts of atrazine into the mixture solution, the sensor response showed 94% recovery of the original sensing response appeared in atrazine solution. Therefore, the lithographic approach to MIP sensing system provides rapid and ultra-sensitive detection in a selective way [29].

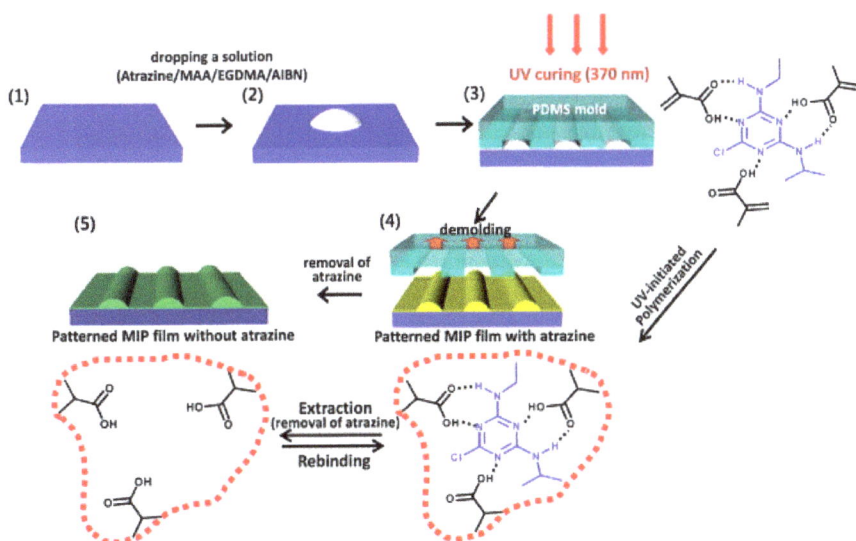

Figure 1. Schematic representation of fabrication process of patterned MIP films via soft lithography and UV polymerization. (**1–2**) Imprinting solution containing the template molecule, functional and cross-linking monomers and initiator was dropped on the substrate. (**3**) The patterned PDMS mold was placed on the droplet of solution under certain pressure to ensure enough physical contact between the PDMS and substrate. Then, UV polymerization was initiated and continued for 7–10 min. (**4**) Following de-molding, the patterned MIP films were dried at 60 °C. (**5**) In the last step, template molecule (atrazine) was removed from the surface (reproduced from [29] with permission).

Voicu et al. [30] reported two dimensional (2D) molecular imprinting technique based on nano-templating and soft lithography techniques. In the basis of 2D molecular imprinting, nano-templating through microcontact imprinting was used. In the first step, by the attachment of a monolayer of the template, the stamp was modified. Then, the modified stamp was brought into contact with the monomer which provides the recognition groups that will interact with the template and then with the anchoring monomer which offers groups able to anchor the whole assembly to the substrate. In the next step, the stamp carrying the recognition and anchoring monomers on top was pressed against the substrate and the polymerization was initiated. In the study, the authors used theophylline as a model template. Several cycles of binding/extraction/rebinding of the template showed the stability of the MIP. Caffeine was used in testing the selectivity of the 2D MIP since caffeine has a very similar structure to theophylline. The results revealed that, even if both caffeine and theophylline were present in the same sample, rebinding of theophylline to the MIP was preferential and demonstrated selectivity towards theophylline. The methodology incorporated the advantages of oriented immobilization of the template and the attachment of the MIP on a substrate surface. By this way, homogeneous recognition sites were obtained on the surface. This system is promising for multi analyte detection at the same time by the immobilization of multiple molecular templates.

Dickert and Hayden [31] used soft lithography to prepare imprinted polymers for selective detection of yeast cells. The yeast imprinted QCM sensor surface showed a regular honeycomb like surface. The authors were able to monitor cell concentrations in the range between 10^4 and 10^9 cells/mL. The technique enabled the authors to generate highly regular imprints by microorganisms. Artificial recognition of microorganisms with MIPs depended on both chemical interactions and a large contact surface between yeast cells and polymer.

Jenik et al. [32] used the same technique to design a QCM sensor for two representatives of picornaviruses, human rhinovirus (HRV) and the foot-and-mouth disease virus (FMDV). In the study, it was shown that in contrast to natural antibodies, the entire surface of the virus particles was recognized by the MIPs because the differences in chemistry between the major and minor groups of the receptor sites which did not have any role in the detection. Weak non-covalent interactions were the main interactions in the recognition mechanism.

2.2.2. Template Immobilization

Template immobilization approach was introduced by Shi et al. in 1999 [33]. In contrary to soft lithography in which template stamp is formed by using self-assembling, in this approach template molecule is immobilized on a solid support via chemical linkages [33,34]. Well-ordered imprinted surfaces can be produced by using this technique. In the general procedure of the technique, template protein is adsorbed onto a mica and a disaccharide layer is formed around the proteins. Then, radio-frequency glow-discharge plasma deposition is used to form polymeric thin films around the proteins coated with disaccharide molecules. The resulting plasma film is fixed to a glass support and mica is peeled-off in the next step. Then, the sample is soaked in a proper solution for dissolution and extraction of the template protein. By this way, nanocavities are created on the surface of the imprint with a shape complementary to that of the protein as shown in Figure 2 [33]. The disaccharides are covalently attached to the polymer film and they create polysaccharide like cavities which exhibit high selectivity for a variety of proteins including albumin, immunoglobulin G, lysozyme, ribonuclease and streptavidin [33,35–37].

Yılmaz et al. introduced "the use of immobilized templates" as a new approach in 2000 [34]. This procedure involves use of a silica based solid support to which the template is immobilized. It provides several advantages to these systems. For the insoluble templates in the polymerization cocktail, this approach might be used as a good alternative. The aggregation of templates in pre-polymerization mixtures which might be seen in some occasions can be prevented by this approach. In contrast to classical MIPs, no porogen was used for the imprinting of the immobilized template molecule in this technique. The pore structure of the obtained polymers was generated by dissolving the silica gel backbone. Because of this, all imprinted sites were located at or close to the surface of the pores which is an important advantage in facilitating the diffusion of the analyte to the binding sites.

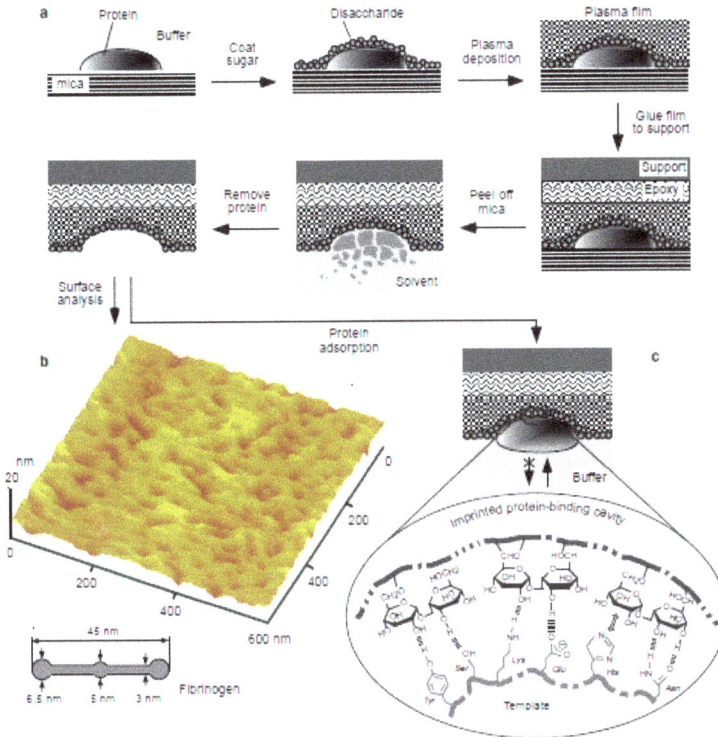

Figure 2. Protocol for template imprinting of proteins; (**a**) Template protein was adsorbed onto a mica and this surface was spin-coated with disaccharide to form a sugar overlayer. Plasma deposition was conducted to form a plasma film on top. The resulting plasma film was fixed to a glass cover slip. Then, mica was peeled-off and the protein was extracted. By this way, a nano-imprint with a shape complementary to the template protein was created on the imprint surface. (**b**) AFM image of the surface of a fibrinogen imprint, (**c**) Mechanism for the specific recognition of the imprinted surface. The main force for the selectivity was steric hindrance and an overall strong interactions which is due to many cooperative weak interactions including hydrogen bonds, Van der Waals forces and hydrophobic interactions (reproduced from [33] with permission).

Shiomi et al. used covalently immobilized hemoglobin (Hb) to create Hb-specific recognition cavities on silica. Covalent immobilization was achieved by forming imine bonds between amino groups on the protein surface and aldehyde groups on silica. After immobilization of the template, the organic silane monomers 3-aminopropyltrimethoxysilane (3-APTMS) and propyltrimethoxy-silane (PTMS) were polymerized on the Hb imprinted silica surface. Then, Hb was removed from the imprinted surface. The resulting protein imprinted silica showed good thermo-stability and mechanical strength [38]. Recently, several new variations of template immobilization including microcontact imprinting method have been introduced.

Microcontact Imprinting

Microcontact imprinting method was first described by Chou et al. in 2005 [39]. They prepared the microcontact imprints between two glass surfaces. In the general procedure of their method, template protein was immobilized onto the cover slip and this was treated with functional monomer to provide site-specific organization of the functional monomer by the template, thereby forming a complex between template and functional monomer(s). In the next step, a cross-linking monomer together with initiator were applied on the other glass and the micro-contact imprints were formed by bringing into contact with the cover slip that has the protein and monomer on top and the support glass carrying the cross-linker. Two functionalized surfaces were brought in contact under UV radiation. Then, the cover slip with the immobilized template molecules was removed from the surface and the excess polymer, non-reacted monomers and any leaking template molecule on the support were washed away. The quantification of re-binding was performed by using ELISA. The imprinted surfaces showed good selectivity for the template molecule (C-reactive protein, CRP) both in single binding and also in competition with human serum albumin (HSA). When the method is used to imprint HSA, the imprints demonstrated significantly greater affinity for the native template HSA than CRP this time. These promising results demonstrated the usability of the method to imprint relatively large proteins. Same authors used the same technique for several model antigens including lysozyme, ribonuclease A and myoglobin [40].

Microcontact imprinting technique allows for rapid and parallel synthesis of MIPs with different compositions [40,41]. Only a small amount of template is used to form a monolayer recognition surface that is an advantage for the templates which might be very expensive or available only in limited amounts. A few microliters of monomer solution are enough and dozens of samples can be polymerized at the same time using the same polymerization batch. Using this technique, potential solubility problems of macromolecular templates including proteins are also avoided. Since the template molecule is immobilized onto a glass support instead of being present in the monomer solution, the conformational stability of the template protein is preserved and aggregation of protein molecules is prevented [42].

Osman et al. [43] used microcontact imprinting technique to prepare SPR biosensor for myoglobin detection. They synthesized myoglobin imprinted poly (hydroxyethyl methacrylate-N-methacryloyl-L-tryptophan methyl ester) [poly (HEMA-MATrp)] nanofilm on the surface of a SPR sensor. The detection limit was found as around 88 ng/mL in the study. Functional monomer MATrp used in this study is monomer conjugated with amino acid which interacts with the functional groups on the template. Therefore, selective binding sites and structural memories for myoglobin are formed during the polymerization. In the selectivity studies, it was demonstrated that the imprinted nano-film recognized myoglobin preferentially over some competing proteins.

Lin et al. [40] reported the effect of cross-linking monomers for the microcontact imprinting of proteins. Four different types of cross-linking monomers: (ethyleneglycol dimethacrylate-EGDMA, tetraethyleneglycol dimethacrylate-TEGDMA, polyethyleneglycol 400 dimethacrylate-PEG400DMA, polyethyleneglycol 600 dimethacrylate-PEG600DMA) were tested and lysozyme rebinding on polyEGDMA, polyTEGDMA, polyPEG400DMA and polyPEG600DMA was measured by quantitative ELISA. The highest rebinding efficiency occurred on the lysozyme imprinted polyTEGDMA polymer. It was suggested that the size of the protein template may be correlated with the optimal number of EG-repeating units in the cross-linking monomers. More effective protein binding and imprinting was obtained with a cross-linking monomer containing more ethylene glycol repeating units. Longer cross-linkers are more flexible and this can lead to allow the polymeric matrix to ensure the protein's conformational stability. The microcontact imprinting technique allows studies on interactions between the polymer surfaces and template molecules.

Sener et al. [42] developed a microcontact imprinted SPR biosensor for real-time detection of procalcitonin which is used as a biomarker to identify sepsis and to give an estimation of the severity of disease. The LOD value was calculated as approximately 3 ng/mL. The authors suggested that more sensitive detection may be possible by decreasing the polymer film thickness below 200 nm on the surface of the sensor chip. Selectivity results showed that the reflectivity change in sensor response in respect to the competing proteins human serum albumin (HSA), myoglobin and cytochrome c (cyt c) was almost the same for both the imprinted and non-imprinted SPR sensor. These results also indicate that the interactions between the imprinted surface and the competing proteins occur in a non-specific manner. Authors also tested the applicability of the sensor for procalcitonin (PCT) detection from simulated blood plasma which comprises HSA, immunoglobulins and fibrinogen at their typical concentrations in human blood plasma. When a comparison was made between the results obtained from analyzing the PCT samples with SPR and ELISA, it was shown that the agreement between two methods was approximately 99%.

The first microcontact imprinting study for the detection of a model analyte, BSA, with the capacitive biosensor was reported by Ertürk et al. [44]. In the study, a capacitive biosensor with an automated flow injection system was used for the detection of the analyte. In the detection principle of capacitive biosensors, the change in dielectric properties on the sensor surface when an analyte interacts with the biorecognition element is measured [45,46]. This measurement is very sensitive and selective, carried out in real-time and directly without using any labeling reagents. In the reported study, the authors prepared microcontact-BSA imprinted capacitive gold electrodes using MAA and PEGDMA as the functional monomer and cross-linker, respectively. After optimization of the conditions, BSA detection was carried out within the concentration range of 1.0×10^{-20}–1.0×10^{-8} M with a LOD value of 1.0×10^{-19} M. The results showed that the BSA imprinted electrodes were selective for BSA compared to HSA and IgG. The cross-reactivity ratios were 5% for HSA and 3% for IgG. A total of 80 assays during a period of 2 months were carried out with the same electrode without any significant differences in the detection performance. The developed method was promising with the advantages of high sensitivity, selectivity and operational stability.

In their next study, the same authors used the same technology for ultra-sensitive detection of an important biomarker, prostate specific antigen (PSA) for the early diagnosis of prostate cancer. A microcontact-PSA imprinted SPR sensor chip was prepared as shown schematically in Figure 3 and PSA detection was performed with standard PSA solutions in the concentration range of 0.1–50 ng/mL with a LOD value of 91 pg/mL. The LOD value obtained from the study was comparable to the LOD values reported in previous studies [47–50]. Because the cut-off value between normal and possible pathological levels of PSA in human serum is 4.0 ng/mL, the developed system with this LOD value meets the requirements of PSA analysis from clinical samples. In the final step, PSA detection was also carried out from prostate cancer patients' serum samples and the results were compared with the ELISA results. The agreement between two methods, SPR and ELISA, was found as approximately 98% which established the reliability and the sensitivity of the method for PSA detection. All the results revealed that the system could be considered as a promising tool and used as an alternative to ELISA for PSA detection in standard PSA solutions and clinical samples. When the authors compared the sensitivity from analysis done with a SPR system using microcontact-PSA imprinted sensor surface with results from a capacitive system with microcontact-PSA imprinted electrodes, the capacitive system was approximately 1000 times more sensitive than the SPR system [51]. Capacitive biosensor technology with the microcontact imprinting method might be used successfully for real-time detection of various analytes even in very low concentrations [52]. Microcontact imprinting has also been used for developing assays of whole cells. Idil et al. [53] monitored *E. coli* cells using a capacitive measuring approach. The sensor chip was essentially prepared according to the procedures used for sensor chips against proteins. The assay was selective against the imprinting microorganism, but there was still some cross-reactivity against competing bacteria. A limit of detection was 70 CFU/mL. The approach on assaying cells via exploitation of microcontact imprinting is promising, but there is still room for improvements [53].

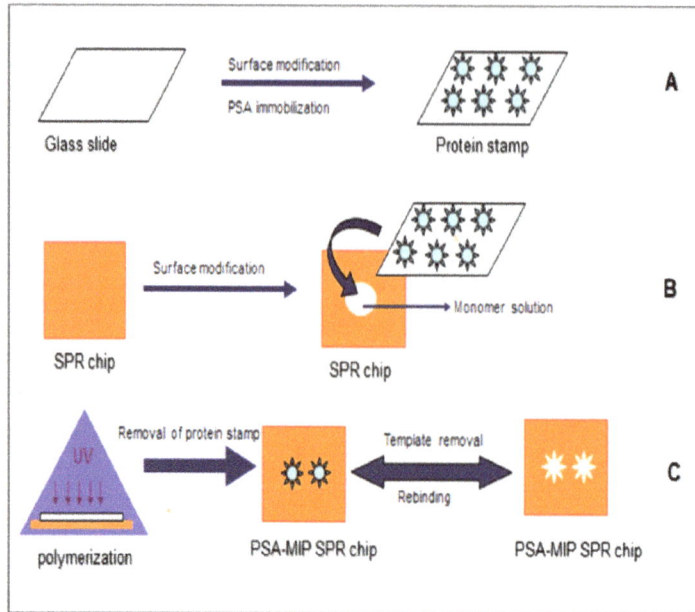

Figure 3. Schematic representation of microcontact imprinting of PSA onto the SPR biosensor surface: (**A**) Preparation of glass cover slips (protein stamps); (**B**) preparation of SPR chips; (**C**) microcontact imprinting of PSA onto the SPR biosensor chip surface via UV-polymerization (reproduced from [52] with permission).

2.2.3. Surface Imprinting via Grafting

In surface grafting, template molecule is adsorbed or attached with the polymeric functional groups which are already grafted on the surface of the support. In other words, contrary to template immobilization method, template molecule is present during the polymerization step in this method [9]. The advantages of the method are improved affinity interactions because of faster mass transfer as a result of higher analyte mobility, better control over polymer shape and morphology.

A molecularly imprinted polymer for domoic acid (DA) was synthesized by Lotierzo et al. [54] by direct photo-grafting onto the SPR gold chip surface. Self-assembly of 2-mercaptoethylamine (2-MEA) was used for the surface functionalization of the SPR gold chip. Then, carbodiimide chemistry was performed for the covalent attachment of the photo initiator to the surface. By using a photo-initiator with symmetrical carboxylic acid group at each arm, covalent attachment of the initiator to the amino-functionalized gold surface was possible by using carbodiimide chemistry. 2-(diethylamino)ethylmethacrylate and EGDMA were used as functional monomer and cross-linker, respectively and thin polymeric film formed only on the surface. The measured MIP film thickness was 40 nm since the immobilization of the photo-initiator to the gold surface prior to being treated with pre-polymerization mixture resulted in the polymerization reaction took place to the close vicinity of the gold surface. The developed system had approximately three times higher detection limit compared to that of monoclonal antibody immobilized system.

BSA surface-imprinted thermosensitive magnetic composite microspheres were prepared via surface grafting co-polymerization method. Temperature sensitive N-isopropylacrylamide (NIPAm) and the functional monomer methacrylic acid (MAA) were used as co-monomers and methylene bis-acrylamide (MBA) as the cross-linking agent. The adsorption-desorption of template molecule was regulated by changing the system temperature due to the thermo-sensitive imprinting layer [55].

An interfacial organic-inorganic hybridization concept was used for the preparation of the spherical imprinted materials. In this surface imprinting study, model template BSA was covalently immobilized by forming peptide bonds with the functional amine groups of biopolymer chitosan [56]. Then, two different kinds of organic siloxanes 3-aminopropyltrimethoxysiloxane (3-APTMS) and tetraethoxysiloxane (TEOS) were assembled. In the next step, polymerization was performed on the polysaccharide-protein surface via sol-gel process. In the last step, template protein BSA was removed from the surface and cavities complementary to the template in size, shape and orientation of the functional groups were created on the surface as shown schematically in Figure 4. Cytochrome c, transferrin, beta-amylase and lysozyme were used as competing proteins. Compared to the imprinted material, the control, non-imprinted material showed poor adsorption performance. The grafting of the imprinted layer through interfacial organic-inorganic hybridization improved stability and reproducibility properties of the final material.

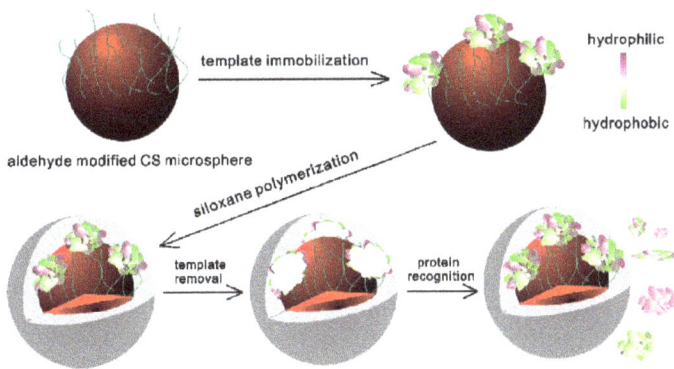

Figure 4. Schematic representation of synthesis of protein imprinted polymers on CS microsphere using immobilized protein as a template. The synthesis involved three steps; Firstly, template BSA was covalently immobilized on the polysaccharide core by forming imine bonds. The second step involved siloxanes polymerization on the polysaccharide–protein surface. It resulted in a polymeric network molded around the template. The template protein was removed in the third step. Cavities complementary to the template protein in shape, size, and functional group orientation were created (reproduced from [56] with permission).

The first report of the automated synthesis of imprinted polymer nanoparticles (nanoMIPs) with size, specificity and solubility characteristics for industrial manufacturing was published by Poma et al. [57]. The protocol developed for the automated synthesis and purification of MIP nanoparticles (Figure 5) was as follows [58]:

(1) In the first step, monomer/initiator mixture was dissolved in an appropriate solvent and then loaded onto a temperature controlled column reactor. This column reactor consisted of the template immobilized onto a solid support.
(2) Then, in the next step, when the optimum reaction conditions were obtained, polymerization was initiated by UV-irradiation of the reactor.
(3) After polymerization, the column reactor was washed with a solvent which resulted in the elution of unreacted monomers and other low molecular weight materials together with low-affinity polymer nanoparticles. For the other high-affinity bound particles, collection was managed by increasing the column temperature or by addition of auxiliary reagents like formic acid.

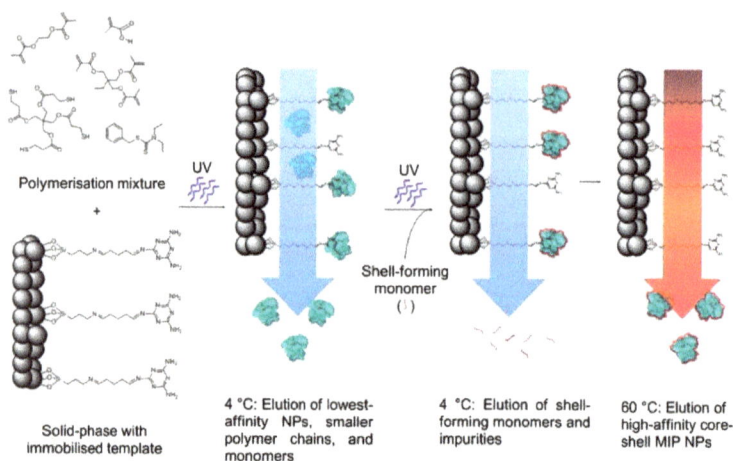

Figure 5. Schematic representation of core-shell MIP NPs. The template molecule was immobilized onto a solid support. A polymerization mixture (containing functional and cross-linking monomers, initiators) were loaded on this solid support. Polymerization was initiated via UV light. Un-reacted monomers and low-affinity particles were eluted from the solid phase at low temperature where the high affinity particles remained bound to the template. The high-affinity particles were subsequently grafted with a secondary monomer and an application of UV-light. After polymerization, high-affinity core-shell MIP NPs were eluted from the shell by increasing the temperature up to 60 °C (reproduced from [58] with permission).

The advantages of this approach are: uniform binding properties, elimination of contamination of the product with template because immobilized templates are used, re-usability of the template in further applications, ease of the procedure, obtaining pure final product which eliminates the need for post-purification steps. Poma and co-workers indicated that by using this technique, high-quality MIP nanoparticles similar to monoclonal antibodies can be produced and these MIP nanoparticles offer a lot of advantages over antibodies for use in assays and sensors in the future.

2.2.4. Emulsion Polymerization

Another surface imprinting strategy is emulsion polymerization. In the general procedure of the technique, core particles are synthesized before they are covered with imprinted shells.

Tan et al. [59] used surface imprinting strategy based on covalently immobilized templates to prepare BSA-imprinted sub-micrometer particles via two-stage core shell mini-emulsion polymerization. Through the encapsulation of Fe_3O_4 nanoparticles, the MIP particles gained magnetic properties which increases their potential applications in various fields including magnetic bioseparation, cell labeling and bioimaging. Imprinted particles with non-immobilized (free) template molecules (fMIP) did not display the expected template recognition and this illustrates the importance of the template immobilization for a successful surface imprinting. The MIP prepared according to this strategy displayed high specific recognition for BSA without any adsorption of the competitive proteins. Therefore the technique has potential to be used for the preparation of highly selective biosensor surfaces in further applications.

Thin MIP coatings on magnetic Fe_3O_4 nanoparticles (NPs) with a uniform core-shell structure were developed in a study [60] (Figure 6). BSA, bovine hemoglobin, RNase and lysozyme were used as the templates. A combination of surface imprinting and sol-gel techniques was used for synthesizing the magnetic protein selective MIPs. Bovine hemoglobin imprinted Fe_3O_4 NPs showed the best template recognition and highest adsorption capability compared to other MIPs. To determine the applicability of Fe_3O_4 for binding bovine hemoglobin-MIPs from complex real samples, selective capture of bovine hemoglobin from bovine blood samples was tested. It was shown that, BHb was almost completely removed from the blood sample after treatment with BHb-MIPs which suggests the potential of MIPs in practical applications including the recognition and enrichment of proteins. Because of the great potential of the technique to capture the target analyte almost completely from the blood sample, the technique is promising to develop biosensors especially for selective detection of analytes which are found in complex real samples.

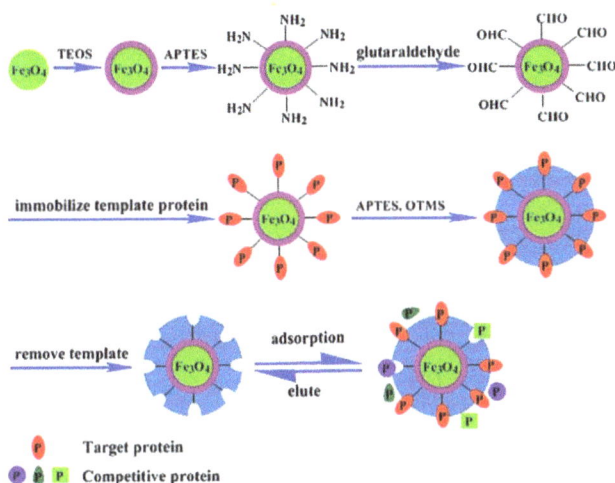

Figure 6. General method for synthesis of molecularly imprinted magnetic Fe_3O_4@MIP NPs using an immobilized template protein. In the protocol, the surface of Fe_3O_4 NPs was transformed to a silica shell using tetraethoxysilane (TEOS) to form Fe_3O_4@SiO$_2$. Then, the surface of Fe_3O_4@SiO$_2$ was reacted with 3-aminopropyltriethoxysilane (APTES) to introduce amino groups on the surface which were reacted with glutaraldehyde (GA) in the following step to form aldehyde modified Fe_3O_4@SiO$_2$ NPs. In the next step, template protein was covalently immobilized on the aldehyde modified Fe_3O_4@SiO$_2$ NPs. In the last step, siloxane co-polymerization by APTES and octyltrimethoxysilane resulted in a polymeric network around the template protein on the surface of the Fe_3O_4@SiO$_2$ – protein complex. Then, the template was removed from the surface and imprinted cavities which were complementary to the template in size, shape and functional groups were obtained (reproduced from [60] with permission).

The tendency of bacteria to assemble at oil-water interfaces was used in a study to create bacteria-imprinted recognition sites on the surface of polymer beads. The authors demonstrated that MIPs for bacterial recognition can be prepared from bacteria based pre-polymer networks which are also known as Pickering emulsions [61]. The basic principle of the technique is the partitioning of solid particles between two immiscible liquids. In this study, living microorganisms were used as particle stabilizers in Pickering emulsion. The general procedure of the technique involves three steps:

(1) In the first step, negatively charged bacteria were assembled with a positively charged pre-polymer which contained vinyl groups,
(2) In the next step, bacteria-pre polymer complexes were used as the particle stabilizer to form emulsion of a cross-linking monomer (the oil phase) in water. The cross-linking monomer was polymerized by free radical polymerization and by this way the pre-polymer was covalently fixed to the core of the polymer beads.
(3) After template removal, bacteria-imprinted sites were obtained on the surface of the polymer beads.

N-acrylylchitosan (NAC) was used as the vinyl-containing pre-polymer. The binding experiments of the groups of bacteria proved that the bacterial recognition on the imprinted polymer beads was dependent on the nature of the polymers and the target bacteria.

Micro-emulsion polymerization was used to construct glutathione peroxidase-like active sites on polystyrene nanoparticles [62]. Two functional monomers were used in the model system. One of them was introduced on the surface of the nanoparticle and acted as a catalytic center while the other was designed as a binding site for the complexation of the carboxyl group of the substrate. The imprinted polymer showed very high catalytic activity and substrate specificity. When the advantages of the glutathione peroxidase mimics including high catalytic activity, good water solubility and substrate specificity are taken into consideration, these imprints might have potential applications in medicine.

2.3. Epitope Imprinting

Rachkov and Minoura [63] proposed an alternative method to the use of whole molecule as templates. In their method, a domain with the same sequence as one of the terminal chains of the target protein was used as template. In this "epitope approach", the peptide epitope was covalently attached to a glass/silicon surface on which the monomers were polymerized in the next step. By this way, after removal of glass/silicon with the target molecules, a molecularly imprinted polymer (MIP) film was produced. In the following steps, the MIP was used to bind and capture the target protein from protein mixtures.

In the epitope imprinting approach, more specific and stronger interactions can be obtained by using a small part or fragment of a macromolecule. Therefore, non-specific binding can be minimized and the affinity can be improved. The imprinted polymer has the ability to recognize the template as well as the entire protein when the sequence is exposed.

"C termini" of proteins are selected as "epitopes" for the imprinting process because these sites are generally less frequently prone to post-translational modifications and the minimum length of peptide necessary to create "unique" recognition for the whole target protein has been reported to be around 9 amino acids according to Nishino et al. [64]. The short epitopes target the primary structure of the peptide rather than the more complex secondary and tertiary structures. Therefore, this approach is also promising for the capture of target proteins based only on genomic information.

Tai et al. [65] used a 15-mer peptide, which is a linear epitope of the dengue virus NS1 protein, as a template to prepare molecularly imprinted QCM sensors. The polymers imprinted with a peptide recognized both the template and other proteins which possess the same epitope part in the structure. The system is ideal as an in vitro cellular assay for quantitatively recognizing proteins.

Ertürk et al. [6] developed an immunoglobulin G sensor via imprinting the F_{ab} fragments of the antibody onto a SPR sensor chip surface. F_{ab} fragments are smaller than the IgG molecule and they are the functional components of the whole molecule. The developed system had showed high affinity to the IgG molecule and the system was also suitable to detect IgG from human plasma with approximately 99% precision in the studied concentration range. Correlation between the developed method and ELISA was found to be approximately 99%. The selectivity coefficients showed that the F_{ab} imprinted sensor's selectivity for IgG was 21 and 14 times higher than that of BSA and F_c fragments, respectively. By using the epitope imprinting approach in the study, it was aimed to overcome some of the drawbacks of macromolecule imprinting. The results revealed that the sensor could detect IgG in a highly sensitive and specific way even from human plasma with a high agreement with results from conventional ELISA method.

Gp41 (HIV type-1 related glycoprotein 41) is a transmembrane protein which plays an important role in the early diagnosis of immunodeficiency syndrome (AIDS). Therefore, ultra-sensitive detection of this protein is especially important in early diagnosis and monitoring of pathogenic conditions. Lu et al. [66] developed a biomimetic sensor for the detection of gp41 using epitope imprinting approach. The authors used dopamine as functional monomer and cross-linking agent as shown in Figure 7. The advantages of using dopamine were high hydrophobicity and biocompatibility of the prepared sensor surface. During the polymerization of dopamine on QCM sensor chip, the template peptide was embedded in it. It was indicated that the thickness of poly-dopamine film would affect the distribution of the imprinted sites as well as the rebinding capacity. Therefore the authors demonstrated that when 5 mg/mL dopamine was used for the polymerization, the MIP coated QCM chip reached maximum template binding capacity. A thicker film would lead into a decrease in rebinding which would mean lower sensitivity. The LOD value for HIV-1 gp41 protein was reported as 2 ng/mL. When the sensor was used for the detection of the protein in human urine sample, the recovery was found in the range of 86.5%–94.1%.

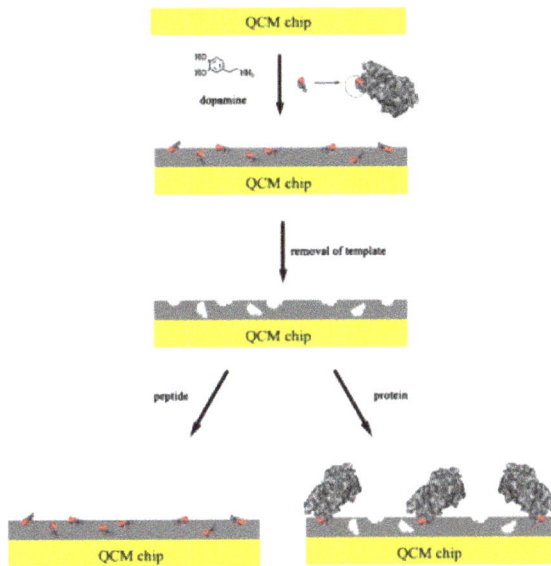

Figure 7. Schematic diagram of epitope imprinting. Dopamine was used as functional monomer and cross-linker. The synthetic peptide was embedded in the polydopamine film on the QCM sensor chip. After template removal step, the recognition sites which were complementary to the template peptide and able to recognize the whole protein were formed on the surface (reproduced from [66] with permission).

3. Concluding Remarks

The area of molecular imprinting is in a rapid development phase at present. What earlier was regarded as impossible is now certainly reachable with new technology. This is illustrated in the present review, by e.g., imprinting of proteins and microbial cells.

The development of nanoparticles with imprinted structures opens new possibilities both in analysis and in the medical area. Synthesis of bi-functional MIPs may be a reality in the future, and then the field is open for e.g., sandwich assays.

As the MIPs are now approaching the sizes of immunoglobulins one can imagine that stable MIPs may replace antibodies in many applications in harsh environments and when repeated assays are to be performed. It also opens up for more frequent on-line assays based on binding reactions since then exchange of the affinity binder is far less frequent. There is still a lack of good information on molecular properties of MIPs and then also on means to improve stability features of structures that are regarded as already very stable.

Acknowledgments: This project was supported by The KK-Foundation for research, project number 20150086.

Conflicts of Interest: The authors declare no conflict of interest.

References

1. Alexander, C.; Andersson, H.S.; Andersson, L.I.; Ansell, R.J.; Kirsch, N.; Nicholls, I.A.; O'Mahony, J.; Whitcombe, M.J. Molecular imprinting science and technology: A survey of the literature for the years up to and including 2003. *J. Mol. Recognit.* **2016**, *19*, 106–180. [CrossRef] [PubMed]
2. Asliyuce, S.; Uzun, L.; Rad, A.Y.; Unal, S.; Say, R.; Denizli, A. Molecular imprinting based composite cryogel membranes for purification of anti-hepatitis B surface antibody by fast protein liquid chromatography. *J. Chromatogr. B* **2012**, *889*, 95–102. [CrossRef] [PubMed]
3. Zhang, H.; Ye, L.; Mosbach, K. Non-covalent molecular imprinting with emphasis on its application in separation and drug development. *J. Mol. Recognit.* **2016**, *19*, 248–259.
4. Kempe, M.; Mosbach, K. Molecular imprinting used for chiral separations. *J. Chromatogr. A* **1995**, *694*, 3–13. [CrossRef]
5. Vidyasankar, S.; Arnold, F.H. Molecular imprinting: Selective materials for separations, sensors and catalysis. *Curr. Opin. Biotechnol.* **1995**, *6*, 218–224. [CrossRef]
6. Ertürk, G.; Uzun, L.; Tümer, M.A.; Say, R.; Denizli, A. F ab fragments imprinted SPR biosensor for real-time human immunoglobulin G detection. *Biosens. Bioelectron.* **2011**, *28*, 97–104. [CrossRef] [PubMed]
7. Uzun, L.; Say, R.; Ünal, S.; Denizli, A. Production of surface plasmon resonance based assay kit for hepatitis diagnosis. *Biosens. Bioelectron.* **2009**, *24*, 2878–2884. [CrossRef] [PubMed]
8. Brüggemann, O.; Haupt, K.; Ye, L.; Yilmaz, E.; Mosbach, K. New configurations and applications of molecularly imprinted polymers. *J. Chromatogr. A* **2000**, *889*, 15–24. [CrossRef]
9. Mujahid, A.; Iqbal, N.; Afzal, A. Bioimprinting strategies: From soft lithography to biomimetic sensors and beyond. *Biotechnol. Adv.* **2013**, *31*, 1435–1447. [CrossRef] [PubMed]
10. Lieberzeit, P.A.; Gazda-Miarecka, S.; Halikias, K.; Schirk, C.; Kauling, J.; Dickert, F.L. Imprinting as a versatile platform for sensitive materials–nanopatterning of the polymer bulk and surfaces. *Sens. Actuators B Chem.* **2005**, *111*, 259–263. [CrossRef]
11. Latif, U.; Mujahid, A.; Afzal, A.; Sikorski, R.; Lieberzeit, P.A.; Dickert, F.L. Dual and tetraelectrode QCMs using imprinted polymers as receptors for ions and neutral analytes. *Anal. Bioanal. Chem.* **2011**, *400*, 2507–2515. [CrossRef] [PubMed]
12. Rao, T.P.; Kala, R.; Daniel, S. Metal ion-imprinted polymers—novel materials for selective recognition of inorganics. *Anal. Chim. Acta* **2006**, *578*, 105–116. [CrossRef] [PubMed]
13. Whitcombe, M.J.; Chianella, I.; Larcombe, L.; Piletsky, S.A.; Noble, J.; Porter, R.; Horgan, A. The rational development of molecularly imprinted polymer-based sensors for protein detection. *Chem. Soc. Rev.* **2011**, *40*, 1547–1571. [CrossRef] [PubMed]
14. Hillberg, A.; Tabrizian, M. Biomolecule imprinting: Developments in mimicking dynamic natural recognition systems. *IRBM* **2008**, *29*, 89–104. [CrossRef]
15. Zahedi, P.; Ziaee, M.; Abdouss, M.; Farazin, A.; Mizaikoff, B. Biomacromolecule template-based molecularly imprinted polymers with an emphasis on their synthesis strategies: A review. *Polym. Adv. Technol.* **2016**, *27*, 1124–1142. [CrossRef]
16. Hu, X.; Li, G.; Huang, J.; Zhang, D.; Qiu, Y. Construction of Self-Reporting Specific Chemical Sensors with High Sensitivity. *Adv. Mater.* **2007**, *19*, 4327–4332. [CrossRef]
17. Lin, T.-Y.; Hu, C.-H.; Chou, T.-C. Determination of albumin concentration by MIP-QCM sensor. *Biosens. Bioelectron.* **2004**, *20*, 75–81. [CrossRef] [PubMed]
18. Qin, L.; He, X.-W.; Yuan, X.; Li, W.-Y.; Zhang, Y.-K. Molecularly imprinted beads with double thermosensitive gates for selective recognition of proteins. *Anal. Bioanal. Chem.* **2011**, *399*, 3375–3385. [CrossRef] [PubMed]
19. Ge, Y.; Turner, A.P. Too large to fit? Recent developments in macromolecular imprinting. *Trends Biotechnol.* **2008**, *26*, 218–224. [CrossRef] [PubMed]
20. Lv, Y.; Tan, T.; Svec, F. Molecular imprinting of proteins in polymers attached to the surface of nanomaterials for selective recognition of biomacromolecules. *Biotechnol. Adv.* **2013**, *31*, 1172–1186. [CrossRef] [PubMed]
21. Kunitake, T.; Mujahid, A.; Dickert, F.L. Detection of Cells and Viruses Using Synthetic Antibodies. In *Handbook of Molecular Imprinting: Advanced Sensor Applications*; Pan Stanford Publishing: Stanford, CA, USA, 2012; pp. 527–567.

22. Xia, Y.; Whitesides, G.M. Soft lithography. *Annu. Rev. Mater. Sci.* **1998**, *28*, 153–184. [CrossRef]
23. Whitesides, G.M.; Ostuni, E.; Takayama, S.; Jiang, X.; Ingber, D.E. Soft lithography in biology and biochemistry. *Annu. Rev. Biomed. Eng.* **2001**, *3*, 335–373. [CrossRef] [PubMed]
24. Bain, C.D.; Whitesides, G.M. Modeling organic surfaces with self-assembled monolayers. *Angew. Chem. Int. Ed. Engl.* **1989**, *28*, 506–512. [CrossRef]
25. Dickert, F.L.; Hayden, O.; Bindeus, R.; Mann, K.-J.; Blaas, D.; Waigmann, E. Bioimprinted QCM sensors for virus detection—screening of plant sap. *Anal. Bioanal. Chem.* **2004**, *378*, 1929–1934. [CrossRef] [PubMed]
26. Dickert, F.L.; Lieberzeit, P.; Gazda-Miarecka, S.; Halikias, K.; Mann, K.-J. Modifying polymers by self-organisation for the mass-sensitive detection of environmental and biogeneous analytes. *Sens. Actuators B Chem.* **2004**, *100*, 112–116. [CrossRef]
27. Lieberzeit, P.A.; Glanznig, G.; Jenik, M.; Gazda-Miarecka, S.S.; Dickert, F.L.; Leidl, A. Soft lithography in chemical sensing–analytes from molecules to cells. *Sensors* **2005**, *5*, 509–518. [CrossRef]
28. Hayden, O.; Lieberzeit, P.A.; Blaas, D.; Dickert, F.L. Artificial antibodies for bioanalyte detection—Sensing viruses and proteins. *Adv. Funct. Mater.* **2006**, *16*, 1269–1278. [CrossRef]
29. Yang, J.C.; Shin, H.-K.; Hong, S.W.; Park, J.Y. Lithographically patterned molecularly imprinted polymer for gravimetric detection of trace atrazine. *Sens. Actuators B Chem.* **2015**, *216*, 476–481. [CrossRef]
30. Voicu, R.; Faid, K.; Farah, A.A.; Bensebaa, F.; Barjovanu, R.; Py, C.; Tao, Y. Nanotemplating for two-dimensional molecular imprinting. *Langmuir* **2007**, *23*, 5452–5458. [CrossRef] [PubMed]
31. Dickert, F.; Hayden, O. Bioimprinting of polymers and sol-gel phases. Selective detection of yeasts with imprinted polymers. *Anal. Chem.* **2002**, *74*, 1302–1306. [CrossRef] [PubMed]
32. Jenik, M.; Schirhagl, R.; Schirk, C.; Hayden, O.; Lieberzeit, P.; Blaas, D.; Paul, G.; Dickert, F.L. Sensing picornaviruses using molecular imprinting techniques on a quartz crystal microbalance. *Anal. Chem.* **2009**, *81*, 5320–5326. [CrossRef] [PubMed]
33. Shi, H.; Tsai, W.-B.; Garrison, M.D.; Ferrari, S.; Ratner, B.D. Template-imprinted nanostructured surfaces for protein recognition. *Nature* **1999**, *398*, 593–597. [PubMed]
34. Yilmaz, E.; Haupt, K.; Mosbach, K. The use of immobilized templates—A new approach in molecular imprinting. *Angew. Chem. Int. Ed.* **2000**, *39*, 2115–2118. [CrossRef]
35. Ratner, B.D.; Shi, H. Recognition templates for biomaterials with engineered bioreactivity. *Curr. Opin. Sol. State Mater. Sci.* **1999**, *4*, 395–402. [CrossRef]
36. Ratner, B.D. The engineering of biomaterials exhibiting recognition and specificity. *J. Mol. Recognit.* **1996**, *9*, 617–625. [CrossRef]
37. Shi, H.; Ratner, B.D. Template recognition of protein-imprinted polymer surfaces. *J. Biomed. Mater. Res.* **2000**, *49*, 1–11. [CrossRef]
38. Shiomi, T.; Matsui, M.; Mizukami, F.; Sakaguchi, K. A method for the molecular imprinting of hemoglobin on silica surfaces using silanes. *Biomaterials* **2005**, *26*, 5564–5571. [CrossRef] [PubMed]
39. Chou, P.-C.; Rick, J.; Chou, T.-C. C-reactive protein thin-film molecularly imprinted polymers formed using a micro-contact approach. *Anal. Chim. Acta* **2005**, *542*, 20–25. [CrossRef]
40. Lin, H.-Y.; Hsu, C.-Y.; Thomas, J.L.; Wang, S.-E.; Chen, H.-C.; Chou, T.-C. The microcontact imprinting of proteins: The effect of cross-linking monomers for lysozyme, ribonuclease A and myoglobin. *Biosens. Bioelectron.* **2006**, *22*, 534–543. [CrossRef] [PubMed]
41. Ertürk, G.; Mattiasson, B. From imprinting to microcontact imprinting—A new tool to increase selectivity in analytical devices. *J. Chromatogr. B* **2016**, *1021*, 30–44. [CrossRef] [PubMed]
42. Sener, G.; Ozgur, E.; Rad, A.Y.; Uzun, L.; Say, R.; Denizli, A. Rapid real-time detection of procalcitonin using a microcontact imprinted surface plasmon resonance biosensor. *Analyst* **2013**, *138*, 6422–6428.
43. Osman, B.; Uzun, L.; Beşirli, N.; Denizli, A. Microcontact imprinted surface plasmon resonance sensor for myoglobin detection. *Mater. Sci. Eng. C* **2013**, *33*, 3609–3614. [CrossRef] [PubMed]
44. Ertürk, G.; Berillo, D.; Hedström, M.; Mattiasson, B. Microcontact-BSA imprinted capacitive biosensor for real-time, sensitive and selective detection of BSA. *Biotechnol. Rep.* **2014**, *3*, 65–72. [CrossRef]
45. Erlandsson, D.; Teeparuksapun, K.; Mattiasson, B.; Hedström, M. Automated flow-injection immunosensor based on current pulse capacitive measurements. *Sens. Actuators B Chem.* **2014**, *190*, 295–304. [CrossRef]
46. Erlandsson, D.; Hedstrom, M.; Mattiasson, B.; Larsson, J. Method of Measuring a Capacitance. U.S. Patent 9,304,096, 5 April 2016.

47. Lee, J.U.; Nguyen, A.H.; Sim, S.J. A nanoplasmonic biosensor for label-free multiplex detection of cancer biomarkers. *Biosens. Bioelectron.* **2015**, *74*, 341–346. [CrossRef] [PubMed]
48. Liang, J.; Yao, C.; Li, X.; Wu, Z.; Huang, C.; Fu, Q.; Lan, C.; Cao, D.; Tang, Y. Silver nanoprism etching-based plasmonic ELISA for the high sensitive detection of prostate-specific antigen. *Biosens. Bioelectron.* **2015**, *69*, 128–134. [CrossRef] [PubMed]
49. Jolly, P.; Formisano, N.; Tkáč, J.; Kasák, P.; Frost, C.G.; Estrela, P. Label-free impedimetric aptasensor with antifouling surface chemistry: A prostate specific antigen case study. *Sens. Actuators B Chem.* **2015**, *209*, 306–312. [CrossRef]
50. Song, H.Y.; Wong, T.I.; Sadovoy, A.; Wu, L.; Bai, P.; Deng, J.; Guo, S.; Wang, Y.; Knoll, W.; Zhou, X. Imprinted gold 2D nanoarray for highly sensitive and convenient PSA detection via plasmon excited quantum dots. *Lab Chip* **2015**, *15*, 253–263. [CrossRef] [PubMed]
51. Ertürk, G.; Hedström, M.; Tümer, M.A.; Denizli, A.; Mattiasson, B. Real-time prostate-specific antigen detection with prostate-specific antigen imprinted capacitive biosensors. *Anal. Chim. Acta* **2015**, *891*, 120–129.
52. Ertürk, G.; Özen, H.; Tümer, M.A.; Mattiasson, B.; Denizli, A. Microcontact imprinting based surface plasmon resonance (SPR) biosensor for real-time and ultrasensitive detection of prostate specific antigen (PSA) from clinical samples. *Sens. Actuators B Chem.* **2016**, *224*, 823–832. [CrossRef]
53. Idil, N.; Hedström, M.; Denizli, A.; Mattiasson, B. Whole cell based microcontact imprinted capacitive biosensor for the detection of *Escherichia coli. Biosens. Bioelectron.* **2017**, *87*, 807–815. [CrossRef] [PubMed]
54. Lotierzo, M.; Henry, O.; Piletsky, S.; Tothill, I.; Cullen, D.; Kania, M.; Hock, B.; Turner, A.P. Surface plasmon resonance sensor for domoic acid based on grafted imprinted polymer. *Biosens. Bioelectron.* **2004**, *20*, 145–152.
55. Li, X.; Zhang, B.; Li, W.; Lei, X.; Fan, X.; Tian, L.; Zhang, H.; Zhang, Q. Preparation and characterization of bovine serum albumin surface-imprinted thermosensitive magnetic polymer microsphere and its application for protein recognition. *Biosens. Bioelectron.* **2014**, *51*, 261–267. [CrossRef] [PubMed]
56. Li, F.; Li, J.; Zhang, S. Molecularly imprinted polymer grafted on polysaccharide microsphere surface by the sol–gel process for protein recognition. *Talanta* **2008**, *74*, 1247–1255. [CrossRef] [PubMed]
57. Poma, A.; Guerreiro, A.; Whitcombe, M.J.; Piletska, E.V.; Turner, A.P.; Piletsky, S.A. Solid-Phase Synthesis of Molecularly Imprinted Polymer Nanoparticles with a Reusable Template–"Plastic Antibodies". *Adv. Funct. Mater.* **2013**, *23*, 2821–2827. [CrossRef] [PubMed]
58. Moczko, E.; Poma, A.; Guerreiro, A.; Sansalvador, I.P.D.; Caygill, S.; Canfarotta, F.; Whitcombe, M.J.; Piletsky, S. Surface-modified multifunctional MIP nanoparticles. *Nanoscale* **2013**, *5*, 3733–3741.
59. Tan, C.J.; Chua, H.G.; Ker, K.H.; Tong, Y.W. Preparation of bovine serum albumin surface-imprinted submicrometer particles with magnetic susceptibility through core-shell miniemulsion polymerization. *Anal. Chem.* **2008**, *80*, 683–692. [CrossRef] [PubMed]
60. Gao, R.; Kong, X.; Wang, X.; He, X.; Chen, L.; Zhang, Y. Preparation and characterization of uniformly sized molecularly imprinted polymers functionalized with core–shell magnetic nanoparticles for the recognition and enrichment of protein. *J. Mater. Chem.* **2011**, *21*, 17863–17871. [CrossRef]
61. Shen, X.; Bonde, J.S.; Kamra, T.; Bülow, L.; Leo, J.C.; Linke, D.; Ye, L. Bacterial imprinting at Pickering emulsion interfaces. *Angew. Chem. Int. Ed.* **2014**, *53*, 10687–10690. [CrossRef] [PubMed]
62. Huang, X.; Liu, Y.; Liang, K.; Tang, Y.; Liu, J. Construction of the active site of glutathione peroxidase on polymer-based nanoparticles. *Biomacromolecules* **2008**, *9*, 1467–1473. [CrossRef] [PubMed]
63. Rachkov, A.; Minoura, N. Towards molecularly imprinted polymers selective to peptides and proteins. The epitope approach. *Biochim. Biophys. Acta (BBA)-Protein Struct. Mol. Enzymol.* **2001**, *1544*, 255–266. [CrossRef]
64. Nishino, H.; Huang, C.S.; Shea, K.J. Selective protein capture by epitope imprinting. *Angew. Chem. Int. Ed.* **2006**, *45*, 2392–2396. [CrossRef] [PubMed]
65. Tai, D.-F.; Lin, C.-Y.; Wu, T.-Z.; Chen, L.-K. Recognition of dengue virus protein using epitope-mediated molecularly imprinted film. *Anal. Chem.* **2005**, *77*, 5140–5143. [CrossRef] [PubMed]
66. Lu, C.-H.; Zhang, Y.; Tang, S.-F.; Fang, Z.-B.; Yang, H.-H.; Chen, X.; Chen, G.-N. Sensing HIV related protein using epitope imprinted hydrophilic polymer coated quartz crystal microbalance. *Biosens. Bioelectron.* **2012**, *31*, 439–444. [CrossRef] [PubMed]

sensors

MDPI

Review
Capacitive Biosensors and Molecularly Imprinted Electrodes

Gizem Ertürk [1,*] and Bo Mattiasson [1,2]

[1] CapSenze Biosystems AB, Lund 223 63, Sweden; bo.mattiasson@biotek.lu.se
[2] Department of Biotechnology, Lund University, Lund 222 40, Sweden
[*] Correspondence: ge@capsenze.se

Academic Editor: Nicole Jaffrezic-Renault
Received: 14 December 2016; Accepted: 8 February 2017; Published: 17 February 2017

Abstract: Capacitive biosensors belong to the group of affinity biosensors that operate by registering direct binding between the sensor surface and the target molecule. This type of biosensors measures the changes in dielectric properties and/or thickness of the dielectric layer at the electrolyte/electrode interface. Capacitive biosensors have so far been successfully used for detection of proteins, nucleotides, heavy metals, saccharides, small organic molecules and microbial cells. In recent years, the microcontact imprinting method has been used to create very sensitive and selective biorecognition cavities on surfaces of capacitive electrodes. This chapter summarizes the principle and different applications of capacitive biosensors with an emphasis on microcontact imprinting method with its recent capacitive biosensor applications.

Keywords: capacitive biosensors; affinity biosensors; microcontact imprinting

1. Introduction

Affinity biosensors can be divided into two main groups: those that measure direct binding between the target molecule and the affinity surface on the sensor, and those biosensors which are adopted to binding assays using labelled reagents [1]. Biosensors operating with labelled affinity-reagents are variations of conventional immunoassay technology in which fluorescent markers, active enzymes, magnetic beads, radioactive species or quantum dots are generally used as labelling agents to label the target molecules [2,3].

Labelling is generally used to significantly facilitate the signal generation and to confirm the interaction between the probe and target molecules [4]. An important feature in using labelled reagents is the amplification of the registered signals and thereby also the sensitivity one can reach. Assays based on the use of labelled reagents are time consuming and the labelled reagents are often expensive. Furthermore, assays with labelled reagents are usually multistep processes and that limit their application for real-time measurements.

The design of label-free affinity based biosensors is the objective of much current research. The aims are to establish alternative methods to the commercial ELISA-based immunoassays. The most attractive features of these types of biosensors are that they allow the monitoring of the analytes directly and in real-time [5].

Biosensors can be divided into four main groups according to transducer types [6]. These main groups involve: electrochemical transducers which involve potentiometric, voltammetric, conductometric, impedimetric and field-effect transistors; optical transducers which include surface plasmon resonance (SPR) biosensors; piezoelectric transducers to which quartz crystal microbalance (QCM) biosensors can be given as an example; and thermometric transducers which measure the amount of heat with a sensitive thermistor to determine the analyte concentration.

Among the different types of label-free biosensors, electrochemical biosensors have received particular attention owing to their properties [4]. These biosensors can also be miniaturized which is very important for many applications that need portable integrated systems. Miniaturization not only allows use at point of care, in a clinic, doctor's office or at home but also reduces the cost of the diagnostic assays. Electrical biosensors fulfil these purposes as being fast, cheap, portable, miniaturized and label-free devices. Electrical biosensors can be classified as amperometric, voltametric impedance or capacitive sensors.

This review deals with capacitive biosensors, different applications of capacitive biosensors developed for both detection of various targets and by using molecular imprinting technology with an emphasis on microcontact imprinting method. In this aspect, this is a novel review which includes the two technologies, capacitive biosensors and molecular imprinting; at the same time with lots of examples from published reports. The reports in molecular imprinting section were selected mainly from capacitive biosensors developed by using microcontact imprinting method.

2. Capacitive Biosensors

Capacitive biosensors belong to the sub-category of impedance biosensors [3]. Capacitive biosensors measure the change in dielectric properties and/or thickness of the dielectric layer at the electrolyte-electrode interface when an analyte interacts with the receptor which is immobilized on the insulating dielectric layer [1].

The electric capacitance between the working electrode (an electrolytic capacitor/the first plate) and the electrolyte (the second plate) is given by Equation (1) [7]:

$$C = (\varepsilon_0 \varepsilon A)/d \tag{1}$$

where ε is the dielectric constant of the medium between plates, ε_0 is the permittivity of the free space (8.85×10^{-12} F/m), A is the surface area of the plates (m^2) and d is the thickness of the insulating layer (m).

According to the given equation above, when the distance between the plates increases, the total capacitance decreases. In other words, in the assaying principle of this type of capacitive biosensors when a target molecule binds to the receptor, displacement of the counter ions around the capacitive electrode results in a decrease in the capacitance. The higher the amount of target molecules bound to the receptor is, the greater is the achieved displacement and the decrease in the registered capacitance [8].

The assaying principle of capacitive biosensors which are developed according to this rule is shown in Figure 1.

The Equation (1) can be represented by two capacitors in series where the inner part includes the dielectric layer (C_{dl}) and the outer one corresponds to the biomolecule layer (C_{bm}). Then, the total capacitance (C_t) can be described as Equation (2) [7].

$$\frac{1}{C_t} = \frac{1}{C_{dl}} + \frac{1}{C_{bm}} \tag{2}$$

The electrochemical capacitors which are described based on the above-mentioned equation are known as constant phase element (CPE). The presence of CPE indicates that the observed capacitance of the system is frequency dependent.

Capacitance can also be defined as Equation (3) [7].

$$Z = \frac{1}{\omega C} \tag{3}$$

where Z is the impedance and ω is the radial frequency expressed in rad·s^{-1}. This model implies that all of the measured current is capacitive.

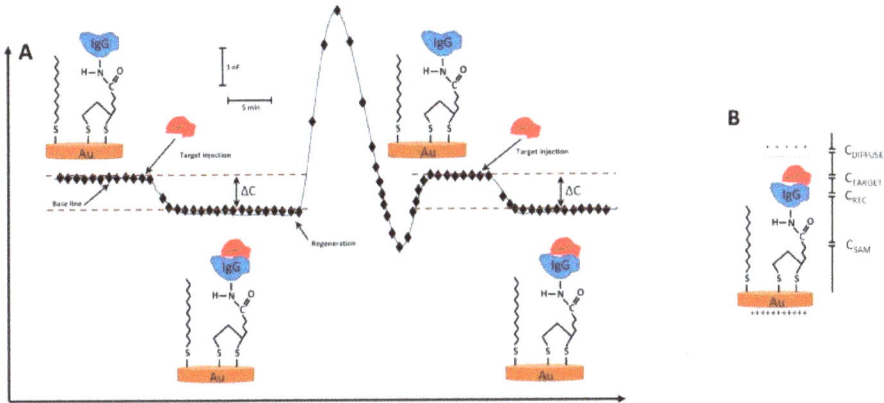

Figure 1. (**A**) Schematic diagram showing the change in capacitance (ΔC) as a function of time when the analyte (IgG) interacts with the receptor molecule (Protein A) immobilized on the surface of the electrode. Subsequent rise in signal is due to the dissociation after the injection of the regeneration solution. In an ideal sensorgram, the baseline should turn back to the original level after regeneration of the surface; (**B**) Immobilization of the receptor molecule on the transducer surface via a self-assembled monolayer (SAM) of alkylthiols. When the target molecule interacts with the receptor, this creates a double layer of counter ions around the gold transducer which results in a change in the capacitance. (Reproduced from Reference [8] with permission).

3. Different Applications of Capacitive Biosensors

Different applications of capacitive biosensors developed for different targets are summarized in Table 1.

Table 1. Different applications of capacitive biosensors developed for different targets.

	Target	Sensor Preparation Method	Dynamic range (M)	Limit of Detection (M)	Selectivity	Stability	Ref.
Proteins	Cholera toxin (CT)	Immobilization of anti-CT antibodies on self-assembled monolayer (SAM) of lipoic acid and 1-Ethyl-3-(3-dimethylaminopropyl) carbodiimide (EDC)	1.0×10^{-13}–1.0×10^{-10}	1.0×10^{-14} uiu	N/D	N/D	[9]
	Cholera toxin (CT)	Immobilization of anti-CT on gold nanoparticles incorporated on a poly-tyramine layer	0.1×10^{-18}–10×10^{-12}	9.0×10^{-20}	N/D	Up to 36 times with an RSD of 2.5%	[10]
	HIV-1 p24 antigen	Immobilization of anti-HIV 1 p24 antigen on gold nanoparticles incorporated on a poly-tyramine layer	10.1×10^{-20}–10.1×10^{-17}	3.32×10^{-20}	N/D	N/D	[11]
	VEGF	Immobilization of anti-VEGF aptamer first capturing the VEGF protein then, sandwiching with antibody-conjugated magnetic beads	13×10^{-14}–2.6×10^{-11}	N/D	N/D	N/D	[12]
Nucleic acids	25-mer oligo C	Covalent attachment of 25-mer oligo C on poly-tyramine modified electrode	10^{-8}–10^{-11}	10^{-11}	Oligo-T was used as the competing agent, when the temperature was increased from RT to 50 °C, the ΔC value decreased from 48 nF·cm^{-2} to 3 nF·cm^{-2}	N/D	[13]
	ssDNA	Thiol modified oligonucleotides were immobilized on Au and 3-glycidoxypropyl-tri-methoxy silane (GOPTS)	0.5×10^{-6}–1.0×10^{-3}	N/D	N/D	GOPTS functionalized surfaces were more stable at 4 °C. Ten-fold decrease in fluorescence intensity after 1 week even when the substrates were stored at 4 °C.	[14]
	Nampt	Immobilization of ssDNA aptamers on SAM of mercaptopropionic acid (MPA)	0–45×10^{-10}	1.8×10^{-11}	N/D	N/D	[15]
	Target DNA	Immobilization of pyrrolidinyl peptide nucleic acid probes (acpcPNA)	1.0×10^{-11}–1.0×10^{-10}	6–10×10^{-12}	Complementary DNA provided a much higher ΔC compared to single and double mismatched DNA	Could be reused for 58–73 times with an average residual activity of $\geq 98\%$	[16]

Table 1. *Cont.*

Target		Sensor Preparation Method	Dynamic range (M)	Limit of Detection (M)	Selectivity	Stability	Ref.
Cells	Total bacteria	Based on the interaction between *E. coli* and concanavalin A immobilized on a modified gold surface	12 CFU·mL^{-1}–1.2 × 10^{-6} CFU·mL^{-1}	12 CFU·mL^{-1}	N/D	For the first 35 cycles, the residual activity was 95% ± 3% (RSD = 3.2%). After 35 cycles, it was 85%.	[17]
	E. coli	*E. coli* cells immobilized on SAM of Mercaptopropionic acid (MPA)	8 × 10^5 CFU·mL^{-1}–8 × 10^7 CFU·mL^{-1}	N/D	N/D	N/D	[18]
Heavy metals	Hg(II), Cu(II), Zn(II), Cd(II)	Immobilization of metal resistance and metal regulatory proteins on gold electrode	10^{-15}–10^{-3}	N/D	N/D	N/D	[19]
	Cu(II), Cd(II), Hg(II)	1. Immobilization of whole bacterial cell to emit a bioluminescent/fluorescent signal in the presence of heavy metal ions	0–200 × 10^{-6}	1.0 × 10^{-6}	N/D	84% of the activity loss within 6 days	[20]
		2. Immobilization of heavy metal binding proteins	10^{-15}–10^{-1}			Stable over 16 days	
Saccharides	Glucose	Immobilization of ConA on gold nanoparticles incorporated on the tyramine modified gold electrode	1.0 × 10^{-6}–1.0 × 10^{-2}	1.0 × 10^{-6}	Small sugars including D-fructose, D-mannose, D-maltose, methyl-α-D-glucopyranoside, methyl-α-D-mannopyranoside also bound instead of glucose	A neglectable loss in sensitivity after 10 cycles (7.5%)	[21]
	Glucose	Immobilization of ConA and replacement of small glucose with the large glucose polymer	1.0 × 10^{-5}–1.0 × 10^{-1}	1.0 × 10^{-6}	Small molecules and high molecular weight dextran also bound instead of glucose	N/D	[22]
	Metergoline	Immobilization of molecularly imprinted spherical beads on modified gold electrode	1–50 × 10^{-6}	1.0 × 10^{-6}	Cross reactant contribution was maximum 1.3 nF	N/D	[23]
Small molecules	Aflatoxin B1	Bioimprinting	3.2 × 10^{-6}–3.2 × 10^{-9}	6.0 × 10^{-12}	Competing agents' binding was significantly lower than aflatoxin B1	Little variation over 28 injections with non-reduced Schiff's bases	[24]
	Ochratoxin A (OTA)	Monoclonal anti-OTA immobilization on Si$_3$N$_4$ substrate combined with magnetic nanoparticles (MNPs)	2.47–49.52 × 10^{-12}	4.57 × 10^{-12}	Differences for ochratoxin B and aflatoxin B1 were not significant	N/D	[25]

3.1. Protein Detection

Labib et al. [9] developed a sensitive method for detection of cholera toxin (CT) using a flow-injection capacitive immunosensor based on self-assembled monolayers. Monoclonal antibodies against the subunit of CT (anti-CT) were immobilized on the gold electrode surface which was modified by lipoic acid and 1-Ethyl-3-(3-dimethylaminopropyl)carbodiimide (EDC). The immunosensor showed linear response to CT concentrations in the concentration range between 1.0×10^{-13} M and 1.0×10^{-10} M under optimized conditions. Limit of detection (LOD) value was 1.0×10^{-14} M. The LOD value obtained from capacitive immunosensor was compared with the LOD values that were obtained from sandwich ELISA and SPR based immunosensors. The ELISA had a LOD of 1.2×10^{-12} M whereas SPR had a LOD value of 1.0×10^{-11} M. The results proved that the method is more sensitive than the other two techniques used in this study.

In another study by the same research group, a label-free capacitive immunosensor was developed for direct detection of CT present at sub-attomolar level. Gold nanoparticles (AuNPs) were incorporated on a polytyramine modified gold electrode and anti-CT antibody was immobilized on this surface. Tyramine provides free amino groups which are very useful to immobilize the affinity ligand to the transducer. At the same time very thin and uniform films can be formed on the electrode surfaces during the electro-polymerization of tyramine. After the immobilization step, the formation of antigen-antibody complexes resulted in a change in capacitance and by this way the concentration of CT was determined. The dynamic range was between 0.1 aM and 10 pM where the LOD value was 9×10^{-20} M (0.09 aM). The electrode could be regenerated with a good reproducibility for up to 36 times with a relative standard deviation (RSD) value of 2.5%. Real sample analyses were performed from water samples collected from a local stream and matrix effect was eliminated with a 10.000 times dilution prior to analysis. The developed system had potential to be used as a portable electrochemical analyser for field conditions [10].

The same strategy was used to develop a capacitive biosensor for sensitive detection of HIV-1 p24 antigen [11]. Following polytyramine electro-polymerization on the gold electrode surface, gold nanoparticles were incorporated onto the electrode and then, anti-HIV-1 p24 monoclonal antibodies were immobilized on top. HIV-1 p24 antigen was detected from standard p24 solutions in the concentration range of 10.1×10^{-20} to 10.1×10^{-17} M. The reasons for extreme sensitivity of the capacitive biosensors were explained by the authors in two ways. Firstly, the capacitive technique is very convenient to detect the size and position of the electrical double layer formed at the interface. However, in other electrochemical techniques such as the amperometric technique, the detection is based on the measurement of only the transport of electrons. The second factor is the increased surface area on the electrode via the use of immobilized AuNPs. The gold nanoparticles also significantly contributed to the decrease in the interfacial resistance which facilitated the electron transfer at the electrode surface [11].

Qureshi et al. [12] developed a capacitive aptamer based sensor for detection of vascular endothelial growth factor (VEGF) in human serum. Systematic evolution of ligands by exponential enrichment (SELEX) process was utilized to select the highly specific and selective anti-VEGF aptamer to bind VEGF to the aptasensor surface. When a sandwich assay was tested by forming sandwich complex with anti-VEGF aptamer+VEGF+anti-VEGF antibody, the generated signal was enhanced by 3–8 folds compared to the direct assay. The developed sensor showed a dynamic detection range from 13×10^{-14} M to 2.6×10^{-11} M of VEGF protein in human serum. The results showed that the developed system could be successfully used in clinical diagnosis to detect biomarkers in real samples in a convenient and sensitive way.

3.2. Nucleic Acid Detection

Mahadhy et al. [26] used in a model study the capacitive sensor to monitor the capture of complementary single-stranded nucleic acids. A 25-mer oligo-C was immobilized onto the polytyramine modified gold electrode surface. Temperature was raised to 50 °C to reduce non-specific hybridization in order to increase the selectivity and hybridization was used in order to amplify the signal by using longer nucleic acid molecules.

Later, Mahadhy et al. [13] developed a promising ultrasensitive, automated flow-based and portable gene sensor. The PCR-free biosensor proved the possibility for powerful detection of foodborne pathogens in diagnostic situations and multi-drug resistant bacteria in the near future. Rapid detection of foodborne pathogens is crucial before many people are infected while early detection of multi-drug resistant bacteria is important to isolate the infected patients earlier and to reduce the risk of spreading.

Two functionalization layers; gold (Au) and 3-glycidoxypropyl-tri-methoxy silane (GOPTS) were used to immobilize thiol modified oligonucleotides on silicon surfaces. GOPTS showed better performance as a functionalization layer because the hybridization efficiency was higher, the stability over time was better and regeneration of the surface after analyte binding was easier. Therefore for the development of microcantilever or micro-membrane based biosensors, GOPTS might be a more promising alternative [14].

A single stranded DNA (ssDNA) aptamer was developed to bind to nicotinamide phosphoribosyl transferase (Nampt) through SELEX and implemented in a capacitive biosensor [15]. The LOD for Nampt was 1.8×10^{-11} M with a dynamic detection range in serum of up to 9×10^{-10} M. Nampt is an important biomarker for obesity-related metabolic diseases, some types of cancers and chronic diseases. Normal level of Nampt in human plasma is around 15 ng·mL^{-1} (27×10^{-11} M). Therefore, the developed system has potential as a diagnostic tool and in point-of-care applications.

Pyrrolidinyl peptide nucleic acid probes were immobilized onto the self-assembled monolayer (SAM) modified gold electrode surface to develop a DNA capacitive biosensor [16]. Four different alkanethiols with various chain lengths were used as a SAM to determine the influence of the length and the terminating head group of blocking thiols on the sensitivity and specificity. In the study, the blocking thiol which had an equal length to the –OH terminating head group gave the highest sensitivity and high binding specificity.

Thus far, there are no good examples on use of MIPs in connection to monitoring of specific DNA sequences. Since the potential is large, one can expect such developments to happen in the near future.

3.3. Cell Detection

Jantra et al. [17] developed a label-free affinity biosensor for detecting and enumerating total bacteria based on the interaction between *E. coli* and Concanavalin A (Con A) immobilized on a modified gold surface. The analyses were completed in less than 20 min with both SPR and impedimetric capacitive biosensor. Compared to SPR (LOD: 6.1×10^7 CFU·mL^{-1}), the capacitive system showed much higher sensitivity (LOD: 12 CFU·mL^{-1}). The developed system might be used successfully for total bacteria analysis from water sources.

Rydosz et al. [27] developed a new type of label-free microwave sensor in a form of interdigitated capacitor for bacterial lipopolysaccharide detection. The sensor surface was coated with T4 phage gp37 adhesin. The adhesin molecule bound *E. coli* by recognizing its bacterial host lipopolysaccharide (LPS). The binding was highly specific and irreversible. Recognition between the phage adhesion and bacterial LPS was based on the recognition of saccharide determinants of LPS which means very specific determination of bacterial strain or its endotoxins within the genus and the species. The selectivity experiments showed that the response for specific LPSs was significantly different from the reference measurements and the response for the non-specific LPSs was very close to the reference values. The developed method was promising for label-free LPS detection and could be used as an alternative for fiber-optic, electrochemical and classic biochemical and immunochemical methods. In their other study [28], same authors used bacteriophage-adhesin-coated long-period gratings for recognition of bacterial lipopolysaccharides. Long-period gratings (LPG) bio-functionalization methodology was based on coating the LPG surface with nickel ions which were capable of binding of gp37-histidine tag. The advantage of using adhesins for the bio-functionalization of the biosensor was to give ability for low-cost bio-sensitive molecule exchange and surface regeneration. In this work, for the first time, adhesion has been applied for bacteria and their endotoxin detection. T4 phage adhesin bound *E. coli* B LPS in its native or denatured form in a highly specific and irreversible way.

Rocha et al. [29] used alternating current electrokinetics (ACEK) capacitive sensing to detect and quantify the microbial cell abundance in aquatic systems. Microbial abundance was detected by measuring the electrical signal. Three different microbial cell cultures including *Bacillus subtilis*, *Alcanivorax borkumensis* and *Microcystis aeruginosa* were detected by using the developed system. The results showed that the sensor is capable of reliably detecting microbial cells even though they have major physiological differences between Gram-positive (*B. subtilis*), Gram-negative (*A. borkumensis*) and cyanobacteria (*M. aeruginosa*). The system is promising to detect and estimate microorganism population sizes in batch cultures, environmentally sourced seawater and groundwater systems.

For expanded use and commercialization of nanotechnology products, toxicity determination is an important field application. For this purpose, Qureshi et al. [18] developed a whole-cell based capacitive biosensor to determine the biological toxicity of nanoparticles (NPs). They used iron oxide (Fe$_3$O$_4$) nanoparticles as models in the study. The living *E. coli* cells were immobilized on the capacitive sensor chips. Then, these chips were interacted with different sizes of Fe$_3$O$_4$ NPs (5, 20 and 100 nm). The smallest Fe$_3$O$_4$ NPs resulted in a maximum capacitance change because they were able to interact with *E. coli* cells on the sensor chip very efficiently. The morphological changes on the surface of *E. coli* cells after interacting with Fe$_3$O$_4$ NPs were examined with SEM.

3.4. Heavy Metal Detection

Two metal binding proteins were over-expressed in *E. coli*, purified and immobilized on a thiol-modified capacitive sensor surface. Capacitive sensor was used to monitor conformational changes following heavy metal binding including copper, cadmium, mercury and zinc. Metal ion detection could be done in down to femtomolar concentrations with the developed system [30].

The same group expressed and purified metal resistance and metal regulatory proteins from bacterial strains and immobilized these proteins on the capacitive biosensor surface for heavy metal detection [19]. The system allowed the detection of heavy metals including Hg(II), Cu(II), Zn(II) and Cd(II) in pure solutions down to 10^{-15} M concentrations.

Corbisier at al. [20] established two different biosensor technologies for detection of several heavy metal ions in environmental samples. The principle of the first approach was to develop whole cell bacterial biosensors which emitted a bioluminescent/fluorescent signal in the presence of heavy metal ions. In the second approach, direct interaction between metal binding proteins and heavy metal ions was used as the detection principle in the capacitive biosensors. In the study, the main advantage of the whole cell based sensors was their ability to react only to biologically available metal ions, whereas the latter one (protein based sensors) was more sensitive towards metal ions.

MIPs selective for heavy metal ions have been presented in connection to separation technology [31–36]. It is an obvious development that also sensors for heavy metal ions will be developed based on selective MIPs.

3.5. Saccharide Detection

By using capacitive biosensors, Labib et al. [21] used the capacitive biosensor to detect glucose based on gold nanoparticles which were fixed on a poly-tyramine modified gold electrode surface. Dextran (MW: 39 kDa) was used as a regeneration agent by utilizing a competitive assay for glucose in the study. The dynamic range for glucose detection was between 1.0×10^{-6} and 1.0×10^{-2} M with a LOD value of 1.0×10^{-6} M.

By using capacitive biosensors, the same authors [22] developed a technique based on the competition between a small molecular mass analyte and a large analyte-carrier conjugate. In the basis of the technique, when a large glucose polymer binds to the biorecognition molecule (Con A) immobilized on the electrode surface, it results in a decrease in the capacitance. Then, in the next step, when the low molecular mass analyte (glucose) is introduced to the system, the effect is reverse, the small glucose molecule will replace the large glucose polymer which is bound on the immobilized Con A as shown in Figure 2. By measuring the shift-back in capacitance, the glucose concentration could be determined by the technique. The authors used this technique to measure IgG as a glycoconjugate and detect its aggregation using immobilized Con A. When the glycoconjugate (IgG) was injected, the decrease in capacitance was measured to determine its concentration. In the second step, when concentrated glucose was injected into the system, the increase in capacitance was employed to determine the glucose concentration. The results showed that this technique is promising for monitoring small molecules with high sensitivity and broad detection range.

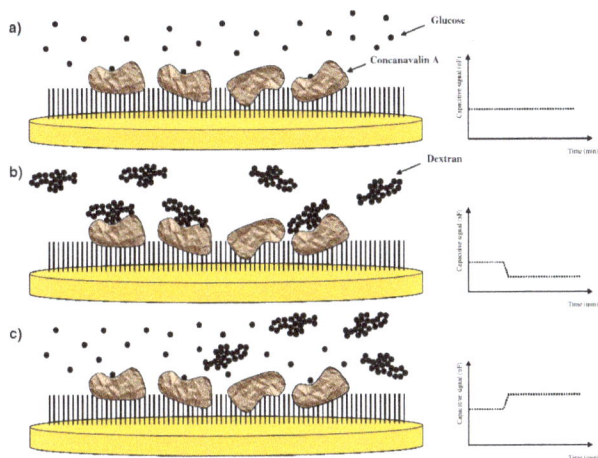

Figure 2. Schematic representation of the competitive glucose binding assay. (**a**) When glucose is injected into the capacitive system, it binds to the immobilized Concanavalin A (ConA) on the surface. However, this binding does not make any change in the capacitance level, as shown in the graph on the right, due to the small size of the glucose molecule; (**b**) When a glucose polymer (dextran) is injected into the system, binding of this big polymer to ConA results in a decrease in the capacitance signal; (**c**) When glucose is injected into the system again, displacement of dextran with glucose results in the capacitance turn back to the original baseline level. (Reproduced from Reference [22] with permission).

The radio frequency (RF) detection method is one of the promising methods for glucose detection out of several detection methods available [37]. When an analyte is injected into the RF biosensor, changes will occur owing to the inductive and capacitive effects [38]. These changes will cause losses and considerable shifts in the resonance frequency of the device. The change in capacitance is proportional to the dielectric constant and the distance between the biomolecule layer and the dielectric layer. A reusable robust RF biosensor was developed by Kim et al. [38] to monitor real-time glucose level in human serum. The resonance behaviour of the system was analysed with human serum samples containing different glucose concentrations ranging from 148–268 mg·dL^{-1}, 105–225 mg·dL^{-1} and at a deionized water glucose concentration in the range of 25–500 mg·dL^{-1}. The response time for glucose was measured as 60 s with a LOD value of 8.01 mg·dL^{-1}. A total of 21 different experiments for each concentration of serum and D-glucose solution were analysed for reusability and the relative standard deviation (RSD) was less than 1% for each concentrations of serum samples and aqueous D-glucose solutions.

In the area of MIPs used for bioseparation, much work has been done concerning carbohydrates [39–41]. It is obvious that one can make good MIPs with high efficiency in binding target molecules or fragments thereof. Based on the observations from affinity chromatography, one can foresee that such systems will also soon be presented for capacitive biosensors [42–48] (Table 2).

Table 2. Molecularly imprinted polymers (MIPs) produced with high binding efficiency for affinity chromatography applications.

Template	Method	Matrix	Comments	Ref.
Benzo[a]pyrene (BAP)	BAP-imprinted poly (2-hydroxyethylmethacrylate-N-methacryloyl-(L)-phenylalanine composite cryogel cartridge	Aqueous solutions	• Preconcentration on BAP with HPLC equipped with a fluorescence detector (HPLC-FLD)	[42]
Melamine	Melamine imprinted monolithic cartridges	Water + milk	• MIP-solid phase extraction • Extraction and enrichment of melamine	[43]
Cholesterol	Cholesterol imprinted polymeric nanospheres	Gastrointestinal mimicking solution	• Cholesterol adsorption	[44]
Catalase	Iron chelated poly (2-hydroxyethylmethacrylate-N-methacryloyl-(L)-glutamic acid cryogel discs	Rat liver	• Catalase purification from rat liver	[45]
L-phenylalanine (L-Phe)	L-Phe imprinted cryogel cartridges	Aqueous solutions	• Chiral separation of l-phenylalanine with FPLC (fast protein liquid chromatography)	[46]
Triazine	Triazine imprinted monolithic columns	Aqueous solutions	• Separation of triazine with capillary electro-chromatography (CEC)	[47]
Cytochrome c	Surface imprinted bacterial cellulose nanofibers	Rat liver	• Cytochrome c purification from rat liver	[48]

3.6. Small Organic Molecules

Small molecules such as pesticides, herbicides, and antibiotics are widely discarded and encountered in naturally flowing waters. These pollutants in environment have an impact on communities and eco-systems. Therefore, detection of these molecules in a sensitive, cheap, robust and fast way is crucial. Lenain et al. [23] chose metergoline as a model compound representingsmall organic molecules such as pharmaceutical residues. Emulsion polymerization was used to produce small, uniformly sized, spherical MIPs. These MIP beads were attached to the poly-tyramine modified gold surface. Scanning electron microscope (SEM) images of the electrode surface are shown in Figure 3. Working range for metergoline was from 1.0×10^{-6} M to 50×10^{-6} M with a LOD value of 1.0×10^{-6} M. In cross reactivity analysis, even though the structural analogs showed binding, this contribution was only around 1.3 nF. The sensor response was more stable at higher ionic strength but the extent of capacitance change for different concentrations of analyte was less pronounced compared to lower electrolyte concentrations.

Bioimprinting is a technology used to mimic specific sites for modification of biological molecules. The process consists of four steps, as shown in Figure 4 [24].

(1) Unfolding the conformation of the starting protein under acidic conditions;
(2) Addition of template molecule and allow interaction between the template molecule and the denatured protein in order to form new molecular configurations;

(3)   Cross-linking of the protein to stabilize the new molecular protein conformation; and

(4)   Dialysis to remove the template molecule.

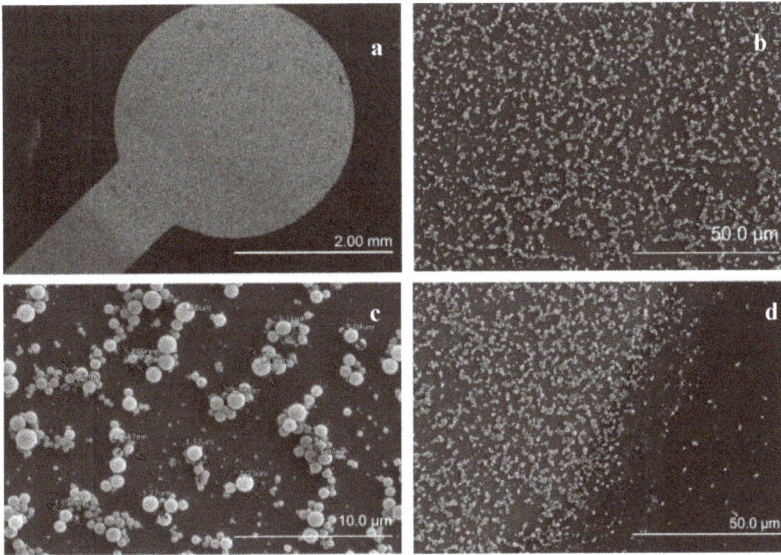

Figure 3. Scanning electron microscope (SEM) pictures of the electrode surface after functionalization with imprinted polymers. From left to right, top to bottom: (**a**) SEM picture of electrode surface; (**b,c**) SEM pictures of centre of the electrode; and (**d**) SEM picture of the border between the gold layer and wafer. (Reproduced from Reference [23] with permission).

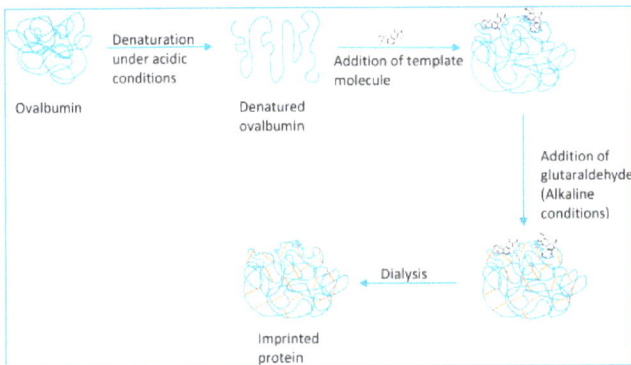

Figure 4. Schematic representation of bio-imprinting process. (Reproduced from Reference [24] with permission).

Gutierrez et al. [24] used bioimprinting to develop a capacitive biosensor for aflatoxin detection. Aflatoxins are natural food contaminants with a high risk for human health. Ovalbumin was used as platform for bioimprinting of aflatoxin because when bovine serum albumin (BSA) was used, there was no change in capacitance owing to the high hydrophobicity of sites of BSA. Three competitive mycotoxins were used in the cross-reactivity analysis and the changes in capacitance were significantly lower than that registered from aflatoxin solution.

Silicon nitride substrate (Si_3N_4) combined with magnetic nanoparticles (MNPs) was used to develop a capacitive immunosensor for ochratoxin A (OTA) detection. Silicon nitride allows an easy control of the film composition and thickness and also prevents the undesirable impurities. These are the main advantages of the substrate used in the study. Magnetic nanoparticles comprised of a conductive core and a carboxylic acid modified shell which was used to immobilize OTA antibodies. The LOD value was calculated as 4.57×10^{-12} M in the study and the selectivity results against ochratoxin B and aflatoxin G1 showed that the potential difference was not so significant when compared to the difference for OTA detection [25].

4. Molecular Imprinting

During the early 1970s, Wulff et al. [49] and Klotz et al. [50] introduced the molecular imprinting to imprint templates in organic polymers. Then, Mosbach et al. [51] reported the use of molecularly imprinted polymers (MIPs) in biosensors instead of antibodies which was a breakthrough.

The formation of MIPs involves three steps:

(1) Pre-complexation of functional monomers around the template molecule in solution either by forming covalent bonds or by self-assembling with non-covalent bonds;
(2) Polymerization of the resulting complex in the presence of cross-linking monomers and suitable solvents/ionic liquids as porogens; and
(3) Removal of template molecule from the synthesized polymer.

The resulting MIP contains recognition cavities capable of selective recognition of compounds that fit these cavities with respect to shape, size, position and orientation of the recognition sites [52]. How MIPs can mimic natural recognition units in different applications are shown schematically in Figure 5.

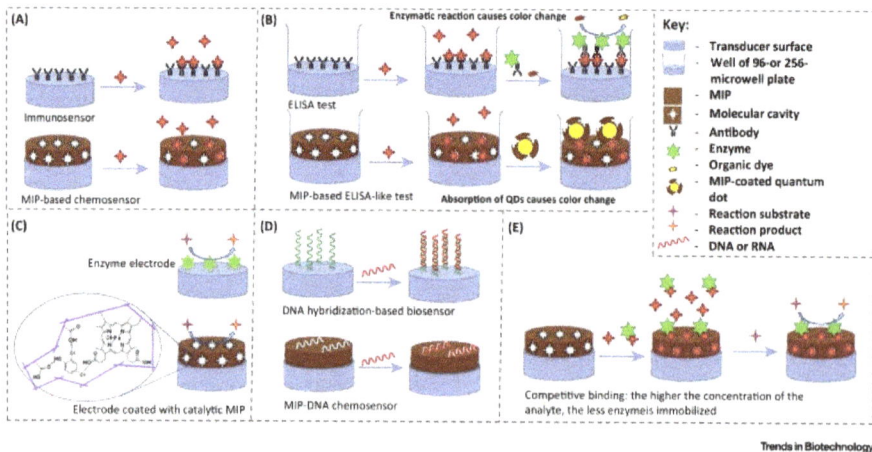

Figure 5. Different applications of MIPs in: (**A**) immunosensors; (**B**) enzyme-linked immunosorbent assay (ELISA); (**C**) enzyme electrodes, reaction rate and analyte concentration of enzyme electrodes and catalytic MIP-coated electrodes can be estimated by electroactive substrate/product consumption/production during the catalytic reaction or electron transfer from the electrode surface to the active centre of enzyme/MIP; (**D**) DNA chips; and (**E**) enzyme immobilization and competitive binding of the analyte. (Reproduced from Reference [52] with permission).

MIP technology has successfully been used for imprinting of low molecular weight templates. However there are still some difficulties of molecular imprinting technique when it is used for macromolecular templates including proteins. Due to this, many researchers have focused on the alternative techniques including imprinting the template directly onto a substrate or immobilizing the template protein on a glass support and use it as a protein stamp. The latter is called microcontact imprinting.

5. Microcontact Imprinting

Microcontact imprinting technique was first introduced by Chou et al. [53]. In the study, the authors formed the microcontact imprints between two cleaned glass surfaces. Template protein was immobilized on the cover slip and then, functional monomer was added on top in order to allow site-specific organization of the functional monomer by the template. In the next step, a drop of solution including cross-linker and initiator was dropped on the pre-modified support glass. Support glass was brought into contact with the cover slip to provide the two functionalized surfaces to contact. The glass assemblies were placed in a UV reactor to initiate the polymerization and it continued for 17 h. After polymerization, the cover slip was removed from the surface with forceps and the support was washed with various solutions. Polymer film thickness measurements showed it to be around 10 μm. The imprinted polymers showed better selectivity for their native templates than competing proteins. By this study, it was reported that microcontact imprinting technique might be used successfully for relatively large proteins in biosensor applications for detection and quantification [53].

There are lots of advantages of the technique over conventional molecular imprinting technique, as described in previous reports [54,55]. Only a few microliters of monomer solution are enough to polymerize dozens of samples at the same time in the same polymerization batch. Therefore, the method is useful to imprint templates which are very expensive or available in limited amounts. The technique also avoids potential solubility and conformational stability problems encountered with macromolecular targets, especially proteins. This is because immobilized templates are used in the process rather than adding the template to the monomer solution which otherwise is the standard procedure. This is also an important advantage for the ease of the template removal step after polymerization [54,55].

6. Applications of Microcontact Imprinting Method with Capacitive Biosensors

Ertürk et al. used microcontact imprinting method to prepare capacitive biosensors for various applications, please see Table 3. In their first application, the authors used BSA as the model protein to prepare microcontact BSA imprinted capacitive biosensor [56]. In the first step, the authors prepared glass cover slips with immobilized protein. This entity was called protein stamp. The cover slips were first modified with 3-aminopropyl-triethoxysilane (APTES) to introduce amino groups on the surface and then with glutaraldehyde to modify these amino groups. Then, the cover slips were immersed in BSA solution overnight at 4 °C. The electrode surface was on the other hand modified with polytyramine and then acryloyl chloride to introduce reactive groups on the surface which would be involved in the subsequent polymerization process. For the microcontact imprinting of BSA onto the gold electrode, the monomer solution containing methacrylic acid (MAA) as the functional monomer and poly-ethyleneglycoldimethacrylate (PEGDMA) as the cross-linker was prepared and the initiator was added into this solution. The modified gold electrode was treated with this monomer solution and then the protein stamp was brought into contact with this monomer treated electrode. Polymerization was initiated under UV light and ended in 15 min. After polymerization, the cover slip was removed from the surface with forceps and the microcontact BSA imprinted electrode was rinsed with water. The measurements were performed with a capacitive sensor involving the microcontact imprinted electrode inserted in an automated flow injection system developed by Erlandsson et al. [57].

Table 3. Capacitive biosensors developed for different targets using microcontact imprinting method.

Target	Biosensing Method	Monomers	Dynamic Range	LOD	Selectivity	Stability	Ref.
Bovine Serum Albumin (BSA)	Capacitive biosensor with current pulse method	Methacrylic acid (MAA); Poly ethyleneglycol-dimethacrylate (PEGDMA)	1.0×10^{-20} M–1.0×10^{-8} M	1.0×10^{-19} M	For human serum albumin (HSA): 5%; For IgG: 3%	>70 assays during 2 months	[56]
Prostate specific antigen (PSA)	Capacitive biosensor with current pulse method	MAA; EGDMA	2.0×10^{-17} M–2.0×10^{-10} M	16×10^{-17} M	Selectivity coefficient (k) = 2.27 for HSA, k = 2.02 for IgG	About same level during 50 injections	[58]
E. coli	Capacitive biosensor with current pulse method	HEMA; (2-Hydroxyethyl methacrylate), N-methacryloyl-L-histidine methyl ester (MAH), EGDMA	1.0×10^{2}–1.0×10^{7} CFU·mL^{-1}	70 CFU·mL^{-1}	K = 3.14 for *B. subtilis*, k = 3.32 for *S. aureus*, k = 2.98 for *S. paratyphi*	About same level during 70 injections	[59]
Trypsin	Capacitive biosensor with current pulse method	N-isopropylacrylamide (NIPAm), N,N-methylenebisacryl, amide (MBAAm), Acrylamide, Hydroxymethylacrylamide	1.0×10^{-13} M–1.0×10^{-7} M	3.0×10^{-13} M	K = 733.1 for chymotrypsin (chy), k = 10.56 for BSA, k = 6.50 for lysozyme (Lyz), k = 3.46 for cytochrome c (cyt c)	The loss in performance was about 2% after 80 analyses	[60]

Capacitive biosensors based on a potential pulse have been used in many applications [1,11,61,62] including microcontact imprinting as seen in Table 3. In this type of capacitive biosensors, a small potential pulse is applied to the working electrode and the capacitance is measured. This potentiometric pulse concept is sensitive to external electronic disturbances. Therefore, the system is prone to inaccurate measurements and poor baseline stability. The sharp potential pulse may induce damage to the surface of the working electrode after a while which results in decreasing response of the electrode and eventually to replacement of the working electrode with a new electrode [57]. Another alternative way is to use a current pulse method to measure the capacitance at the electrode/solution interface. Erlandsson et al. [57] developed a new concept to measure capacitance based on a constant current pulse to the biosensor transducer. In the basis of the principle, the system could be described as a simple resistor-capacitor (RC) circuit model. The schematic representation of the capacitance measurement via current pulse method is shown in Figure 6. The system consists of:

(1)   A current source;
(2)   An electro-chemical flow-cell which includes three electrodes: the working electrode which is a thin gold film coated with an insulating layer which functions as a bio-recognition layer to immobilize the ligand, the auxiliary and reference electrodes which are made from a platinum wire;
(3)   A potential differential amplifier; and
(4)   A processor which converts the analogue potential to digital signal.

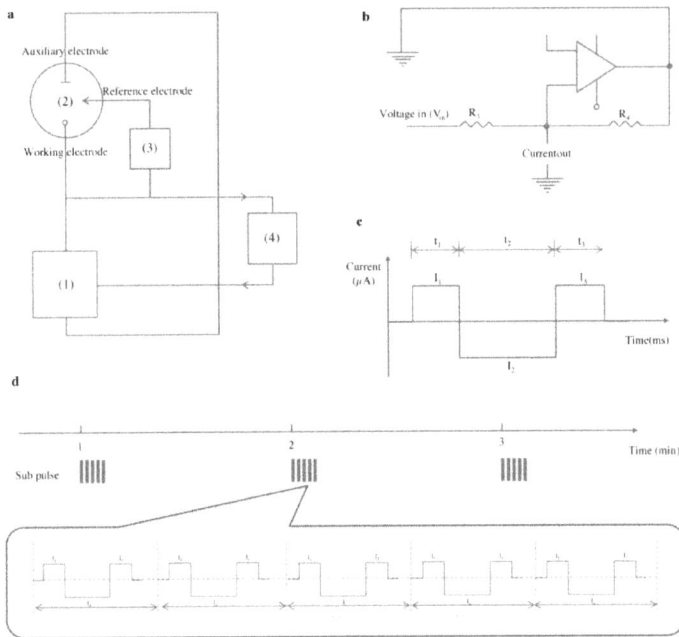

Figure 6. (**a**) Schematic representation of the capacitive system with current pulse method. The system is comprised of: (1) current source; (2) flow cell which is connected to the working, reference and auxiliary electrodes; (3) potential differential amplifier; and (4) a processor and ADC where the analogue potential is converted to digital signal; (**b**) A schematic view of Howland current pump used for supplying constant current; (**c**) Constant current supply to the sensor during the determined time periods to measure the resistance and capacitance; (**d**) Capacitance is measured every minute and each minute (pulse) contains five sub pulse measurements with 20 ms intervals. (Reproduced from reference [57] with permission).

When using current pulse method, the capacitive measurements were performed with an automated flow-injection system as shown in Figure 7 [63].

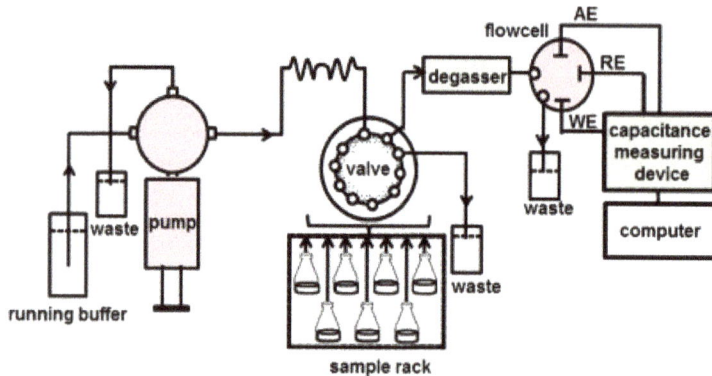

Figure 7. Schematic representation of automated flow injection capacitive system. The components shown in the figure are integrated into a box to make a single, portable unit. (Reproduced from Reference [63] with permission).

In their next study, Ertürk et al. [58] used microcontact imprinting method for detection of an important biomarker, prostate specific antigen (PSA), for early detection of prostate cancer with capacitive biosensors. The standard solutions of different concentrations of PSA (2.0×10^{-17} M–2.0×10^{-10} M) were analysed by the system and the LOD value was calculated as 16×10^{-15} M. HSA and IgG were used as the competing proteins in order to test the selectivity of the system for PSA. These proteins are normally found in the levels of $mg \cdot mL^{-1}$ where PSA is found at approximately $4.0 \ ng \cdot mL^{-1}$ in human serum. Therefore, when the developed system is tested against proteins at concentrations of 1 mg/mL which are normally found in one-million-fold higher concentrations, the system showed around two times more selectivity for PSA compared to HSA and IgG. In the next step, the sensitivity and the selectivity of the MIP system were compared with the performance of the system based on immobilized Anti-PSA antibody. The LOD was determined to be 12×10^{-14} M with the Anti-PSA system. The results showed that the MIP capacitive system was very promising to detect biomarkers which are important for diagnosis of various diseases. The authors compared the sensitivity of capacitive system for PSA detection with the microcontact imprinted surface plasmon resonance (SPR) system [64]. The LOD value was calculated around 91 $pg \cdot mL^{-1}$ (18×10^{-14} M) with the SPR biosensors. This result proves that the capacitive system is approximately 1000 times more sensitive than the SPR system.

Microcontact imprinting method was used for *E. coli* detection by Idil et al. [59]. The authors used a histidine containing specific monomer (N-methacryloyl-amido-histidine, MAH) as a metal chelating ligand. By using a meal-chelate between MAH and copper(II) (MAH-Cu^{2+}), an enhanced selectivity against certain amino acid residues present on the cell wall of *E. coli* was achieved. N-(hydroxyethyl)methacrylate (HEMA) was used as a functional monomer to make a complex with MAH-Cu^{2+} (pHEMA-MAH-Cu^{2+}) and EGDMA was used as cross-linker. The dynamic range for *E. coli* detection was between 1.0×10^2 and 1.0×10^7 $CFU \cdot mL^{-1}$ with a LOD value of 70 $CFU \cdot mL^{-1}$. *Bacillus subtilis*, *Staphylococcus aureus* and *Salmonella paratyphi* strains were used in selectivity experiments as competing strains. The cross-reactivity ratios were between 24% and 58% against pre-mixed suspensions of all bacterial strains. Even though the system showed cross-reactivity against competing strains, the ratio was negligible compared to the response of the system against *E. coli*. River water and apple juice were used to show the detection performance of the system from complex, real samples. Recovery value was found between 81% and 97% for *E. coli* detection from *E. coli* spiked (1.0×10^2 to

1.0×10^4 CFU·mL^{-1}) river water samples. The sensor had potential to monitor *E. coli* in contaminated water or food supplies.

Ertürk et al. [60] used microcontact imprinting method to develop capacitive biosensor for trypsin detection. A monomer solution containing hydroxymethylacrylamide, N-isopropylacrylamide (NIPAm), acrylamide and N,N-methylenebisacrylamide was prepared and N,N,N′,N′-tetramethylethylenediamine (TEMED, 5%, *v/v*) and ammonium persulfate (APS, 10%, *v/v*) were added into this solution. When the modified gold electrode was treated with monomer solution and brought into contact with the protein stamp which was carrying immobilized trypsin on top, the polymerization continued for 3–5 h at room temperature. Schematic representation of microcontact imprinting of trypsin on capacitive electrodes is shown in Figure 8. The dynamic range for trypsin detection was between 1.0×10^{-13} M and 1.0×10^{-7} M with a LOD value of 3.0×10^{-13} M. In order to test the selectivity and cross-reactivity of trypsin imprinted capacitive system for trypsin (MW 23.3 kDa, isoelectric point (pI): 10.1–10.5), chymotrypsin (chy) (MW: 25.6 kDa, pI 8.3), bovine serum albumin (BSA) (MW: 66.5 kDa, pI 4.7), lysozyme (lyz) (MW 14.3 kDa, pI 11.35) and cytochrome c (cyt c) (MW: 12.3 kDa, pI 10.0–10.5) were selected as competing agents. The concentrations of the proteins in the selectivity and cross-reactivity experiments were 1.0 mg·mL^{-1}. If it was tested with the lower concentrations including 1.0 µg·mL^{-1} or 1.0 pg·mL^{-1}, the selectivity results would be better because it would not be possible to detect interfering proteins in these concentration levels. Very low affinity of the system towards chy shows that the system can be used successfully for trypsin detection from pancreatic secretions where chy and try are found together. When the re-usability of the system was tested by monitoring the change in capacitance (-pF·cm^{-2}) at the same concentration of trypsin (10^{-7} M), the loss in the detection performance was around 2% after 80 analyses and there was not any significant difference in the performance after storage for two months at 4 °C. In the last step, trypsin activity measured by capacitive system was compared with the trypsin activity measured by spectrophotometer at 410 nm. One unit of enzyme was defined as the amount of enzyme catalyzing the conversion of one micromole of substrate, N$_\alpha$-benzoyl-DL-arginine 4-nitroanilide hydrochloride (BAPNA), at 25 °C, pH: 8.1 per minute. The trypsin activity measured spectrophotometrically was 9 mU·mL^{-1} where the value was around 8.0 mU·mL^{-1} measured by capacitive system. The results showed that there was correlation between two methods. The advantages of the developed method including detection of trypsin in 20 min, high selectivity towards interfering proteins, high correlation with the spectrophotometer show that the system might be used successfully for detection of proteases and as a point of care system for diagnosis of pancreatic diseases.

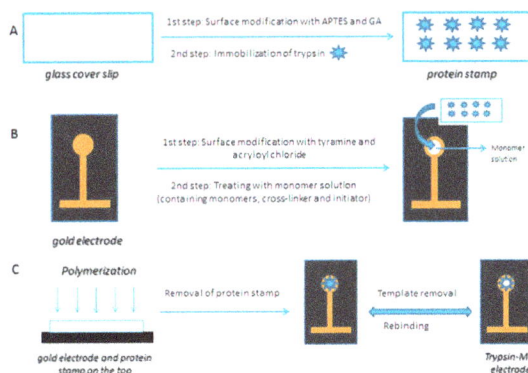

Figure 8. Schematic representation of preparation of trypsin imprinted capacitive electrodes using microcontact imprinting procedure: (**A**) preparation of glass cover slips (protein stamps); (**B**) preparation of capacitive gold electrodes; and (**C**) imprinting of trypsin to the electrode surface via microcontact imprinting method. (Reproduced from Reference [60] with permission).

7. Concluding Remarks

From the above, it is obvious that the combination of capacitive biosensors and MIPs offers great possibilities, both with regard to sensitive and selective assays as well as to stable measurements of extended periods of time. Sensitivity is now at a level where there are no needs to struggle for even more sensitive assays. What could be improved might be selectivity. It should however be mentioned that, with microcontact MIPs, higher selectivity was obtained as compared to that of monoclonal antibodies.

Cells and other particulate targets are still a bit difficult to assay. The number of CFU/mL that can be detected is far better than what can be reached with many of the methods used today. However, selectivity is too low and needs to be improved.

Conflicts of Interest: The authors declare no conflict of interest.

References

1. Berggren, C.; Bjarnason, B.; Johansson, G. Capacitive biosensors. *Electroanalysis* **2001**, *13*, 173–180. [CrossRef]
2. Waggoner, P.S.; Craighead, H.G. Micro- and nanomechanical sensors for environmental, chemical, and biological detection. *Lab Chip* **2007**, *7*, 1238–1255. [CrossRef] [PubMed]
3. Daniels, J.S.; Pourmand, N. Label-free impedance biosensors: Opportunities and challenges. *Electroanalysis* **2007**, *19*, 1239–1257. [CrossRef] [PubMed]
4. Tsouti, V.; Boutopoulos, C.; Zergioti, I.; Chatzandroulis, S. Capacitive microsystems for biological sensing. *Biosens. Bioelectron.* **2011**, *27*, 1–11. [CrossRef] [PubMed]
5. Mattiasson, B.; Teeparuksapun, K.; Hedström, M. Immunochemical binding assays for detection and quantification of trace impurities in biotechnological production. *Trends Biotechnol.* **2010**, *28*, 20–27. [CrossRef] [PubMed]
6. Korotkaya, E.V. Biosensors: Design, classification and applications in the food industry. *Food Raw Mater.* **2014**, *2*, 161–171. [CrossRef]
7. Prodromidis, M.I. Impedimetric immunosensors—A review. *Electrochim. Acta* **2010**, *55*, 4227–4233. [CrossRef]
8. Mattiasson, B.; Hedström, M. Capacitive biosensors for ultra-sensitive assays. *TrAC Trends Anal. Chem.* **2016**, *79*, 233–238. [CrossRef]
9. Labib, M.; Hedström, M.; Amin, M.; Mattiasson, B. A capacitive immunosensor for detection of cholera toxin. *Anal. Chim. Acta* **2009**, *634*, 255–261. [CrossRef] [PubMed]
10. Loyprasert, S.; Hedström, M.; Thavarungkul, P.; Kanatharana, P.; Mattiasson, B. Sub-attomolar detection of cholera toxin using a label-free capacitive immunosensor. *Biosens. Bioelectron.* **2010**, *25*, 1977–1983. [CrossRef] [PubMed]
11. Teeparuksapun, K.; Hedström, M.; Wong, E.Y.; Tang, S.; Hewlett, I.K.; Mattiasson, B. Ultrasensitive detection of HIV-1 p24 antigen using nanofunctionalized surfaces in a capacitive immunosensor. *Anal. Chem.* **2010**, *82*, 8406–8411. [CrossRef] [PubMed]
12. Qureshi, A.; Gurbuz, Y.; Niazi, J.H. Capacitive aptamer–antibody based sandwich assay for the detection of VEGF cancer biomarker in serum. *Sens. Actuators B Chem.* **2015**, *209*, 645–651. [CrossRef]
13. Mahadhy, A. Development of an Ultrasensitive Capacitive DNA-Sensor: A Promising Tool towards Microbial Diagnostics. Ph.D. Thesis, Lund University, Lund, Sweden, 2015.
14. Tsekenis, G.; Chatzipetrou, M.; Tanner, J.; Chatzandroulis, S.; Thanos, D.; Tsoukalas, D.; Zergioti, I. Surface functionalization studies and direct laser printing of oligonucleotides toward the fabrication of a micromembrane DNA capacitive biosensor. *Sens. Actuators B Chem.* **2012**, *175*, 123–131. [CrossRef]
15. Park, J.W.; Kallempudi, S.S.; Niazi, J.H.; Gurbuz, Y.; Youn, B.-S.; Gu, M.B. Rapid and sensitive detection of Nampt (PBEF/visfatin) in human serum using an ssDNA aptamer-based capacitive biosensor. *Biosens. Bioelectron.* **2012**, *38*, 233–238. [CrossRef] [PubMed]
16. Thipmanee, O.; Samanman, S.; Sankoh, S.; Numnuam, A.; Limbut, W.; Kanatharana, P.; Vilaivan, T.; Thavarungkul, P. Label-free capacitive DNA sensor using immobilized pyrrolidinyl PNA probe: effect of the length and terminating head group of the blocking thiols. *Biosens. Bioelectron.* **2012**, *38*, 430–435. [CrossRef] [PubMed]

17. Jantra, J.; Kanatharana, P.; Asawatreratanakul, P.; Hedström, M.; Mattiasson, B.; Thavarungkul, P. Real-time label-free affinity biosensors for enumeration of total bacteria based on immobilized concanavalin A. *J. Environ. Sci. Health Part A* **2011**, *46*, 1450–1460. [CrossRef] [PubMed]

18. Qureshi, A.; Pandey, A.; Chouhan, R.S.; Gurbuz, Y.; Niazi, J.H. Whole-cell based label-free capacitive biosensor for rapid nanosize-dependent toxicity detection. *Biosens. Bioelectron.* **2015**, *67*, 100–106. [CrossRef] [PubMed]

19. Bontidean, I.; Lloyd, J.R.; Hobman, J.L.; Wilson, J.R.; Csöregi, E.; Mattiasson, B.; Brown, N.L. Bacterial metal-resistance proteins and their use in biosensors for the detection of bioavailable heavy metals. *J. Inorg. Biochem.* **2000**, *79*, 225–229. [CrossRef]

20. Corbisier, P.; van der Lelie, D.; Borremans, B.; Provoost, A.; de Lorenzo, V.; Brown, N.L.; Lloyd, J.R.; Hobman, J.L.; Csöregi, E.; Johansson, G. Whole cell-and protein-based biosensors for the detection of bioavailable heavy metals in environmental samples. *Anal. Chim. Acta* **1999**, *387*, 235–244. [CrossRef]

21. Labib, M.; Hedström, M.; Amin, M.; Mattiasson, B. A novel competitive capacitive glucose biosensor based on concanavalin A-labeled nanogold colloids assembled on a polytyramine-modified gold electrode. *Anal. Chim. Acta* **2010**, *659*, 194–200. [CrossRef] [PubMed]

22. Labib, M.; Hedström, M.; Amin, M.; Mattiasson, B. Competitive capacitive biosensing technique (CCBT): A novel technique for monitoring low molecular mass analytes using glucose assay as a model study. *Anal. Bioanal. Chem.* **2010**, *397*, 1217–1224. [CrossRef] [PubMed]

23. Lenain, P.; de Saeger, S.; Mattiasson, B.; Hedström, M. Affinity sensor based on immobilized molecular imprinted synthetic recognition elements. *Biosens. Bioelectron.* **2015**, *69*, 34–39. [CrossRef] [PubMed]

24. Gutierrez, A.; Hedström, M.; Mattiasson, B. Bioimprinting as a tool for the detection of aflatoxin B1 using a capacitive biosensor. *Biotechnol. Rep.* **2016**, *11*, 12–17. [CrossRef]

25. Bougrini, M.; Baraket, A.; Jamshaid, T.; El Aissari, A.; Bausells, J.; Zabala, M.; El Bari, N.; Bouchikhi, B.; Jaffrezic-Renault, N.; Abdelhamid, E. Development of a novel capacitance electrochemical biosensor based on silicon nitride for ochratoxin A detection. *Sens. Actuators B Chem.* **2016**, *234*, 446–452. [CrossRef]

26. Mahadhy, A.; Ståhl-Wernersson, E.; Mattiasson, B.; Hedström, M. Use of a capacitive affinity biosensor for sensitive and selective detection and quantification of DNA—A model study. *Biotechnol. Rep.* **2014**, *3*, 42–48. [CrossRef]

27. Rydosz, A.; Brzozowska, E.; Gorska, S.; Wincza, K.; Gamian, A.; Gruszczynski, S. A broadband capacitive sensing method for label-free bacterial LPS detection. *Biosens. Bioelectron.* **2016**, *75*, 328–336. [CrossRef] [PubMed]

28. Brzozowska, E.; Smietana, M.; Koba, M.; Gorska, S.; Pawlik, K.; Gamian, A.; Bock, W.J. Recognition of bacterial lipopolysaccharide using bacteriophage-adhesin-coated long-period gratings. *Biosens. Bioelectron.* **2015**, *67*, 93–99. [CrossRef] [PubMed]

29. Rocha, A.M.; Yuan, Q.; Close, D.M.; O'Dell, K.B.; Fortney, J.L.; Wu, J.; Hazen, T.C. Rapid Detection of Microbial Cell Abundance in Aquatic Systems. *Biosens. Bioelectron.* **2016**, *85*, 915–923. [CrossRef] [PubMed]

30. Bontidean, I.; Berggren, C.; Johansson, G.; Csöregi, E.; Mattiasson, B.; Lloyd, J.R.; Jakeman, K.J.; Brown, N.L. Detection of heavy metal ions at femtomolar levels using protein-based biosensors. *Anal. Chem.* **1998**, *70*, 4162–4169. [CrossRef] [PubMed]

31. Tekin, K.; Uzun, L.; Şahin, Ç.A.; Bektaş, S.; Denizli, A. Preparation and characterization of composite cryogels containing imidazole group and use in heavy metal removal. *React. Funct. Polym.* **2011**, *71*, 985–993. [CrossRef]

32. Say, R.; Birlik, E.; Ersöz, A.; Yılmaz, F.; Gedikbey, T.; Denizli, A. Preconcentration of copper on ion-selective imprinted polymer microbeads. *Anal. Chim. Acta* **2003**, *480*, 251–258. [CrossRef]

33. Ersöz, A.; Say, R.; Denizli, A. Ni(II) ion-imprinted solid-phase extraction and preconcentration in aqueous solutions by packed-bed columns. *Anal. Chim. Acta* **2004**, *502*, 91–97. [CrossRef]

34. Birlik, E.; Ersöz, A.; Açıkkalp, E.; Denizli, A.; Say, R. Cr(III)-imprinted polymeric beads: Sorption and preconcentration studies. *J. Hazard. Mater.* **2007**, *140*, 110–116. [CrossRef] [PubMed]

35. Esen, C.; Andac, M.; Bereli, N.; Say, R.; Henden, E.; Denizli, A. Highly selective ion-imprinted particles for solid-phase extraction of Pb 2+ ions. *Mater. Sci. Eng. C* **2009**, *29*, 2464–2470. [CrossRef]

36. Say, R.; Ersöz, A.; Denizli, A. Selective separation of uranium containing glutamic acid molecular-imprinted polymeric microbeads. *Sep. Sci. Technol.* **2003**, *38*, 3431–3447. [CrossRef]

37. Lee, H.J.; Lee, J.H.; Jung, H.I. A symmetric metamaterial element-based RF biosensor for rapid and label-free detection. *Appl. Phys. Lett.* **2011**, *99*, 163703. [CrossRef]

38. Kim, N.Y.; Dhakal, R.; Adhikari, K.K.; Kim, E.S.; Wang, C. A reusable robust radio frequency biosensor using microwave resonator by integrated passive device technology for quantitative detection of glucose level. *Biosens. Bioelectron.* **2015**, *67*, 687–693. [CrossRef] [PubMed]

39. Altunbaş, C.; Uygun, M.; Uygun, D.A.; Akgöl, S.; Denizli, A. Immobilization of inulinase on concanavalin A-attached super macroporous cryogel for production of high-fructose syrup. *Appl. Biochem. Biotechnol.* **2013**, *170*, 1909–1921. [CrossRef] [PubMed]

40. Perçin, I.; Yavuz, H.; Aksöz, E.; Denizli, A. Mannose-specific lectin isolation from *Canavalia ensiformis* seeds by PHEMA-based cryogel. *Biotechnol. Prog.* **2012**, *28*, 756–761. [CrossRef] [PubMed]

41. Sarı, M.M.; Armutcu, C.; Bereli, N.; Uzun, L.; Denizli, A. Monosize microbeads for pseudo-affinity adsorption of human insulin. *Colloids Surf. B Biointerfaces* **2011**, *84*, 140–147. [CrossRef] [PubMed]

42. Çorman, M.E.; Armutcu, C.; Uzun, L.; Denizli, A. Rapid, efficient and selective preconcentration of benzo [a] pyrene (BaP) by molecularly imprinted composite cartridge and HPLC. *Mater. Sci. Eng. C* **2017**, *70*, 41–53. [CrossRef] [PubMed]

43. Dursun, E.M.; Üzek, R.; Bereli, N.; Şenel, S.; Denizli, A. Synthesis of novel monolithic cartridges with specific recognition sites for extraction of melamine. *React. Funct. Polym.* **2016**, *109*, 33–41. [CrossRef]

44. Inanan, T.; Tüzmen, N.; Akgöl, S.; Denizli, A. Selective cholesterol adsorption by molecular imprinted polymeric nanospheres and application to GIMS. *Int. J. Biol. Macromol.* **2016**, *92*, 451–460. [CrossRef] [PubMed]

45. Göktürk, I.; Perçin, I.; Denizli, A. Catalase purification from rat liver with iron chelated poly (hydroxyethyl methacrylate-N-methacryloyl-(l)-glutamic acid) cryogel discs. *Prep. Biochem. Biotechnol.* **2015**, *46*, 602–609. [CrossRef] [PubMed]

46. Akgönüllü, S.; Yavuz, H.; Denizli, A. Preparation of imprinted cryogel cartridge for chiral separation of l-phenylalanine. *Artif. Cells Nanomed. Biotechnol.* **2016**, 1–8. [CrossRef] [PubMed]

47. Aşır, S.; Derazshamshir, A.; Yilmaz, F.; Denizli, A. Triazine herbicide imprinted monolithic column for capillary electrochromatography. *Electrophoresis* **2015**, *36*, 2888–2895. [CrossRef] [PubMed]

48. Tamahkar, E.; Kutsal, T.; Denizli, A. Surface imprinted bacterial cellulose nanofibers for cytochrome c purification. *Process Biochem.* **2015**, *50*, 2289–2297. [CrossRef]

49. Wulff, G.; Sarhan, A. *Use of Polymers with Enzyme-Analogous Structures for Resolution of Racemates*, Angewandte Chemie-International ed.; Wiley-VCH Verlag Gmbh: Muhlenstrasse, Berlin, Germany, 1972; pp. 33–34.

50. Takagishi, T.; Klotz, I.M. Macromolecule-small molecule interactions; introduction of additional binding sites in polyethyleneimine by disulfide cross–linkages. *Biopolymers* **1972**, *11*, 483–491. [CrossRef] [PubMed]

51. Vlatakis, G.; Andersson, L.I.; Müller, R.; Mosbach, K. Drug assay using antibody mimics made by molecular imprinting. *Nature* **1993**, *361*, 645–647. [CrossRef] [PubMed]

52. Cieplak, M.; Kutner, W. Artificial Biosensors: How Can Molecular Imprinting Mimic Biorecognition? *Trends Biotechnol.* **2016**, *34*, 922–941. [CrossRef] [PubMed]

53. Chou, P.C.; Rick, J.; Chou, T.C. C-reactive protein thin-film molecularl imprinted polymers formed using a micro-contact approach. *Anal. Chim. Acta* **2005**, *542*, 20–25. [CrossRef]

54. Ertürk, G.; Mattiasson, B. From imprinting to microcontact imprinting—A new tool to increase selectivity in analytical devices. *J. Chromatogr. B* **2016**, *1021*, 30–44. [CrossRef] [PubMed]

55. Lin, H.Y.; Hsu, C.Y.; Thomas, J.L.; Wang, S.E.; Chen, H.C.; Chou, T.C. The microcontact imprinting of proteins: The effect of cross-linking monomers for lysozyme, ribonuclease A and myoglobin. *Biosens. Bioelectron.* **2006**, *22*, 534–543. [CrossRef] [PubMed]

56. Ertürk, G.; Berillo, D.; Hedström, M.; Mattiasson, B. Microcontact-BSA imprinted capacitive biosensor for real-time, sensitive and selective detection of BSA. *Biotechnol. Rep.* **2014**, *3*, 65–72. [CrossRef]

57. Erlandsson, D.; Teeparuksapun, K.; Mattiasson, B.; Hedström, M. Automated flow-injection immunosensor based on current pulse capacitive measurements. *Sens. Actuators B Chem.* **2014**, *190*, 295–304. [CrossRef]

58. Ertürk, G.; Hedström, M.; Tümer, M.A.; Denizli, A.; Mattiasson, B. Real-time prostate-specific antigen detection with prostate-specific antigen imprinted capacitive biosensors. *Anal. Chim. Acta* **2015**, *891*, 120–129. [CrossRef] [PubMed]

59. Idil, N.; Hedström, M.; Denizli, A.; Mattiasson, B. Whole cell based microcontact imprinted capacitive biosensor for the detection of *Escherichia coli*. *Biosens. Bioelectron.* **2017**, *87*, 807–815. [CrossRef] [PubMed]

60. Ertürk, G.; Hedström, M.; Mattiasson, B. A sensitive and real-time assay of trypsin by using molecular imprinting-based capacitive biosensor. *Biosens. Bioelectron.* **2016**, *86*, 557–565. [CrossRef] [PubMed]

61. Berggren, C.; Bjarnason, B.; Johansson, G. An immunological Interleukine-6 capacitive biosensor using perturbation with a potentiostatic step. *Biosens. Bioelectron.* **1998**, *13*, 1061–1068. [CrossRef]

62. Hedström, M.; Galaev, I.Y.; Mattiasson, B. Continuous measurements of a binding reaction using a capacitive biosensor. *Biosens. Bioelectron.* **2005**, *21*, 41–48. [CrossRef] [PubMed]

63. Lebogang, L.; Hedström, M.; Mattiasson, B. Development of a real-time capacitive biosensor for cyclic cyanotoxic peptides based on Adda-specific antibodies. *Anal. Chim. Acta* **2014**, *826*, 69–76. [CrossRef] [PubMed]

64. Ertürk, G.; Özen, H.; Tümer, M.A.; Mattiasson, B.; Denizli, A. Microcontact imprinting based surface plasmon resonance (SPR) biosensor for real-time and ultrasensitive detection of prostate specific antigen (PSA) from clinical samples. *Sens. Actuators B Chem.* **2016**, *224*, 823–832. [CrossRef]

sensors

MDPI

Review

Molecular Imprinting Technology in Quartz Crystal Microbalance (QCM) Sensors

Sibel Emir Diltemiz [1], Rüstem Keçili [2], Arzu Ersöz [1] and Rıdvan Say [1,3,*]

[1] Chemistry Department, Faculty of Science, Anadolu University, 26470 Eskisehir, Turkey; semir@anadolu.edu.tr (S.E.D.); arzuersoz@anadolu.edu.tr (A.E.)

[2] Department of Medical Services and Techniques, Yunus Emre Vocational School of Health Services, Anadolu University, 26470 Eskisehir, Turkey; rkecili@anadolu.edu.tr

[3] Bionkit Co. Ltd., 26470 Eskisehir, Turkey

[*] Correspondence: ridvansayy@gmail.com; Tel.: +90-533-411-7091

Academic Editors: Bo Mattiasson and Gizem Ertürk

Received: 15 December 2016; Accepted: 21 February 2017; Published: 24 February 2017

Abstract: Molecularly imprinted polymers (MIPs) as artificial antibodies have received considerable scientific attention in the past years in the field of (bio)sensors since they have unique features that distinguish them from natural antibodies such as robustness, multiple binding sites, low cost, facile preparation and high stability under extreme operation conditions (higher pH and temperature values, etc.). On the other hand, the Quartz Crystal Microbalance (QCM) is an analytical tool based on the measurement of small mass changes on the sensor surface. QCM sensors are practical and convenient monitoring tools because of their specificity, sensitivity, high accuracy, stability and reproducibility. QCM devices are highly suitable for converting the recognition process achieved using MIP-based memories into a sensor signal. Therefore, the combination of a QCM and MIPs as synthetic receptors enhances the sensitivity through MIP process-based multiplexed binding sites using size, 3D-shape and chemical function having molecular memories of the prepared sensor system toward the target compound to be detected. This review aims to highlight and summarize the recent progress and studies in the field of (bio)sensor systems based on QCMs combined with molecular imprinting technology.

Keywords: molecularly imprinted polymers (MIPs); quartz crystal microbalance (QCM); biosensors; biomolecular recognition; synthetic receptors

1. Introduction

Biomolecular recognition plays a crucial role in biological systems where the enzyme-substrate, DNA-protein and antibody-antigen interactions are carried out [1]. These interactions and binding phenomena are usually based on lock and key models where receptors and substrates specifically interact with each other. These specific interactions include non-covalent interactions such as hydrogen bonding, metal coordination, hydrophobic interactions, Van der Waals interactions, π-π interactions and electrostatic interactions [2]. This specific molecular recognition phenomenon is commonly used in biosensor applications. For this purpose, antibodies are used as the recognition elements in biosensors since they have high selectivity and sensitivity toward the target compound. However, antibodies display some drawbacks of such as high cost and low stability under extreme conditions (higher pH, temperature and pressure).

Molecularly imprinted polymers (MIPs) also called "artificial antibodies" can overcome these disadvantages of natural antibodies. MIPs are man-made artificial materials that show high affinity and selectivity toward a target compound ("template"). MIPs are prepared by polymerization of an appropriate functional monomer and a cross-linker in the presence of a template as schematically shown in Figure 1. Since MIPs are very selective toward the target compound, they are commonly used as recognition elements in the fabrication of biosensors. MIPs have the ability to bind target compounds not only by their 3-D shape, because the incorporation of specific binding groups into the selective cavities of a polymeric network enhances its affinity and selectivity toward the target analyte [3–10].

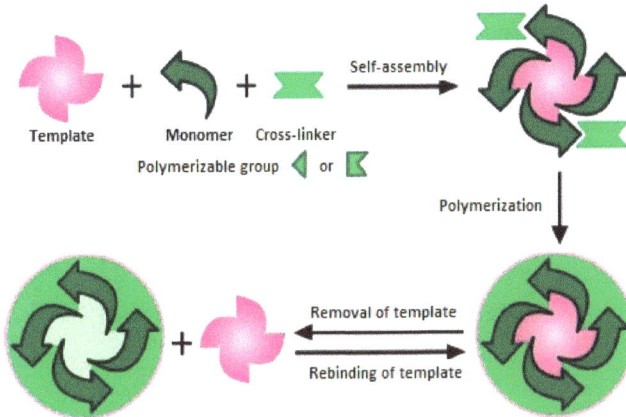

Figure 1. Schematic representation of molecular imprinting (reproduced with permission from [11]).

On the other hand, the Quartz Crystal Microbalance (QCM), is well suited as a transducer element for chemical sensors, being rapid, easy to use, highly stable and portable. Increased mass on the gold surface, associated with the binding reaction, results in a decrease of the frequency. QCM-based sensors have been used in the detection of several analytes even in very different matrix environments.

The combination of QCMs and MIPs can be carried out by two main approaches which are immobilization of pre-prepared MIP particles and in-situ polymerization. For the immobilization of MIP particles on the surface of QCM sensors, modification of the gold sensor surface with self-assembled monolayers composed of thiol-containing compounds such as 11-mercaptoundecanoic acid is performed in the first step and then MIP particles are attached to the modified sensor surface.

In in-situ polymerization, a chemical, thermal or photochemical initiator is used [12]. Photochemical initiation has some advantages such as easy control and polymerization at room temperature. In addition, MIP layers can also be formed by electropolymerization on the surface of the QCM sensor [13]. In this strategy, the thickness of the MIP layer can easily be controlled by changing the applied voltage [14]. However, there may be some difficulties such as adhesion problems during the washing step of the prepared MIP layer on the surface. Thus, some special pre-treatment processes should be applied to increase the adhesion of MIP layers on the surface of the sensor system. In this review, we provide an overview of the recent progress and applications of MIP based-QCM sensors.

2. MIP Based-Quartz Crystal Microbalance (QCM) Sensors

There are many reported studies on the applications of QCM sensors based on molecular imprinting technology. Some examples from the literature are briefly described in the following sections.

2.1. MIP-Based QCM Sensors for Biological Applications

In the last decade, one of the most promising technical applications based on the use of MIPs has been QCM. QCM sensors have been developed for the detection of various targets such as proteins (e.g., enzymes) and cells.

In one of the first reported studies performed by Dickert and Hayden [15], a MIP-based QCM sensor for selective recognition of yeast cells was prepared by surface imprinting (Figure 2). The prepared sensor exhibited recognition ability toward target yeast cells in growth media.

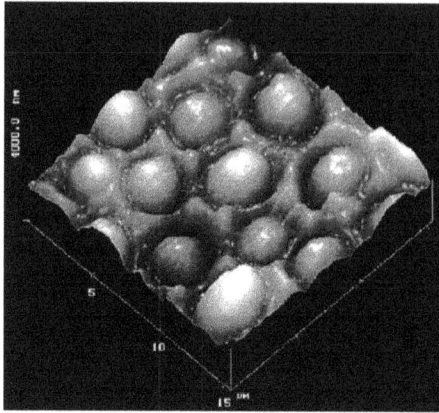

Figure 2. AFM image of a *Saccharomyces cerevisiae cell* imprinted sensor surface (reproduced with permission from [15]).

QCM sensors modified with selective MIP layers also have the ability to recognize their target compounds, even in complex biological samples such as blood. Dickert and Hayden have reported another interesting study [16]. They developed a QCM sensor coated with MIP for selective recognition of erythrocytes in blood samples. The functional monomer 1-vinyl-2-pyrrolidone and template erythrocytes were polymerized in the presence of the cross-linker N,N'-methylene-bis(acrylamide) under UV-light. Then, the prepared MIP-based QCM sensor was used for the recognition of erythrocytes. The results obtained from the performance tests showed that the prepared QCM sensor has affinity and selectivity toward erythrocytes.

Due to their size, viruses cannot be recognized using conventional optical techniques. Microbiological assays are therefore used for the detection of viruses. In this case, QCM sensors combined with MIPs prepared by surface imprinting can efficiently be used for selective detection of viruses. For example, Tai et al. have reported the first study of MIP-based QCM sensors for successful recognition of dengue virus [17]. In their study, a selective MIP for the recognition of nonstructural protein 1 was prepared on the surface of a QCM sensor. The prepared sensor was successfully applied for the recognition of nonstructural protein and the obtained results showed that the developed assay can be used for the recognition of various flaviviruses such as four different types of dengue viruses. The authors reported that additional experiments are needed to determine the diagnostic accuracy of the prepared MIP-based QCM sensor systems for the detection of acute phase dengue virus infection.

In another important study, a MIP film-coated QCM sensor for the detection of anthrax protective antigen was developed by the same researh group a few years later [18]. The results showed that the prepared MIP based-QCM sensor display high affinity toward the target epitope of anthrax protective antigen in a picomolar concentration range. This sensor platform is a fast and selective assay that can be efficiently applied for the detection of other bacterial antigens.

Liu et al. [19] have produced a QCM sensor for detection of staphylococcal enterotoxin B (SEB). This QCM sensor was coated by molecularly imprinted sol-gel thin film. They firstly mixed organosilanes with SEB and then this combination was coated on the sensor surface (Figure 3). Their results showed that the prepared sensor was successfully applied in the working range of 1.0×10^{-1} to 1.0×10^3 $\mu g \cdot mL^{-1}$ and detection limit was 6.1 $ng \cdot mL^{-1}$. Selectivity studies were done with staphylococcal enterotoxin A (SEA), staphylococcal enterotoxin C1 (SEC1), bovine serum albumin

(BSA) and ovalbumin (OVA) and the QCM sensor system exhibited high selectivity toward the templates' analogues. These results showed that MIP-QCM combination systems are very effective tools for the determination of SEB. The studies included a comparison between the MIP-QCM combination and immunochips for the detection of SEB. Thus, they developed an alternative method to expensive immunochip systems. In the light of the obtained data the authors stated that they had developed a simple, rapid, low cost and sensitive MIP-coated QCM sensor.

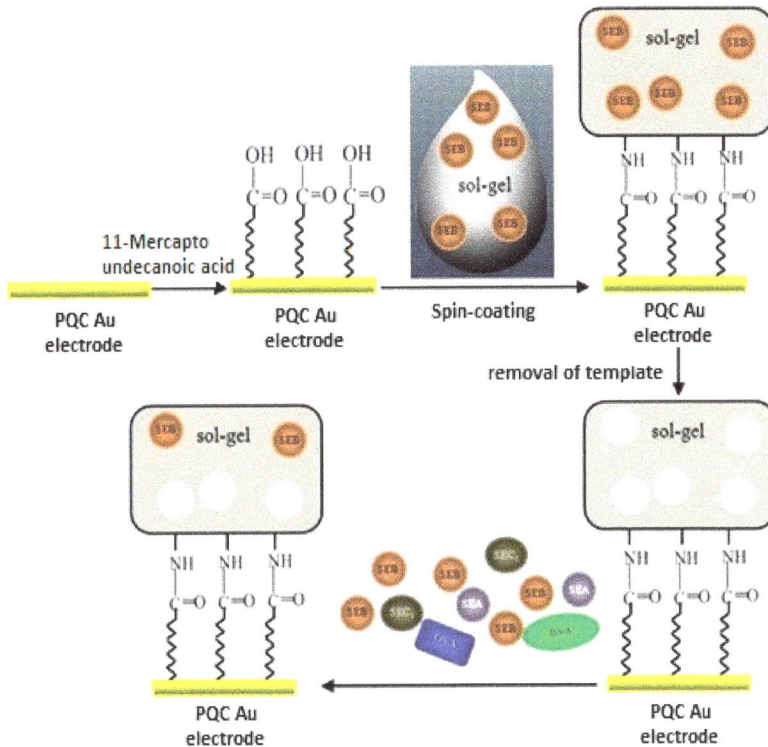

Figure 3. Schemes of the preparation of sol-gel imprinted thin film on the surface of the piezoelectric quartz crystal (PQC) Au-electrode for the detection of staphylococcal enterotoxin B (SEB) (reproduced with permission from [19]).

In another study, Lu et al. [20] used a biomimetic sensor system for the determination of glycoprotein 41, gp41. Glycoprotein 41 is a protein related with human immunodeficiency virus type 1 (HIV-1). In their study the epitope-imprinting technique was used. Figure 4 shows a schematic diagram of epitope-imprinting. They also used dopamine as the functional monomer; a peptide with 35 amino acid residues of gp41 as template molecule and their report was the first study to describe the application of polydopamine in epitope-imprinting. The QCM sensor surface was coated with the hydrophilic MIP film and it was seen that this film is very selective for the template peptide and also gp41 protein. The dissociation constant (Kd) was found to be 3.17 nM, a value is very close to that obtained with monoclonal antibodies. They also investigated the analytical performance of the sensor by the imprinting effect, selectivity and real sample analysis. The presented study is very important in terms of biomolecule analysis in that it shows that small peptide groups of large biomolecule structures can be used as template molecules.

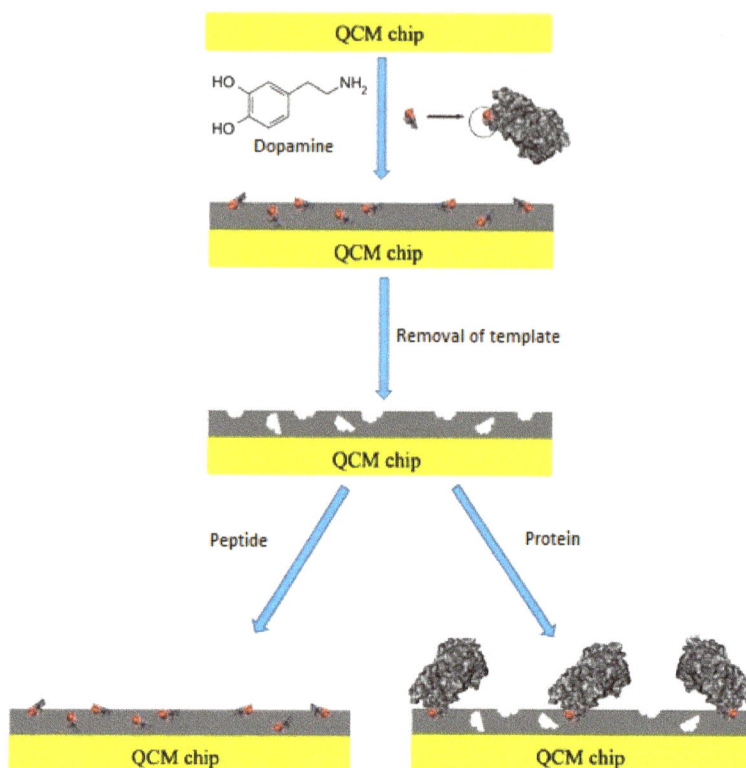

Figure 4. Schematic representation of epitope-imprinting (reproduced with permission from [20]).

Liu et al. [21] have developed MIP-QCM-based sensors for the detection of staphylococcal enterotoxins (SEs), the cause of the most common type of food poisoning known as staphylococcal food poisoning. They used real substances for the determination of SEA and SEB. Unlike their previous study, they used QCM-D. They demonstrated that, QCM-D is more stable than QCM 200 for detection of SEA and SEB from real substances. QCM-D device is more advantageous due to the fact that the dissipation shift (ΔD) can also be measured. TEOS, APTES, OTES and template molecule- containing sol gel was coated on a QCM electrode by a spin coater. The prepared electrode was evaluated by repeated measurements of spiked milk samples. This study showed that MIP-based QCM sensors have very beneficial properties when applied in different areas such as food safety, environmental and biological applications.

Another QCM-MIP based study was published by Phan et al. [22] in 2014. This study is very important because the authors investigated the effects of changes in cross-linker and monomer ratio on the sensitivity and response time. They have also examined the effects of different polymer preparation mechanisms such as stamp imprinting, bulk imprinting and solvent effect. For this purpose an albumin imprinted polymer was prepared on a QCM sensor. Studies have shown that a more hydrophilic polymer is obtained when acrylic acid is used instead of methacrylic acid, that when the amount of cross-linker is increased, the sensor response time is reduced and the effect of solvent change is very small. The study is very useful in terms of examining the different parameters.

In 2015, El-Sharif et al. [23] studied different acrylic amides (acrylamide, AA; *N*-hydroxymethyl-acrylamide, NHMA; *N*-isopropylacrylamide, NiPAm) and showed their effects on MIP selectivity. Their studies are composed of two parts: spectrophotometric and QCM sensor systema. The selectivity of protein-specific MIP and the control NIP was also compared towards the template molecule bovine haemoglobin (BHb). One other important result is that they achieve more selectivity and recognition capacity through the use of hydrophilic NHMA.

Boronic acids can reversibly interact with diols, α-hydroxyacids and α-amino alcohols. Due to their high binding ability toward diol-containing compounds through this feature, they are commonly used as the recognition unit of the sensor systems for the detection of carbohydrates. Immunoglobulin M (IgM) has high mannose content. Considering this, Diltemiz et al. [24] developed mannose-imprinted QCM sensors for the selective detection of mannose and IgM. The synthesis of the functional monomer methacryloylamidophenylboronic acid (MAPBA) and its use for the preparation of MIPs were performed for the first time in the literature. For the MIP film preparation on the sensor surface, a MAPBA-mannose pre-organized system was prepared. The QCM electrode surface was coated with a molecularly imprinted film by a spin-coater and the polymerization was completed under UV light. The binding affinity was evaluated by using the Langmuir isotherm and the prepared electrode has high affinity toward the analyte. This study also showed that the imprinting of mannose has the capability of determining IgM.

Latif et al. [25] used a bulk imprinting technique for the detection of the estradiols which are a kind of endocrine disrupting chemicals (EDCs). To achieve highly sensitive and selective detection even in the presence of very similar compounds that have nearly same structure, they used 17β-estradiol (E2) as a template. The interesting point of this study was the successful imprinting of bacterial cells. Therefore, both small molecules and analytes such as bacteria with their larger size could be imprinted on QCM electrode via a polyurethane layer.

L-Nicotine was also used as a template molecule in MIP-QCM studies in the literature. Alenus et al. have published two different studies on the detection of L-nicotine. One of the studies is in aqueous solution [26] while the other one is in a biological sample [27]. Although many researchers have studied L-nicotine template molecules in aqueous solution, Alenus et al. studied the detection of L-nicotine in saliva and urine. They used bulk-polymerized L-nicotine MIPs for this purpose in both studies. The monomer solutions were prepared using MAA as the functional monomer, EDMA as the cross-linker, AIBN as the initiator and L-nicotine as the template molecule, then these solutions were polymerized at 60 °C for 72 h. The obtained polymers were ground, washed to extract excess L-nicotine and then coated on a quartz crystal microbalance-dissipation (QCM-D) electrode with PVC. They demonstrated that MIPs bind 4.03 times more L-nicotine than NIPs in water and 1.99 times more in PBS at pH 9. L-Nicotine-spiked saliva and urine samples were diluted in water and PBS solution. L-Nicotine was successfully determined in samples of patients' saliva after using nicotine gums and smokeless tobacco. This study shows that even though the L-nicotine concentration is in the micromolar range, it could be detected directly in saliva and urine samples.

Another study for the determination of L-nicotine was accomplished by Croux et al. [28]. They used a different approach and application with a MIP-based QCM sensor and prepared a multi-channel system. Their four channel system contains two NIP/MIP pairs. The mixture of MAA, EGDM, AIBN and template molecule L-nicotine was placed on the QCM electrode via a PDMS mold (Figure 5). Finite element analysis (FEA) was used for this study to investigate the flow inside the system. In this way turbulence causing concentration problems was prevented.

Electropolymerization is another crucial approach for the preparation of MIPs on the surface of QCM sensors. In situ polymerization strategies can be used for this purpose. The first attempts using phenolic functional monomers were carried out to investigate the effect of electropolymerization on MIP film preparation, but the obtained MIP film layers were thick and non-specific interactions had to be blocked using other compounds [29,30]. Considerable progress has been achieved in this area by depositing thin film layers at a certain point of the sensor surface [31]. One of the main advantages of the electropolymerization technique is that the thickness of the MIP film layer can be changed by applying different voltage values. Lenain and co-workers [32] first prepared an electrode composed of molecularly imprinted sub-micron spherical particles for the recognition of metergoline. Hybrid structures can also be prepared by using MIPs with proteins or a self-assembled monolayer. The first combination of an enzyme with a MIP film on the sensor surface for the detection of electroactive proteins was reported by the research group of Yarman [33].

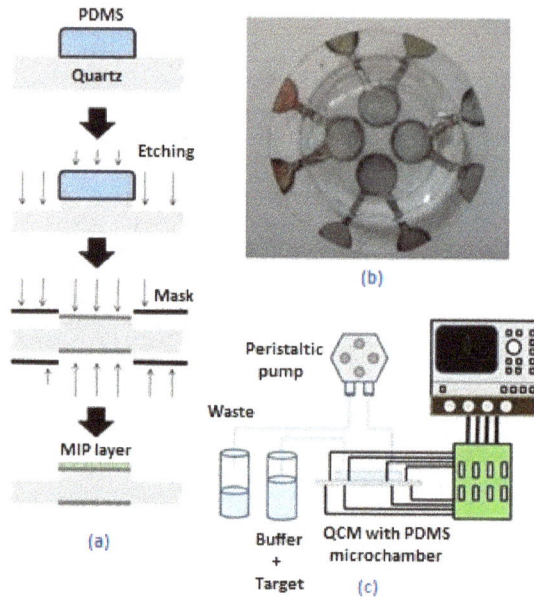

Figure 5. (**a**) Schematic demonstration of MIP-based QCM sensor; (**b**) Prepared MIP based-QCM sensor with multi-channel; (**c**) Schematic depiction of the prepared sensor system combined with impedance analyzer (reproduced with permission from [28]).

Apodaca et al. [34] have developed a MIP-based QCM sensor for the selective recognition of folic acid. MIP film using bisterthiophene dendrons and folic acid as the template was prepared by electropolymerization on the sensor surface. The analytical performance of the prepared sensor was investigated in the presence of pteroic acid, caffeine and theophylline. The obtained results showed that the prepared MIP-based QCM sensor exhibited high affinity and selectivity toward target compound folic acid. The detection limit was found to be as 15.4 μM.

It is also possible to detect target proteins in biological samples by using MIP-based QCM sensors. If the template is a large biomolecule such as an enzyme or protein, the surface imprinting strategy is used. In another remarkable study, a MIP-based QCM sensor selective to ribonuclease A was developed by Liu and his research group [35].

In their previous attempts, they prepared calcium carbonate nanoparticles as the porogen for the preparation of MIP thin layers and the obtained results showed that the synthesized polymeric film exhibited high porous effects. Considering this, a MIP film responsive to ribonuclease A was prepared on a gold sensor surface using a surface imprinting approach by polymerization of ribonuclease A template and $CaCO_3$ nanoparticles in the presence of methacryloylamidohistidine (MAH) and trimethylolpropane trimethacrylate (TRIM) as the functional monomer and cross-linker, respectively. Figure 6 shows the schematic representation of the prepared QCM sensor coated with ribonuclease A MIP film. The results obtained from the experiments on the performance of the prepared QCM sensor toward ribonuclease A showed that the prepared sensor exhibits high selectivity toward the target protein ribonuclease A in the presence of lysozyme as the control protein. The selectivity factor was calculated to be 3.3 at the protein concentration of $1-10^{-6}$ g·mL^{-1}.

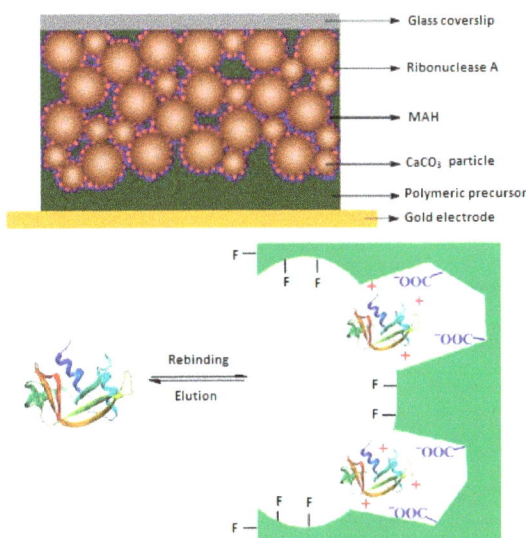

Figure 6. Preparation of MIP film coated QCM sensor for Ribonuclease A (Reproduced with permission from Reference [35]).

The chelating properties of various metal ions with the desired compounds are also used for the preparation of the selective MIPs. Considering this feature, MIPs prepared by using metal-chelate functional monomers were designed for the construction of MIP-based QCM sensors toward biological compounds in real samples. For example, Ersöz et al. [36] have prepared a QCM sensor composed of 8-hydroxy-2'-deoxyguanosine (8-OHdG) imprinted film for the detection of 8-OHdG levels in biological samples. For this purpose, MIP film selective to 8-OHdG was synthesized by using a photo-graft surface polymerization technique. Methacryloylamidoantipyrine-iron, N-N'-methylenebisacrylamide and 8-OHdG were used as the complex functional monomer, cross-linker and template, respectively. The preparation of a MIP-based QCM sensor toward 8-OHdG is schematically represented in Figure 7. The results obtained from the performance studies of the prepared sensor showed that the prepared MIP-based QCM sensor displays high affinity and selectivity toward 8-OHdG. The affinity constant (K_a) of the sensor was found to be 48,510 M^{-1}.

In a study reported by Lee et al. [37] MIP film-coated QCM sensors were used for the selective recognition of lipase, amylase and lysozyme which are digestive enzymes found in saliva. For this purpose, the sensor surface was coated with a mixture of target protein and poly(ethylene-co-vinyl alcohol). Then, a MIP thin film was prepared applying the thermally induced phase separation approach. The prepared sensors were applied to recognize lipase, amylase and lysozyme in real samples. The obtained results showed that the prepared MIP-based QCM sensors have the ability to recognize target proteins at lower concentrations and the limit of detection values were found to be 7 pM, 2.5 pM and 3.5 pM for lysozyme, lipase and amylase, respectively.

A QCM sensor coated with lysozyme imprinted macroporous film was developed by Zhou and co-workers [38]. In their study, methyl methacrylate (MMA) and TRIM were used as the functional monomer and cross-linker, respectively. In addition, $CaCO_3$ nanoparticles were used as porogen to form interconnected macropores in the MIP film on the QCM sensor surface (Figure 8). The results of experiments for the selective recognition of target protein lysozyme by using the prepared MIP-based QCM sensor showed that the prepared sensor displays high affinity and selectivity toward lysozyme in the presence of myoglobin. The selectivity factors were calculated as 3.1, 4.0 and 3.6 at protein concentrations of 5×10^{-5} g·mL^{-1}, 20×10^{-5} g·mL^{-1} and 50×10^{-5} g·mL^{-1}, respectively.

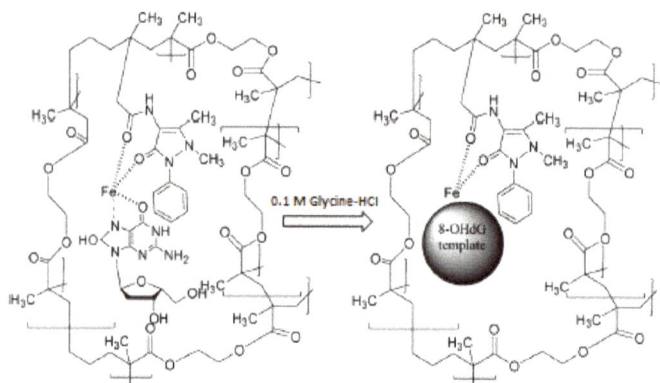

Figure 7. Preparation of MIP for 8-OHdG (reproduced with permission from [36]).

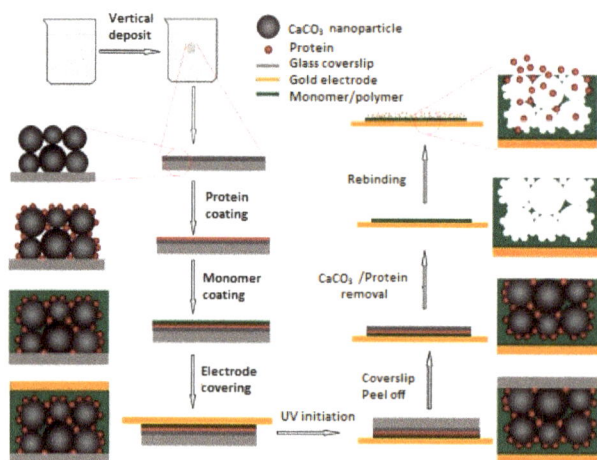

Figure 8. Preparation of QCM sensor toward lysozyme (reproduced with permission from [38]).

Mirmohseni et al. [39] have reported the preparation of a MIP-based QCM sensor for the selective detection of phenylalanine. MIP film sensitive toward phenylalanine on the surface of the sensor was synthesized by co-polymerization of acrylonitrile (AN) and acrylic acid (AA) [poly(AN-co-AA)] in the presence of the target amino acid phenylalanine. The prepared MIP-based QCM sensor exhibited a linear response toward phenylalanine in the concentration range of 50–500 mg·L^{-1}. The limit of detection was found to be 45 mg·L^{-1}.

In another study, a thymine-imprinted thin film-coated QCM sensor was developed by Diltemiz et al. [40]. For this purpose, methacryloylamidoadenine (MA-Ade) and thymine were used as the functional monomer and template compound, respectively. Polymerization was carried out under UV-light to form allyl-based self-assembled monolayers (SAMs) which have rougher surfaces compared to traditional monolayers formed by thiol modification. This approach provides an efficient recognition of DNA. The prepared MIP-based QCM sensor was used for the detection of thymine in the presence of uracil nucleobase. The Langmuir binding isotherm was applied to investigate the binding behavior of the prepared QCM sensor. The obtained results showed that the prepared sensor has homogeneous binding sites and displays high affinity toward the target compound thymine and K_a value was calculated as 1.0×10^5 M^{-1}. This reported approach offers a cost-effective, easy and sensitive MIP-based sensor system based on mimicking of DNA. The prepared allyl-based SAMs are great choice for the optimization of QCM sensors coated with MIP layers which exhibit high potential to become a commercial products.

In 2015, Jha and Hayashi have prepared a MIP coated-QCM sensor for the recognition of aldehyde compounds in body odor [41]. Gas chromatography-mass spectrometer (GC-MS) was applied for the characterization of odor samples. The functional monomer polyacryclic acid (PAA) with presence of a template aldehyde was polymerized on the sensor surface. The obtained results showed that the prepared MIP-based QCM sensor for heptanal exhibits high sensitivity, fast response and reproducibility compared to other prepared sensors toward hexanal and nonanal.

2.2. MIP-Based QCM Sensors for Food and Beverage Applications

Food and beverage industries need sensitive and reliable analytical techniques for the quality control of their products. During the production process, continuous monitoring should be carried out for the food safety. Traditional approaches such as enzyme assays and immuno techniques based on natural antibodies are commonly used for the analysis of food samples, but these techniques have some drawbacks such as high cost, low stability. QCM sensors combined with MIPs as plastic antibodies can be an alternative approach and overcome these drawbacks of traditional approaches.

For example, Sun et al. [42] have reported the preparation of MIP film-coated QCM sensors for the selective recognition of quinine and saccharine in bitter drinks. The functional monomer MAA and cross-linker (EDMA) were polymerized in the presence of template compounds quinine and saccharine to obtain selective MIP films on the sensor surface. Then, recognition performance of the prepared sensors toward target compounds was investigated. Other possible interfering compounds such as vanillin, caffeine, citric acid, sodium benzoate, sodium bicarbonate and sucrose were used to test selectivity of the prepared MIP-based QCM sensor. The obtained results from the experiments showed that the prepared sensor systems have high sensitivity (2.04 mg·L^{-1} and 32.8 mg·L^{-1} for quinine and saccharine, respectively). The MIP-coated QCM sensor systems were developed to provide a sensitive, facile and fast method for the determination of quinine and saccharine in tonic water at a practical concentration range with fast sample throughput and sufficient repeatability. MIP film-coated QCM technology thus provides a promising methodology for the taste application in flavor forecast and quality control of experimental, intermediate and final products for food, drinks and beverages.

In another important study, Iqbal et al. [43] have developed QCM sensors with polystyrene-based MIP membranes for the determination of different terpenes (limonene, α-pinene, β-pinene, estragole, eucalyptol and terpinene) in fresh and dried herb samples. The prepared sensor systems were succesfully applied to recognize target terpenes in rosemary, basil and sage. The sensitivity of the prepared sensors was <20 ppm of target compounds and the linear concentration range of the sensor response was <20 ppm to 250 ppm. Are the following really food and beverage applications?

The research group of Dickert [44] has developed a QCM sensor coated with MIP film for selective detection of tobacco mosaic virus in tobacco plant sap. MIP film on the sensor surface was prepared by polymerization of the functional monomer MAA, cross-linker EDMA and template virus. The prepared sensor was successfully used for the recognition of the target virus in real samples. The results showed that the prepared MIP-based QCM sensor exhibits high recognition ability toward tobacco mosaic virus in the concentration range of 10 ng·mL^{-1} to 1.0 mg·mL^{-1} and that qualitative detection of virus in infected leaf sap was successfully achieved. In this study, the authors also showed that the prepared sensor based on biomimetic polymers in combination with QCM is a new approach for monitoring of plant viruses directly in the plant sap within minutes. Ebarvia et al. [45] has prepared a MIP-based QCM sensor for the selective detection of chloramphenicol. In their work, selective MIP toward chloramphenicol was prepared by using precipitation polymerization and MAA, TRIM and chloramphenicol were used as the functional monomer, cross-linker and the template, respectively. MIP suspension in polyvinylchloride-tetrahydrofuran solution was coated onto the 10 MHz AT-cut quartz crystal by spin-coating. The obtained results showed that the prepared MIP based QCM sensor exhibits high affinity toward chloramphenicol.

In another study published in 2015 by Eren et al. [46], a MIP based-QCM sensor for selective determination of lovastatin (LOV) in red yeast rice was developed by formation of allylmercaptane monolayer on the sensor surface. For this purpose, LOV imprinted poly(2-hydroxyethyl methacrylate-methacryloylamidoaspartic acid) [p(HEMA-MAAsp)] nanofilm was prepared on the surface of a QCM sensor. The obtained results from the performance tests of the prepared sensor showed that the linearity range was 0.10–1.25 nM and the limit of detection was found as 0.030 nM.

In 2014, Dai et al. [47] developed a novel material based on MIP-QCM sensor for histamine in food samples. Histamine (HA) is a critical marker of food quality, being an indicator of bacterial contamination. The obtained results showed that the sensor exhibited linear behavior for HA concentrations of 0.11×10^{-2} to 4.45×10^{-2} mg·L^{-1}, a detection limit of 7.49×10^{-4} mg·kg^{-1} (S/N = 3), and high selectivity for HA (selectivity coefficient > 4) compared with structural analogues, good reproducibility, and long-term stability were observed. Also, the sensor was used to determine the concentration of HA in spiked fish products and the recovery values were found be 93.2%–100.4%. These outstanding detection limits are highly suitable for on-line monitoring of small amounts of histamine.

In another study published by Yan et al. [48], a QCM sensor was coated with MIP film for the determination of daminozide, which is a potential carcinogen in apple samples. For the daminozide sensor preparation, a MIP film using MAA as the functional monomer and EDMA as the cross-linker was prepared in the presence of the target molecule. The obtained results from the performance studies of the prepared MIP based-QCM sensor showed that the sensor has high selectivity and affinity toward the target compound daminozide in real samples. The limit of detection was found to be 5.0×10^{-8} mg·mL^{-1}.

Sun and Fung [49] prepared MIP based-QCM sensor systems for pirimicarb residues in vegetables which are potentially mutagenic and carcinogenic. For this purpose, they synthesized three different MIP particles using the functional monomer MAA by bulk and precipitation polymerization techniques. Then, the prepared MIP particles were coated on the surface of the QCM sensor. The obtained results showed that the nanoscale MIP particles prepared by precipitation polymerization were the best coating material for the sensor surface. The recognition performance of the prepared MIP based-QCM sensor toward pirimicarb in the presence of other interfering compounds such as atrazine, carbaryl, carbofuran and aldicarb was investigated. The prepared sensor showed high selectivity toward pirimicarb in the linear working range of 5.0×10^{-6} to 4.7×10^{-3} M and the limit of detection was calculated as 5.0×10^{-7} M.

Avila et al. [50] developed a QCM sensor coated with MIP for the detection of vanillin in vanilla sugar samples. Other similar compounds such as 4-hydroxybenzyl alcohol, 4-hydroxy-3-methoxy-benzyl alcohol and 4-hydroxybenzaldehyde as potential interferences were used for the selectivity experiments. The obtained results showed that the prepared sensor has high affinity and selectivity toward vanillin. The linear concentration range for the sensor response was 5 to 65 µM.

In another study, a MIP based-QCM sensor for citrinin was prepared by Fang and co-workers [51]. They developed a novel 3D-composite QCM sensor composed of MIP/Au nanoparticles/mesoporous carbon. The prepared sensor was successfully applied for the detection of citrinin in several cereals such as rice, white rice vinegar and wheat. The limit of detection was found as 1.8×10^{-9} M.

2.3. MIP-Based QCM Sensors for Environmental Applications

Environmental pollution is one of the crucial and challenging problem in the world today. Many compounds found in water and soil such as heavy metals, phenolics, pesticides, herbicides and pharmaceuticals, etc. are hazardous for humans, plants, animals and may cause serious health problems. Therefore, it is important to detect and remove these compounds from environmental samples. Conventional techniques such as HPLC, GC and CE have been applied for the analysis of these compounds in environmental samples [52–55]. However, these approaches are quite expensive, time consuming and require experienced researchers. These disadvantages of conventional approaches can be overcome by design and contruction of MIP-based QCM sensor systems that have high affinity and selectivity toward desired analyte in complex matrices.

One of the first examples of a MIP-based QCM sensor for environmental applications was reported by Percival et al. [56]. In their study, a selective QCM sensor coated with MIP thin film sensitive toward L-menthol in aqueous solutions was prepared. To obtain a selective MIP for L-menthol, MAA, L-menthol and EDMA were used as the functional monomer, template and cross-linker, respectively. The prepared thin MIP film was coated on the surface of QCM by using a sandwich casting approach. The limit of detection was found to be 200 ppb within a response range of 0 to 1.0 ppm.

An interesting work regarding the use of MIPs in QCM sensors was performed by Iglesias and his colleagues in 2009 [57]. They prepared a hybrid system composed of a polyurethane based-MIP coated QCM sensor for selective recognition and separation in chromatographic systems. The prepared hybrid system was successfully applied for selective recognition and separation of benzene, toluene, ethylbenzene, and xylenes in gasoline vapors.

In another interesting study reported by Dickert et al. [58], a QCM sensor coated with a polyurethane based-MIP film was developed for detection of degradation products in automotive engine oils.

These reported examples show that QCM sensors combined with MIPs selective to target compound/s can be succesfully used for the desired analyte/s in complex matrices in liquid and gas phases.

The recognition of chemical nerve agents has also been successfully achieved by using MIP-based QCM sensors. In 2012 Vergara et al. [59] developed an electrochemically MIP polythiophene film QCM surface for selective and sensitive detection of pinacolyl methylphosphonate (PMP). The lectrochemistry-QCM (EC-QCM) technique was used for the deposition of the MIP film onto the electrode. Figure 9 shows the schematic depiction of the MIP-based QCM sensor toward PMP. The synthesis was performed using cyclic voltammetry (CV) techniques applying apotential in the range of 0 to 1100 mV.

Organophosphate pesticides are able to cause neurological diseases and are considered toxic compounds. Paraoxon is the most commonly used organophosphate pesticide. Novel MIPs-based QCM biosensors for paraoxon recognition have been investigated by Özkütük et al. [60]. In their study, chitosan-Cd(II) (TCM-Cd(II)) modified with thiourea and epichlorohydrin were used as the functional monomer and cross-linker, respectively. The obtained results showed that the prepared QCM sensor-coated MIP film exhibited high affinity toward paraoxon in the concentration range of 0.02 to 1 μM and the limit of detection was found to be 0.02 μM. In another work [61], they used N-(2-aminoethyl)-3-aminopropyltrimethoxysilane–Cu(II) (AAPTS–Cu(II)) as a new metal–chelating monomer and tetraethoxysilane (TEOS) crosslinking agent for the polymerization.

The MIP-based QCM biosensor technique was applied to the detection of some drugs by Eslami et al. [62] and Kim et al. [63] in 2015. The nanostructured conducting MIP film was synthesized by CV method on the QCM electrode. The electrode was successfully applied for selective detection of ibuprofen in sample solutions.

In another study, a QCM sensor was coated with nano-MIP film for the determination of naproxen (NAP) [64]. MIP film preparation was performed on the surface of gold electrode by polymerization of pyrrole in the presence of template compound NAP as schematically shown in Figure 10. Scanning electron microscopy (SEM), infrared spectroscopy (FT-IR), cyclic voltammetry (CV) and electrochemical impedance spectroscopy (EIS) were used for the characterization of the prepared sensor. The obtained results from performance tests of the MIP-based QCM sensor toward NAP showed that the prepared sensor shows high affinity and selectivity toward NAP. The limit of detection was calculated to be 0.04 μM.

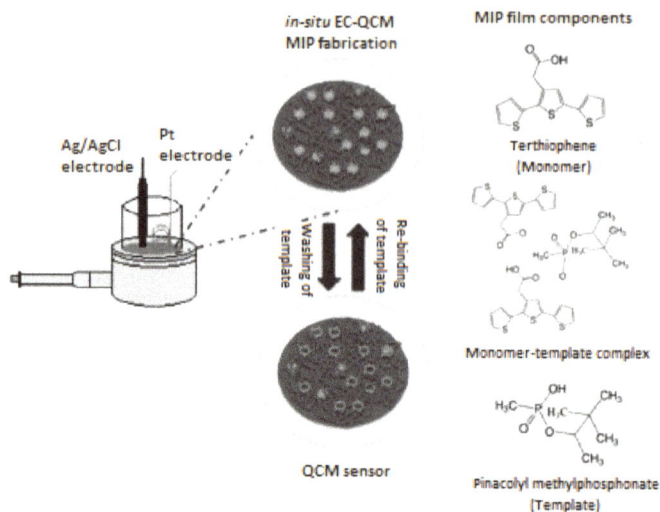

Figure 9. MIP-based QCM sensor toward pinacolyl methylphosphonate (reproduced with permission from [59]).

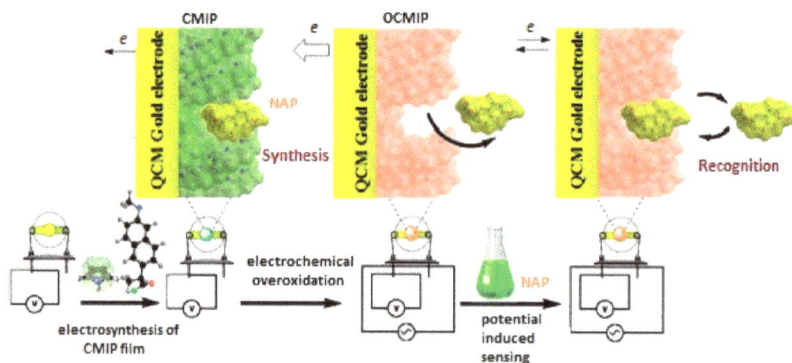

Figure 10. MIP-based QCM sensor for naproxen (reproduced with permission from [64]).

Gao et al. developed QCM sensors coated with ultra-thin MIP films for the detection in tap water of profenofos, which is an organophosphorus pesticide [65]. The sensor systems were prepared by using entrapment and in-situ self-assembly approaches. The results showed that the best recognition performance was obtained by a MIP-based QCM sensor prepared by in-situ self-assembly. Selectivity studies were also carried out in the presence of other potentially interfering compounds such as chlorpyrifos, parathion dichlorvos and omethoate. The prepared MIP based-QCM sensor displayed high affinity and selectivity toward profenofos in real samples. The limit of detection was calculated as 2.0×10^{-7} mg·mL^{-1}.

Another important study was performed by He and co-workers [66]. In their study, MIP-based QCM sensors were prepared for selective recognition of microcystin-LR (MC-LR) (a toxic peptide produced by some types of algae) in lake water. For this purpose, MAA, MC-LR and EDMA were used as the functional monomer, template and cross-linker, respectively. Selectivity of the prepared MIP film coated sensor was investigated in the presence of MC-RR, MC-RY and nodularin as interfering compounds. The obtained results showed that the prepared sensor has high selectivity toward target

MC-LR in the presence of interfering compounds in lake water. The limit of detection was found as 0.04 nM. In this study, the investigation of binding performance of the prepared sensor toward target in real samples shows the potential practical use of the prepared sensor.

In another study, a MIP coated-QCM sensor was prepared for the detection of atrazine in wastewater samples [67]. For the preparation of MIP film on the sensor surface, HEMA, atrazine and EDMA were used as functional monomer, template and cross-linker respectively. The prepared MIP based-QCM sensor showed high affinity and sensitivity toward target compound atrazine. The obtained linear concentration range was 0.08 to 1.5 nM and the limit of detection was calculated as 0.028 nM. Recently reported studies of MIP-based QCM sensors in different applications are given in Table 1.

Table 1. Recent reported studies of MIP-based QCM sensor in different applications.

Reference	Composition of QCM Sensor	Target	Sample
	**Applications to Environmental Samples**		
[68]	MIP film prepared by using functional monomer MAA on the sensor surface	Propranolol	Aqueous solutions
[69]	MIP film prepared by using functional monomer MAA on the sensor surface	Cu^{2+} and Ni^{2+} ions	Aqueous solutions
[70]	MIP film prepared by using functional monomer MAA on the sensor surface	Cu^{2+}	Wastewater
[71]	MIP film prepared by using functional monomer pyrrole on the sensor surface	Trichloroacetic acid	Drinking water
[72]	MIP film prepared by using functional monomer MAA on the sensor surface	Methomyl	Natural water
[73]	MIP film prepared by using functional monomer 3-thiophene acetic acid (3-TAA) on the sensor surface	Melphalan	Aqueous solutions
[74]	Cyclodextrin-modified poly(L-lysine) based- MIP film on the sensor surface	Bisphenol A	Aqueous solutions
	**Applications to Biological Samples**		
[75]	MIP film prepared by using functional monomer 1-Vinyl-2-pyrrolidone on the sensor surface	Heparin	Human plasma
[76]	MIP film prepared by using functional monomer MAA and poly(amidoamine) dendrimer on the sensor surface	Methimazole	Human urine
[77]	MIP film prepared by using functional monomer methacryloylamido tryptophan on the sensor surface	Bilirubin	Human plasma and Urine
[78]	MIP film prepared by using functional monomer methacryloylamido histidine on the sensor surface	Cholic acid	Human serum and Urine
[79]	MIP film prepared by using 3-dimethylaminopropyl methacrylamide as the functional monomer on the sensor surface	Albumin	Human serum
[80]	MIP film prepared by using functional monomer MAA on the sensor surface	D-Methamphetamine	Human urine
	**Applications to Food and Beverage Samples**		
[81]	MIP microsphere modified QCM sensor	Endosulfan	Drinking water and milk
[82]	MIP/poly(o-aminothiophenol) membrane/Au nanoparticles composite on the sensor surface	Ractopamine	Swine feed
[83]	MIP film prepared by using functional monomer methacryloylamido antipyrine on the sensor surface	Caffeic acid	Tea, apple and potato
[84]	MIP film on the surface of the alkanethiol modified-gold electrode	Thiacloprid	Celery Juice
[85]	Gold electrode coated with molecularly imprinted nanoparticles prepared by using functional monomer methacryloylamido histidine	Lysozyme	Chicken egg white
[86]	MIP film prepared by using methacryloylamidoaspartic acid as the monomer on the sensor surface	Kaempferol	Orange and apple juice

3. Conclusions and Future Trends

The examples described in this review highlight the recent progress and applications of MIP based-QCM sensors over the past years. The growing number of reported studies in which QCM sensor systems based on the molecular imprinting technique have been used in various application areas showed that these sensor systems are promising for selective recognition. QCM sensors with high specificity and sensitivity are commonly used as monitoring tools for target compounds in complex matrices where the selectivity is crucial. The combination of QCM sensors with target molecule memories having MIP thin films through the pre-recognition provides affinity toward the target compound, highly selective binding sites and novel, more sensitive sensing systems based on homogeneity in a larger number of recognition sites in MIPs. This combination has led to the design and development of a next generation of sensor platforms providing useful information for the progress of analytical sciences. On the other hand, MIP-coated QCM sensor platforms can also be potentially applied to process control and monitoring, assistance in the development of new products, as well as to the assessment of synergistic effects of food, drug, artificial enzyme and inhibitors and other innovative products.

Conflicts of Interest: The authors declare no conflict of interest.

References

1. Phongphanphanee, S.; Yoshida, N.; Hirata, F. Molecular recognition explored by a statistical-mechanics theory of liquids. *Curr. Pharm. Des.* **2011**, *17*, 1740–1757. [CrossRef] [PubMed]
2. Cheremisinoff, P. *Advances in Environmental Control Technology: Health and Toxicology*; Gulf Professional Publishing: Houston, TX, USA, 1997.
3. Sellergren, B. *Molecularly Imprinted Polymers: Man-Made Mimics of Antibodies and Their Application in Analytical Chemistry: Techniques and Instrumentation in Analytical Chemistry*; Elsevier Science: Amsterdam, The Netherlands, 2001.
4. Kupai, J.; Razali, M.; Büyüktiryaki, S.; Keçili, R.; Szekely, G. Long-term stability and reusability of molecularly imprinted polymers. *Polym. Chem.* **2017**, *8*, 666–673. [CrossRef]
5. Kryscioa, D.R.; Peppas, N.A. Critical review and perspective of macromolecularly imprinted polymers. *Acta Biomater.* **2012**, *8*, 461–473. [CrossRef] [PubMed]
6. Cheong, W.J.; Yang, S.H.; Ali, F. Molecular imprinted polymers for separation science: A review of reviews. *J. Sep. Sci.* **2013**, *36*, 609–628. [CrossRef] [PubMed]
7. Vasapollo, G.; Del Sole, R.; Mergola, L.; Lazzoi, M.R.; Scardino, A.; Scorrano, S.; Mele, G. Molecularly imprinted polymers: Present and future prospective. *Int. J. Mol. Sci.* **2011**, *12*, 5908–5945. [CrossRef] [PubMed]
8. Chen, Z.; Wang, M.; Fu, Y.; Yu, H.; Di, D. Preparation of quercetin molecularly imprinted polymers. *Des. Monomers Polym.* **2012**, *15*, 93–111. [CrossRef]
9. Wang, J.; Liu, F. Enhanced and selective adsorption of heavy metal ions on ion-imprinted simultaneous interpenetrating network hydrogels. *Des. Monomers Polym.* **2014**, *17*, 19–25. [CrossRef]
10. Long, J.-P.; Chen, Z.-B.; Liu, X.-J.; Du, X.Y. Preparation and adsorption property of solanesol molecular imprinted polymers. *Des. Monomers Polym.* **2015**, *18*, 641–649. [CrossRef]
11. Shen, X.; Zhu, L.; Wang, N.; Ye, L.; Tang, H. Molecular imprinting for removing highly toxic organic pollutants. *Chem. Commun.* **2012**, *48*, 788–798. [CrossRef] [PubMed]
12. Uludağ, Y.; Piletsky, S.A.; Turner, A.P.F.; Cooper, M.A. Piezoelectric sensors based on molecular imprinted polymers for detection of low molecular mass analytes. *FEBS J.* **2007**, *274*, 5471–5480. [CrossRef] [PubMed]
13. Liao, H.; Zhang, Z.; Nie, L.; Yao, S. Electrosynthesis of imprinted polyacrylamide membranes for the stereospecific L-histidine sensor and its characterization by AC impedance spectroscopy and piezoelectric quartz crystal technique. *J. Biochem. Biophys. Methods* **2004**, *59*, 75–87. [CrossRef] [PubMed]
14. Dickert, F.; Tortschano, M.; Bulst, W.E.; Fischerauer, G. Molecularly imprinted sensor layers for the detection of polycyclic aromatic hydrocarbons in water. *Anal. Chem.* **1999**, *71*, 4559–4563. [CrossRef]

15. Dickert, F.L.; Hayden, O. Bioimprinting of polymers and sol-gel phases. Selective detection of yeasts with imprinted polymers. *Anal. Chem.* **2002**, *74*, 1302–1306. [CrossRef] [PubMed]

16. Seifner, A.; Lieberzeit, P.; Jungbauer, C.; Dickert, F.L. Synthetic receptors for selectively detecting erythrocyte ABO subgroups. *Anal. Chim. Acta* **2009**, *651*, 215–219. [CrossRef] [PubMed]

17. Tai, D.F.; Lin, C.Y.; Wu, T.Z.; Huang, J.H.; Shu, P.Y. Artificial receptors in serologic tests for the early diagnosis of dengue virus infection. *Clin. Chem.* **2006**, *52*, 1486–1491. [CrossRef] [PubMed]

18. Tai, D.F.; Jhang, M.H.; Chen, G.Y.; Wang, S.C.; Lu, K.H.; Lee, Y.D.; Liu, H.T. Epitope-cavities generated by molecularly imprinted films measure the coincident response to anthrax protective antigen and its segments. *Anal. Chem.* **2010**, *82*, 2290–2293. [CrossRef] [PubMed]

19. Liu, N.; Zhao, Z.; Chen, Y.; Gao, Z. Rapid detection of staphylococcal enterotoxin B by two-dimensional molecularly imprinted film-coated quartz crystal microbalance. *Anal. Lett.* **2012**, *45*, 283–295. [CrossRef]

20. Lu, C.-H.; Zhang, Y.; Tang, S.-F.; Fang, Z.-B.; Yang, H.-H.; Chen, X.; Chen, G.-N. Sensing HIV related protein using epitope imprinted hydrophilic polymer coated quartz crystal microbalance. *Biosens. Bioelectron.* **2012**, *31*, 439–444. [CrossRef] [PubMed]

21. Liu, N.; Li, X.; Ma, X.; Ou, G.; Gao, Z. Rapid and multiple detections of staphylococcal enterotoxins by two-dimensional molecularly imprinted film-coated QCM sensor. *Sens. Actuators B Chem.* **2014**, *191*, 326–331. [CrossRef]

22. Phan, N.; Sussitz, H.; Lieberzeit, P. Polymerization Parameters Influencing the QCM Response Characteristics of BSA MIP. *Biosensors* **2014**, *4*, 161–171. [CrossRef] [PubMed]

23. EL-Sharif, H.F.; Aizawa, H.; Reddy, S.M. Spectroscopic and quartz crystal microbalance (QCM) characterisation of protein-based MIPs. *Sens. Actuators B Chem.* **2015**, *206*, 239–245. [CrossRef]

24. Diltemiz, S.E.; Hür, D.; Keçili, R.; Ersöz, A.; Say, R. New synthesis method for 4-MAPBA monomer and using for the recognition of IgM and mannose with MIP-based QCM sensors. *Analyst* **2013**, *138*, 1558–1563. [CrossRef] [PubMed]

25. Latif, U.; Qian, J.; Can, S.; Dickert, F. Biomimetic Receptors for Bioanalyte Detection by Quartz Crystal Microbalances—From Molecules to Cells. *Sensors* **2014**, *14*, 23419–23438. [CrossRef] [PubMed]

26. Alenus, J.; Galar, P.; Ethirajan, A.; Horemans, F.; Weustenraed, A.; Cleij, T.J.; Wagner, P. Detection of L-nicotine with dissipation mode quartz crystal microbalance using molecular imprinted polymers. *Phys. Status Solidi* **2012**, *209*, 905–910. [CrossRef]

27. Alenus, J.; Ethirajan, A.; Horemans, F.; Weustenraed, A.; Csipai, P.; Gruber, J.; Peeters, M.; Cleij, T.J.; Wagner, P. Molecularly imprinted polymers as synthetic receptors for the QCM-D-based detection of l-nicotine in diluted saliva and urine samples. *Anal. Bioanal. Chem.* **2013**, *405*, 6479–6487. [CrossRef] [PubMed]

28. Croux, D.; Weustenraed, A.; Pobedinskas, P.; Horemans, F.; Diliën, H.; Haenen, K.; Cleij, T.; Wagner, P.; Thoelen, R.; De Ceuninck, W. Development of multichannel quartz crystal microbalances for MIP-based biosensing. *Phys. Status Solidi Appl. Mater. Sci.* **2012**, *209*, 892–899. [CrossRef]

29. Cheng, Z.; Wang, E.; Yang, X. Capacitive detection of glucose using molecularly imprinted polymers. *Biosens. Bioelectron.* **2001**, *16*, 179–185. [CrossRef]

30. Panasyuk, T.L.; Mirsky, V.M.; Piletsky, S.A.; Wolfbeis, O.S. Electropolymerized molecularly imprinted polymers as receptors layers in capacitive chemical sensors. *Anal. Chem.* **1999**, *71*, 4609–4613. [CrossRef]

31. Özcan, L.; Sahin, Y. Determination of a paracetamol based on electropolymerized-molecularly imprinted polypyrrole modified pencil graphite electrode. *Sens. Actuators B Chem.* **2007**, *127*, 362–369. [CrossRef]

32. Lenain, P.; De Saeger, S.; Mattiasson, B.; Hedström, M. Affinity sensor based on immobilized molecular imprinted synthetic recognition elements. *Biosens. Bioelectron.* **2015**, *69*, 34–39. [CrossRef] [PubMed]

33. Yarman, A.; Dechtrirat, D.; Bosserdt, M.; Jetzschmann, K.J.; Gajovic-Eichelmann, N.; Scheller, F.W. Cytochrome c-derived hybrid systems based on molecularly imprinted polymers. *Electroanalysis* **2015**, *27*, 573–586. [CrossRef]

34. Apodaca, D.C.; Pernites, R.B.; Ponnapati, R.R.; Del Mundo, F.R.; Advincula, R.C. Electropolymerized molecularly imprinted polymer films of a bis-terthiophene dendron: Folic Acid quartz crystal microbalance sensing. *ACS Appl. Mater. Interfaces* **2011**, *3*, 191–203. [CrossRef]

35. Liu, S.; Zhou, D.; Guo, T. Construction of a novel macroporous imprinted biosensor based on quartz crystal microbalance for ribonuclease A detection. *Biosens. Bioelectron.* **2013**, *42*, 80–86. [CrossRef] [PubMed]

36. Ersöz, A.; Diltemiz, S.E.; Özcan, A.A.; Denizli, A.; Say, R. 8-OHdG sensing with MIP based solid phase extraction and QCM technique. *Sens. Actuators B Chem.* **2009**, *137*, 7–11. [CrossRef]

37. Lee, M.H.; Thomas, J.L.; Tseng, H.Y.; Lin, W.C.; Liu, B.D.; Lin, H.Y. Sensing of digestive proteins in saliva with a molecularly imprinted poly(ethylene-co-vinyl alcohol) thin film coated quartz crystal microbalance sensor. *ACS Appl. Mater. Interfaces* **2011**, *3*, 3064–3071. [CrossRef] [PubMed]

38. Zhou, D.; Guo, T.; Yang, Y.; Zhang, Z. Surface imprinted macroporous film for high performance protein recognition in combination with quartz crystal microbalance. *Sens. Actuators B Chem.* **2011**, *153*, 96–102. [CrossRef]

39. Mirmohseni, A.; Shojaei, M.; Farbodi, M. Application of a quartz crystal nanobalance to the molecularly imprinted recognition of phenylalanine in solution. *Biotechnol. Bioprocess Eng.* **2008**, *13*, 592–597. [CrossRef]

40. Diltemiz Emir, S.; Hür, D.; Ersöz, A.; Denizli, A.; Say, R. Designing of MIP based QCM sensor having thymine recognition sites based on biomimicking DNA approach. *Biosens. Bioelectron.* **2009**, *25*, 599–603. [CrossRef] [PubMed]

41. Jha, S.K.; Hayashi, K. Polyacrylic acid polymer and aldehydes template molecule based MIPs coated QCM sensors for detection of pattern aldehydes in body odor. *Sens. Actuators B Chem.* **2015**, *206*, 471–487. [CrossRef]

42. Sun, H.; Mo, Z.H.; Choy, J.T.S.; Zhu, D.R.; Fung, Y.S. Piezoelectric quartz crystal sensor for sensing taste-causing compounds in food. *Sens. Actuators B Chem.* **2008**, *131*, 148–158. [CrossRef]

43. Iqbal, N.; Mustafa, G.; Rehman, A.; Biedermann, A.; Najafi, B.; Lieberzeit, P.A.; Dickert, F.L. QCM-arrays for sensing terpenes in fresh and dried herbs via bio-mimetic MIP layers. *Sensors* **2010**, *10*, 6361–6376. [CrossRef] [PubMed]

44. Dickert, F.L.; Hayden, O.; Bindeus, R.; Mann, K.J.; Blaas, D.; Waigmann, E. Bioimprinted QCM sensors for virus detection—Screening of plant sap. *Anal. Bioanal. Chem.* **2004**, *378*, 1929–1934. [CrossRef] [PubMed]

45. Ebarvia, B.S.; Ubando, I.E.; Sevilla, F.B. Biomimetic piezoelectric quartz crystal sensor with chloramphenicol-imprinted polymer sensing layer. *Talanta* **2015**, *144*, 1260–1265. [CrossRef] [PubMed]

46. Eren, T.; Atar, N.; Yola, M.L.; Karimi-Maleh, H. A sensitive molecularly imprinted polymer based quartz crystal microbalance nanosensor for selective determination of lovastatin in red yeast rice. *Food Chem.* **2015**, *185*, 430–436. [CrossRef] [PubMed]

47. Dai, J.; Zhang, Y.; Pan, M.; Kong, L.; Wang, S. Development and application of quartz crystal microbalance sensor based on novel molecularly imprinted sol-gel polymer for rapid detection of histamine in foods. *J. Agric. Food Chem.* **2014**, *62*, 5269–5274. [CrossRef] [PubMed]

48. Yan, S.; Fang, Y.; Gao, Z. Determination of daminozide in apple sample by mip-coated piezoelectric quartz sensor. *Anal. Lett.* **2007**, *40*, 1013–1021. [CrossRef]

49. Sun, H.; Fung, Y. Piezoelectric quartz crystal sensor for rapid analysis of pirimicarb residues using molecularly imprinted polymers as recognition elements. *Anal. Chim. Acta* **2006**, *576*, 67–76. [CrossRef] [PubMed]

50. Avila, M.; Zougagh, M.; Escarpa, A.; Ríos, A. Supported liquid membrane-modified piezoelectric flow sensor with molecularly imprinted polymer for the determination of vanillin in food samples. *Talanta* **2007**, *72*, 1362–1369. [CrossRef] [PubMed]

51. Fang, G.; Liu, G.; Yang, Y.; Wang, S. Quartz crystal microbalance sensor based on molecularly imprinted polymer membrane and three-dimensional Au nanoparticles@mesoporous carbon CMK-3 functional composite for ultrasensitive and specific determination of citrinin. *Sens. Actuators B Chem.* **2016**, *230*, 272–280. [CrossRef]

52. Bagheri, H.; Mohammadi, A.; Salemi, A. On-line trace enrichment of phenolic compounds from water using a pyrrole-based polymer as the solid-phase extraction sorbent coupled with high-performance liquid chromatography. *Anal. Chim. Acta* **2004**, *513*, 445–449. [CrossRef]

53. Bonwick, G.A.; Sun, C.; Abdullatif, P.; Baugh, P.J.; Smith, C.J.; Armitage, R.; Davies, D.H. Determination of permethrin and cyfluthrin in water and sediment by gas chromatography-mass spectrometry operated in the negative chemical-ionization mode. *J. Chromatogr. A* **1995**, *707*, 293–302. [CrossRef]

54. Hladik, M.L.; Smalling, K.L.; Kuivila, K.M. A multi-residue method for the analysis of pesticides and pesticide degradates in water using HLB solid-phase extraction and gas chromatography-ion trap mass spectrometry. *Bull. Environ. Contam. Toxicol.* **2008**, *80*, 139–144. [CrossRef] [PubMed]

55. Vasas, G.; Szydlowska, D.; Gáspár, A.; Welker, M.; Trojanowicz, M.; Borbély, G. Determination of microcystins in environmental samples using capillary electrophoresis. *J. Biochem. Biophys. Methods* **2006**, *66*, 87–97. [CrossRef] [PubMed]

56. Percival, C.J.; Stanley, S.; Galle, M.; Braithwaite, A.; Newton, M.I.; McHale, G.; Hayes, W. Molecular-imprinted polymer-coated quartz crystal microbalances for the detection of terpenes. *Anal. Chem.* **2001**, *73*, 4225–4228. [CrossRef] [PubMed]

57. Iglesias, R.A.; Tsow, F.; Wang, R.; Forzani, E.S.; Tao, N. Hybrid separation and detection device for analysis of benzene, toluene, tthylbenzene, and xylenes in complex samples. *Anal. Chem.* **2009**, *81*, 8930–8935. [CrossRef] [PubMed]

58. Dickert, F.L.; Forth, P.; Lieberzeit, P.A.; Voigt, G. Quality control of automotive engine oils with mass-sensitive chemical sensors—QCMs and molecularly imprinted polymers. *Fresenius J. Anal. Chem.* **2000**, *366*, 802–806. [CrossRef] [PubMed]

59. Vergara, A.V.; Pernites, R.B.; Pascua, S.; Binag, C.A.; Advincula, R.C. QCM sensing of a chemical nerve agent analog via electropolymerized molecularly imprinted polythiophene films. *J. Polym. Sci. A Polym. Chem.* **2012**, *50*, 675–685. [CrossRef]

60. Özkütük, E.B.; Diltemiz, S.E.; Özalp, E.; Ersöz, A.; Say, R. Silan based paraoxon memories onto QCM electrodes. *J. Ind. Eng. Chem.* **2013**, *19*, 1788–1792. [CrossRef]

61. Özkütük, E.B.; Diltemiz, S.E.; Özalp, E.; Gedikbey, T.; Ersöz, A. Paraoxon imprinted biopolymer based QCM sensor. *Mater. Chem. Phys.* **2013**, *139*, 107–112. [CrossRef]

62. Eslami, M.R.; Alizadeh, N. A dual usage smart sorbent/recognition element based on nanostructured conducting molecularly imprinted polypyrrole for simultaneous potential-induced nanoextraction/ determination of ibuprofen in biomedical samples by quartz crystal microbalance sensor. *Sens. Actuators B Chem.* **2015**, *220*, 880–887. [CrossRef]

63. Kim, J.M.; Yang, J.C.; Park, J.Y. Quartz crystal microbalance (QCM) gravimetric sensing of theophylline via molecularly imprinted microporous polypyrrole copolymers. *Sens. Actuators B Chem.* **2015**, *206*, 50–55. [CrossRef]

64. Alizadeh, N. Nanostructured conducting molecularly imprinted polypyrrole based quartz crystal microbalance sensor for naproxen determination and its electrochemical impedance study. *RSC Adv.* **2016**, *6*, 9387–9395.

65. Gao, N.; Dong, J.; Liu, M.; Ning, B.; Cheng, C.; Guo, C.; Zhou, C.; Peng, Y.; Bai, J.; Gao, Z. Development of molecularly imprinted polymer films used for detection of profenofos based on a quartz crystal microbalance sensor. *Analyst* **2012**, *137*, 1252–1258. [CrossRef] [PubMed]

66. He, H.; Zhou, L.; Wang, Y.; Li, C.; Yao, J.; Zhang, W.; Zhang, O.; Li, M.; Li, H.; Dong, W.F. Detection of trace microcystin-LR on a 20 MHz QCM sensor coated with in situ self-assembled MIPs. *Talanta* **2015**, *131*, 8–13. [CrossRef] [PubMed]

67. Gupta, V.K.; Yola, M.L.; Eren, T.; Atar, N. Selective QCM sensor based on atrazine imprinted polymer: Its application to wastewater sample. *Sens. Actuators B Chem.* **2015**, *218*, 215–221. [CrossRef]

68. Piacham, T.; Josell, A.; Arwin, H.; Prachayasittikul, V.; Ye, L. Molecularly imprinted polymer thin films on quartz crystal microbalance using a surface bound photo-radical initiator. *Anal. Chim. Acta* **2005**, *536*, 191–196. [CrossRef]

69. Latif, U.; Mujahid, A.; Afzal, A.; Sikorski, R.; Lieberzeit, P.A.; Dickert, F.L. Dual and tetraelectrode QCMs using imprinted polymers as receptors for ions and neutral analytes. *Anal. Bioanal. Chem.* **2011**, *400*, 2507–2515. [CrossRef] [PubMed]

70. Yang, Z.P.; Zhang, C.J. Designing of MIP-based QCM sensor for the determination of Cu(II) ions in solution. *Sens. Actuators B Chem.* **2009**, *142*, 210–215. [CrossRef]

71. Suedee, R.; Intakong, W.; Lieberzeit, P.A.; Wanichapichart, P.; Chooto, P.; Dickert, F.L. Trichloroacetic acid-imprinted polypyrrole film and its property in piezoelectric quartz crystal microbalance and electrochemical sensors to application for determination of haloacetic acids disinfection by-product in drinking water. *J. Appl. Polym. Sci.* **2007**, *106*, 3861–3871. [CrossRef]

72. Mirmohseni, A.; Houjaghan, M.R. Measurement of the pesticide methomyl by modified quartz crystal nanobalance with molecularly imprinted polymer. *J. Environ. Sci. Health B* **2013**, *48*, 278–284. [CrossRef] [PubMed]

73. Singh, A.K.; Singh, M. QCM sensing of melphalan via electropolymerized molecularly imprinted polythiophene films. *Biosens. Bioelectron.* **2015**, *74*, 711–717. [CrossRef] [PubMed]

74. Matsumoto, K.; Tiu, B.D.B.; Kawamura, A.; Advincula, R.C.; Miyata, T. QCM sensing of bisphenol A using molecularly imprinted hydrogel/conducting polymer matrix. *Polym. J.* **2016**, *48*, 525–532. [CrossRef]

75. Hussain, M. Ultra-sensitive detection of heparin via a PTT using plastic antibodies on QCM-D platform. *RSC Adv.* **2015**, *5*, 54963–54970. [CrossRef]

76. Pan, M.; Fang, G.; Lu, Y.; Kong, L.; Yang, Y.; Wang, S. Molecularly imprinted biomimetic QCM sensor involving a poly(amidoamine) dendrimer as a functional monomer for the highly selective and sensitive determination of methimazole. *Sens. Actuators B Chem.* **2015**, *207*, 588–595. [CrossRef]

77. Çiçek, Ç.; Yılmaz, F.; Özgür, E.; Yavuz, H.; Denizli, A. Molecularly imprinted quartz crystal microbalance sensor (QCM) for bilirubin detection. *Chemosensors* **2016**, *4*, 21–34. [CrossRef]

78. Gültekin, A.; Karanfil, G.; Sönmezoglu, S.; Say, R. Development of a highly sensitive MIP based-QCM nanosensor for selective determination of cholic acid level in body fluids. *Mat. Sci. Eng. C* **2014**, *42*, 436–442. [CrossRef] [PubMed]

79. Lin, T.Y.; Hub, C.H.; Chou, T.C. Determination of albumin concentration by MIP-QCM sensor. *Biosens. Bioelectron.* **2004**, *20*, 75–81. [CrossRef] [PubMed]

80. Arenas, L.F.; Ebarvia, B.S.; Sevilla, F.B. Enantioselective piezoelectric quartz crystal sensor for d-methamphetamine based on a molecularly imprinted polymer. *Anal. Bioanal. Chem.* **2010**, *397*, 3155–3158. [CrossRef] [PubMed]

81. Liu, N.; Han, J.; Liu, Z.; Qu, L.; Gao, Z. Rapid detection of endosulfan by a molecularly imprinted polymer microsphere modified quartz crystal microbalance. *Anal. Methods* **2013**, *5*, 4442–4447. [CrossRef]

82. Kong, L.J.; Pan, M.F.; Fang, G.Z.; He, X.I.; Yang, Y.K.; Dai, J.; Wang, S. Molecularly imprinted quartz crystal microbalance sensor based on poly(o-aminothiophenol) membrane and Au nanoparticles for ractopamine determination. *Biosens. Bioelectron.* **2014**, *51*, 286–292. [CrossRef] [PubMed]

83. Gültekin, A.; Karanfil, G.; Kuş, M.; Sönmezoğlu, S.; Say, R. Preparation of MIP-based QCM nanosensor for detection of caffeic acid. *Talanta* **2014**, *119*, 533–537. [CrossRef] [PubMed]

84. Bi, X.; Yang, K.L. On-Line monitoring imidacloprid and thiacloprid in celery juice using quartz crystal microbalance. *Anal. Chem.* **2009**, *81*, 527–532. [CrossRef] [PubMed]

85. Şener, G.; Özgür, E.; Yılmaz, E.; Uzun, L.; Say, R.; Denizli, A. Quartz crystal microbalance based nanosensor for lysozyme detection with lysozyme imprinted nanoparticles. *Biosens. Bioelectron.* **2010**, *26*, 815–821. [CrossRef] [PubMed]

86. Gupta, V.K.; Yola, M.L.; Atar, N. A novel molecular imprinted nanosensor based quartz crystal microbalance for determination of kaempferol. *Sens. Actuators B Chem.* **2014**, *194*, 79–85. [CrossRef]

sensors

MDPI

Review

Imprinting Technology in Electrochemical Biomimetic Sensors

Manuela F. Frasco, Liliana A. A. N. A. Truta, M. Goreti F. Sales and Felismina T. C. Moreira *

BioMark-CINTESIS/ISEP, School of Engineering, Polytechnic Institute of Porto, 4200-072 Porto, Portugal; mtbff@isep.ipp.pt (M.F.F.); laant@isep.ipp.pt (L.A.A.N.A.T.); mgf@isep.ipp.pt (M.G.F.S.)
* Correspondance: ftcmo@isep.ipp.pt; Tel.: +351-228340544; Fax: +351-228321159

Academic Editors: Bo Mattiasson and Gizem Ertürk
Received: 21 December 2016; Accepted: 3 March 2017; Published: 6 March 2017

Abstract: Biosensors are a promising tool offering the possibility of low cost and fast analytical screening in point-of-care diagnostics and for on-site detection in the field. Most biosensors in routine use ensure their selectivity/specificity by including natural receptors as biorecognition element. These materials are however too expensive and hard to obtain for every biochemical molecule of interest in environmental and clinical practice. Molecularly imprinted polymers have emerged through time as an alternative to natural antibodies in biosensors. In theory, these materials are stable and robust, presenting much higher capacity to resist to harsher conditions of pH, temperature, pressure or organic solvents. In addition, these synthetic materials are much cheaper than their natural counterparts while offering equivalent affinity and sensitivity in the molecular recognition of the target analyte. Imprinting technology and biosensors have met quite recently, relying mostly on electrochemical detection and enabling a direct reading of different analytes, while promoting significant advances in various fields of use. Thus, this review encompasses such developments and describes a general overview for building promising biomimetic materials as biorecognition elements in electrochemical sensors. It includes different molecular imprinting strategies such as the choice of polymer material, imprinting methodology and assembly on the transduction platform. Their interface with the most recent nanostructured supports acting as standard conductive materials within electrochemical biomimetic sensors is pointed out.

Keywords: molecularly imprinted polymers; electrochemistry; biomimetic sensors

1. Introduction

A biomimetic strategy enabling synthetic materials to mimic molecules ranging from amino acids and drugs to much larger proteins, bacteriophage or microbial cells, is the basis of molecular imprinting technology. The ability of imprinted materials to recognize the target analyte is comparable to natural molecular recognition in terms of affinity and sensitivity. Thus, designing biomimetic systems and synthetic analogues of biological processes are the focus of special attention in several research areas.

Molecularly imprinted polymers (MIPs) have the ability to recognize biological compounds like proteins [1–4], amino acids [5–7], peptides [8–11] or nucleotides [12], and chemicals such as pollutants [13], drugs and food additives [14–16]. Moreover, they can be applied in separation and purification processes [17], chemical sensors [18], catalysis [19] and drug delivery [20].

Synthetic materials as biological analogue receptors have been highlighted in different research areas with great advances in sensor design [21]. MIP production is generally based on some form of template-directed synthesis. The template and functional monomers interact to form a complex during the imprinting polymerization and rebinding [22]. The polymerization starts with addition of the cross-linker and initiator. In the last step, the template is removed and the 3-D functional matrix keeps its geometry and organization guaranteeing the ability of MIPs to rebind the target molecule [23] (Figure 1).

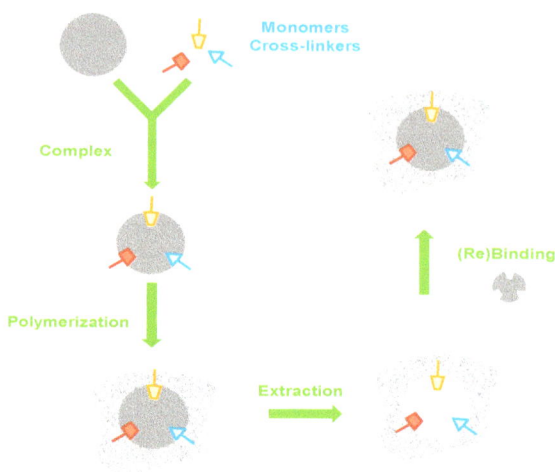

Figure 1. Synthesis of molecularly imprinted polymers.

Electrochemical sensors with MIP as recognition element combine outstanding characteristics. The biomimetic recognition secures their high selectivity, while the electrochemical transduction offers a highly sensitive response, i.e., the lowest possible limit of detection (LOD), in a short time [2,24]. The reusable electrochemical imprinted device produces a signal upon the selective interaction with the analyte. The electrical signal is proportional to the concentration of the target under analysis. Considering MIP construction, both the selection of polymeric materials and suitable imprinting processes for a target molecule are critical steps, especially considering the low detection levels required. Moreover, for combining MIPs with electrochemical sensors, different imprinting approaches (e.g., in situ bulk polymerization, epitope imprinting, surface imprinting, multi-template imprinting) can be employed [25–28].

The successful development of electrochemical sensors entailing the advantageous elimination of sample preparation requires multidisciplinary fields of analytical chemistry. The evaluation of several parameters is crucial to assess the performance of the sensor. These include the sensitivity by analysing the LOD and the linear and dynamic ranges, as well as the selectivity of response in the presence of interfering substances [29,30]. Additional characteristics like response time, reproducibility, stability, reusability and portability are also considered important key factors for evaluating any sensor.

It is possible to achieve higher sensitivity when developing a sensor by introducing some labels that bind to the recognition element or by applying indirect measurements. The main advantage for analyte detection is the signal amplification and the decrease of non-specific binding events [31]. This type of sensor is time consuming since it involves multiple steps, and it is also cost-limited. Thus, progress in label-free sensors able to directly detect the analyte allowing real-time measurements have been achieved [32,33]. However, these sensors lack amplification and need MIPs with very high affinity for the target as well as a very sensitive transducer. Very low LOD is in theory more challenging to achieve.

A general overview of the main polymeric materials and technological approaches for MIP design and fabrication are outlined first. In addition, this review describes the conjugation of MIPs as recognition element with transducer schemes, including immobilization strategies and labelling detection for different electrochemical sensors, providing an outline of electrochemically–based imprinted biomimetic sensors.

2. Polymeric Materials

Polymeric materials holding complimentary features in dimensions and functionality to the template are obtained by different routes. The type of interactions that occur during the phase of imprinting within the matrix-forming material, such as covalent bonds, non-covalent binding or metal-ion mediated imprinting can be customized according to the desired MIP, target bioanalytes and type of sensor, and has been extensively reviewed elsewhere [34–36]. Functional monomers are used to mediate specific chemical recognition processes. In addition, the monomers are held in place by cross-linking agents. The selective artificial recognition cavities are then formed within matrices or impressed on surfaces in appropriate solvents [37–39]. The type of interaction between the template and monomers determines the template removal method, either by a simple extraction or by chemical cleavage. Thus, choosing the starting material or composite with proper recognition ability can deliver molecular cavities complementary to the template in shape, size and function, accomplishing the desired selectivity [4,40].

2.1. Organic Polymers

Most formulations used in molecular imprinting technique for sensing purposes are based on the traditional free radical polymerization, relying on non-covalent interactions, such as electrostatic interactions or hydrogen bonding, between the template and functional groups of the organic monomers [41]. These interactions occur first in the pre-polymerization complex and the polymer is grown afterwards in the presence of a proper cross-linker and solvent type. It is a widespread method because the template has multiple potential binding sites to be targeted, together with the less chemistry involved when compared to the pre-synthesis of covalent adducts [34]. The mild reaction conditions of free radical polymerizations can be implemented using a myriad of vinyl monomers, including ethylene, styrene, methacrylic acid, acrylamide, among many others [34,42]. Methacrylic acid as functional monomer has found broad applicability, mainly because it is a carboxylic acid-based monomer and this group can serve as both a hydrogen bond donor and acceptor, and additionally it can establish ion-pair and dipole-dipole interactions [34]. The design of MIP synthesis should consider a number of variables taking into account the nature and levels of the various components of the mixtures, i.e., template, functional monomer, cross-linker, solvent and initiator, as well as the method of initiation, polymerization time and temperature [42]. A screening of functional vinyl monomers enables the copolymerization of two or more monomers with different functionalities, which is highly desirable to establish multiple types of interactions with the template. MIP synthesis using only non-covalent assembly in polar solvents, namely water, may face some difficulties because the electrostatic and hydrogen bonding interactions are weakened in the polar medium [41,43]. Thus, to achieve good affinity and selectivity, especially for imprinting biological molecules in aqueous medium, adjusting the polarity of the solvent with the template solubility is crucial [40]. Nevertheless, constructing MIPs with improved performance in organic polar solvents or water can be achieved by combining stronger interactions in such media, like hydrophobic and metal-coordination interactions [41,44].

2.2. Hydrogels

Hydrogel substrates are swellable soft materials of considerable interest. Nonetheless, producing chemically and mechanically stable hydrogel MIPs with high specific recognition is challenging [45]. This is due to the lower binding strength of the non-covalent interactions established between the template and monomers, crucial to obtain the imprinting outcome, in aqueous solutions. Among the acrylate family of polymers, polyacrylamide has been successfully used as an imprinting matrix for biological molecules. Along with water solubility, low cost, easy production and engineering are some of polyacrylamide gels' attractive properties concerning molecular imprinting [46]. Studies on polyacrylamide hydrogel-based MIPs have demonstrated sensitivity and selectivity towards several proteins with different biological roles, sizes and electrochemical activities, namely haemoglobin, myoglobin, cytochrome c, bovine serum albumin (BSA) and catalase [47–49]. Further improvements in terms of MIP selectivity have been attempted by including a metal chelating complex comprising bifunctional vinyl groups with the ability to co-polymerize within the polyacrylamide matrix [50]. These advanced imprinted cavities with metal-coding for enhanced selective protein recognition showed higher BSA binding and selectivity. These results may be explained by a better macroporosity and stability of the polymer backbone, as well as additional metal contribution to favorable interactions with the protein [50]. Hydrogel polymers undergo a reversible volume transition between swollen and collapsed phases. Thus, they can be designed to respond with a phase transition as triggered by certain external stimuli, such as temperature, pH, and light, among others. These features of stimuli-responsive hydrogels could benefit MIPs performance by reducing non-specific binding and improving transport of the template [51]. The synthesis of smart hydrogel MIPs that can modulate their affinity for the target macromolecules enabling a switchable capacity for the binding and releasing processes according to external stimuli has attracted considerable research interest and has been comprehensively reviewed [52,53].

2.3. Sol-Gel Materials

The most highlighted properties of sol-gel processing when constructing matrices to detect an analyte with high sensitivity and selectivity are the ability to control the porosity and the nanostructuring while achieving the desired purity and homogeneity [54]. Systems with tailored features were first described using silica gels and since then, growing interest enabled expanding the use of these inorganic materials in MIP design, including hybrid organic-inorganic matrices, for various analytes [55–58]. Likewise, inorganic matrices of titanium oxide sol-gels have enabled the production of imprinted films with specific binding cavities with a variety of functionalities [59,60]. One of the advantages presented by titanium over silane sol-gels is the improved stability when exposed to higher temperatures [54]. Generally, the sol-gel process occurring at mild conditions consists of a system transition from liquid "sol" into solid "gel". Alkoxide precursors are hydrolyzed and react with each other (or with the unhydrolyzed molecules), followed by polycondensation reactions that release water or alcohol when the matrix is cross-linking [54]. As the condensation propagates, there is a gradual increase of the matrix viscosity until originating an interconnected, homogeneous and rigid material. The degree of cross-linking and the network formation can be tuned because there are several parameters that affect the sol-gel reactions. An optimization requires considering the type of solvent and the pH of the solution, the catalyst chosen and its concentration, the reaction temperature and time, and the nature of the (R) group of the alkoxides [61,62]. Thus, when adjusting these variables, fundamental properties (e.g., refractive index, surface area, porosity, and mechanical features) can be established according to the goal of the sol-gel sensor [54]. For example, a silica sol-gel thin film molecularly imprinted was used to coat a piezoelectric sensor, and its ability to determine the presence of amino acids, namely L-histidine, has been described. The imprinted polymer obtained without pre-protection of the amino acid exhibited excellent properties in terms of stereoselectivity, specificity and stability [55]. The flexibility and low density properties are advantageous since the thin film is fabricated using mild polymerization conditions without decomposition problems and guarantying excellent selectivity [55]. Other studies include the optimization of sol-gel based MIPs to recognize molecules like metabolites, proteins, cells, pesticides [61,63–65].

3. Imprinting Technology

MIPs denote polymer matrices designed for highly selective recognition of target molecules. The ideal sensitivity for a certain envisioned application can be achieved with a proper control of the polymer matrix and assuring an optimal design of the imprinting process. An important role is played by the conditions under which the template molecules are introduced to the polymerization mixture, i.e., appropriate procedures for imprinting the affinity sites with memory for the target are critical.

3.1. Bulk Imprinting

In the most usual imprinting approach, the surface of the transducer is coated with a pre-polymer mixture to which typically a small molecule template is added. These components interact in the solution creating a network that is then polymerized. The template that is entrapped within the polymer matrix in the curing process is removed by elution, leaving behind cavities throughout the bulk material that possess shape and size selectivity to recognize the analyte in the subsequent sensor measurements [3]. The binding sites in the imprinted material should retain their shape, thus the polymer needs to meet certain requirements in terms of porosity, allowing molecular diffusion, and also stiffness, obtained via suitable cross-linking [66]. To assist the process of template removal, physical (e.g., increasing the temperature) or (bio)chemical (e.g., washing with solvents, acids/bases, detergents, and digesting enzymes) methods are usually applied [67]. Due to the creation of binding sites within the polymeric bulk, the features of this type of imprinting are most convenient to generate binding cavities to relatively small molecules. The advantages for macromolecules would be related to the formation of 3-D binding sites for the entire structure with easy preparation procedures. However, the drawbacks are linked to the bulk volume, originating poor mass transfer and slow binding kinetics due to long diffusion path lengths and reduced number of effective imprinted sites in the polymer matrices for template rebinding [3,68]. These limitations are reflected in the type of molecules that are imprinted, usually restricted to small molecules [69]. Thus, alternative methods are proposed to place the binding sites mainly on the surface, and thus enlarging the applications with enhanced efficiency.

3.2. Surface Imprinting

In order to extend the use of MIPs to larger molecules, allowing their diffusion and facilitating template removal and rebinding, alternatives have been conceived, such as: (a) post-processing of bulk imprinted material by grinding and sieving to expose the imprinted sites; (b) imprinting hydrogels; (c) adding substances to increase the porosity of the polymer; (d) increase the surface-to-volume ratio by using polymer nanoparticles; (e) use a conducting polymer film deposited by electropolymerization [67,68]. In these approaches, the binding sites that are imprinted are positioned near the polymer surface and allow the synthesis of thin polymer films. Due to the existence of the support, the imprinted assembly is more robust and is easily integrated within electrochemical platforms. The main drawback is related to the available surface that is limiting the extent of binding sites. Nonetheless, these materials offer expedite accessibility to larger target analytes, namely proteins, virus and cells. Different approaches towards the production of surface imprinted materials have been described in the literature and presented next.

3.2.1. Microcontact Imprinting

Included in the soft lithography techniques, microcontact printing is commonly used to transfer molecules onto a surface using a patterned elastomeric mold or stamp (e.g., poly(dimethylsiloxane)—PDMS). This simple and cost-effective technique is very attractive taking also into account the versatility of substrates and materials to be imprinted [70]. Briefly, the stamp, previously molded from a solid master, is coated with biomolecule-containing solution and then brought into contact with the substrate to be patterned; in this stage, it transfers the biomolecules from the microscale raised surfaces in the mold to the substrate [71]. As previously mentioned, the creation of artificial recognition sites by molecular imprinting is comparable to natural receptors with the additional advantage of resisting much harsher conditions. To achieve successful specific molecular recognition, factors like size, shape and chemical functionality have to be addressed. Thus, when considering macromolecules, such as proteins and cells, the use of bulk monoliths to form the imprinted sites may hinder a proper recognition due to diffusion constraints [3,72]. In this context, microcontact printing techniques can be useful to form highly sensitive, low cost and stable thin films with imprinted surface molecular recognition architectures [73]. Biomimetic sensors incorporating microcontact imprinted films for various protein biomarkers (e.g., C-reactive protein, ribonuclease, lysozyme, myoglobin, ovalbumin) have been studied [73,74]. The general procedure of extended standard microcontact printing technique applied to MIP construction can be depicted as (Figure 2): a glass slide is used to make the template stamp by its adsorption on pre-treated glass surface, while the pre-polymer (monomers and cross-linker) mixture solution is deposited in the transducer support serving as MIP substrate; the template stamp is brought into contact with the substrate and the polymerization is initiated; the last step involves removing the stamp and eluting the bound template from the obtained MIP [75].

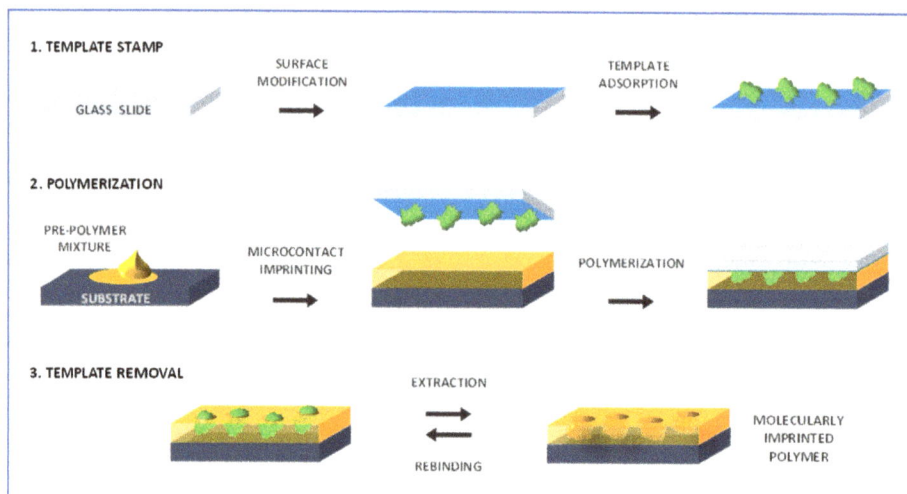

Figure 2. Schematic representation of microcontact imprinted polymer fabrication.

The flat and optically transparent glass used as the adsorption substrate enables the sensing layer to be synthesized by UV polymerization. Also, the glass surface is easily modified with organic compounds (e.g., amino groups) to enhance its affinity for the template biomolecules. In order to increase the site-specific spatial organization of the functional monomer, the template can be first assembled with the monomer in the cover glass slip and only after introduced into the remaining polymer mixture with subsequent polymerization. Once the analyte molecules interact with the sensor surface they can easily rebind to the recognition cavities. This type of procedure has been used to create high affinity MIPs for several biomolecules, such as ribonuclease A (RNase A), lysozyme and myoglobin [75,76]. One study used microcontact imprinting to prepare capacitive gold electrodes for BSA detection as a model protein. BSA was imprinted onto pre-modified gold electrode surface by microcontact with subsequent UV-polymerization and the results showed good selectivity and stability [77]. As for other types of imprinting techniques, the functional monomer(s) chosen, which mimic structures or chemical features of natural receptors, the cross-linking agent and the respective ratios are very important in achieving higher selectivity for the target molecule [73,78]. Moreover, in terms of MIP recognition of protein targets, evidence suggests that conformational changes may occur, along with protein denaturation leading to some degree of aggregation, thus recognition of protein fragments may take place to some extent [76]. An example of such occurrence was described for RNase A when interacting with the functional monomer styrene, that despite providing MIPs with higher affinity for the enzyme could lead to RNase A denaturation. The study of surface morphology and imprinted cavities suggests the presence of large sizes of imprinted cavities fitting with the hypothesis that aggregates rather than individual RNase molecules bind into MIPs cavities [76]. Nonetheless, good recognition is not of concern and the presence of any small structural fragments can work also as recognition elements [75]. Microcontact printing has also been combined with layer-by-layer (LbL) assembly to have an unconventional approach towards the concept of molecular imprinting, simultaneously tackling the current difficulties of slow binding kinetics and low accessibility of binding sites in the bulk polymer [79]. This method has been advanced to introduce new functionalities to nanostructured LbL thin films, namely to fabricate surface molecularly imprinted films [79]. Interesting results have also been reported on the use of this methodology to prepare bacterial stamps and proceed to the imprinting and quantification of bacterial cells [80,81].

3.2.2. Polymer-Brush Imprinting

Polymer-brush imprinting relies on the tethering (grafting) to a solid interface of polymer chains through one end. The template molecules are first attached to a polymer layer with subsequent removal by enzymatic/chemical treatment. A grafted polymer occupies the space neighboring the adsorbed template, and when the target analyte recognizes and rebinds to the substrate, it only occurs in the imprinted cavities [82] (Figure 3).

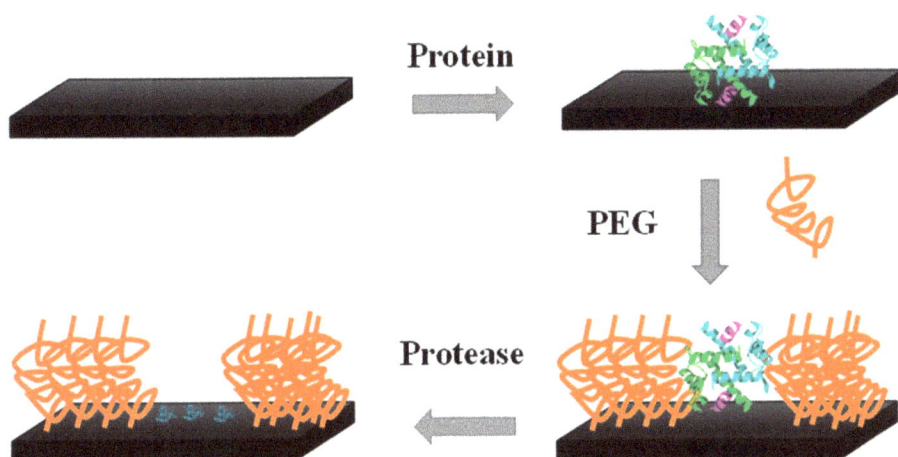

Figure 3. Polymer-brush imprinting.

In the particular case of electrochemical biosensors, the main challenge is to tackle the tendency of abundant proteins present in the complex mixture of compounds that compose biological samples to physically adsorb without specific receptor-recognition interactions. This adsorption is non-specific and it naturally reduces the expected biosensor performance leading to electrode fouling and decreasing the sensitivity when it comes to detect the target biomarkers typically in a very low concentration. Non-specific interactions can be prevented or minimized by using self-assembled monolayers (SAMs), namely of poly(ethylene glycol) (PEG) or oligo(ethylene glycol) (OEG) [82–85].

Zdyrko et al. [86] employed grafting of polymer brushes as an approach to surface protein imprinting. An ultrathin polymer layer of poly(glycidyl methacrylate) served to chemically bind the protein molecules that were later removed by protease treatment. A PEG grafted layer was introduced on the surrounding space of the adsorbed molecules, so that the residual amino acids on the surface after template removal formed nanosized recognition sites. The imprinted substrate was selective, differentiating the bovine serum fibrinogen from BSA [86]. Another method of surface imprinting using SAMs was showed by Wang et al. [87] by producing a thiol film on a gold surface with the co-adsorption of the template biomolecules. The template is removed afterwards, leaving the SAMs matrix with recognition sites. This system was used to detect the carcinoembryonic antigen, a cancer biomarker, in samples containing only the purified biomarker and also using the culture medium of a human colon cancer cell line. More recently, Yang and co-authors used silica particles to anchor proteins and polymer brushes through a reaction of thiol—disulfide exchange. The grafted polymer brushes were temperature-sensitive and in their collapsed state above the lower critical solution temperature the protein molecules were protected what enabled to control their catalytic activity [88].

A device has been reported to detect uric acid (UA) using differential pulse cathodic stripping voltammetric (DPCSV) analysis [89]. The electrodes were prepared of sol-gel modified graphite and worked as support to graft MIP brushes of poly(melamine-co-chloranil). This approach enabled to avoid non-specific binding arising from cross-reactivity and matrix interactions. The DPCSV signals were generated by the electrochemical oxidation of UA with subsequent cathodic stripping, allowing a selective detection within the limits required to diagnose hyperurecemia. In a similar work developed to detect dopamine, Prasad et al. applied DPCSV in combination with a solid-phase microextraction fiber. In this method, MIP brushes prepared using SAMs of low molecular weight were grafted to a sol–gel matrix coupled to an optical fiber [90]. In summary, polymer brushes hold multi-faceted properties, like self-assembly, wettability, switchability, and biocompatibility.

3.2.3. Surface Grafting

Surface grafting imprinting is another effective approach to overcome the difficulty of template mass transfer within MIPs and particularly valuable when considering the imprinting of macromolecular structures such as proteins, polysaccharides or microorganisms [91].

Biomimetic sensor development can highly benefit from surface protein-imprinted nanoparticles. The imprinted material can be obtained through the synthesis of a thin polymer film or by using a substrate (e.g., nanoparticles, nanotubes) that functions as support to attach the template to its surface with following polymerization around it [1,2,92,93]. The imprinted binding sites are positioned nearby the surface of the outer polymer layer, avoiding limited mass transfer and enabling easy template removal [93,94]. The drawback is the limited number of binding sites.

Kryscio and Peppas described several macromolecular imprinting works published through 1994–2010, showing a considerable rise of interest since 2005. This critical review shows that this particularly field has been ruled by surface imprinting that comprises around 60% of the overall published works [3]. Surface imprinting techniques are widely applicable to proteins that have great interest as disease biomarkers. In this context, surface grafting MIPs present some advantageous properties because the presence of a support confers a higher physical robustness and it is more easily integrated in sensory platforms. Still, as the protein is only partially imprinted, some loss in specificity can be a drawback. Some studies using this approach in combination with electrochemical transduction are highlighted below.

Moreira et al. [1] described a MIP on the surface of multiwalled carbon nanotubes (MWCNT-COOH) that functioned as an artificial antibody for the selective detection of troponin T. The protein was covalently linked to MWCNT surface and a polymer prepared with acrylamide and N,N'-methylenebisacrylamide filled the vacant spaces. The protein was removed from the polymeric matrix by acidic treatment with oxalic acid. The electroactive modified nanotubes were incorporated in a PVC/plasticizer mixture forming a membrane, and the performance of the biomimetic sensor was subsequently tested. In general, these simply designed sensors offer excellent properties in terms of short measurement times, high precision, accuracy and throughput, as well as low LOD, and good selectivity [1].

The same authors reported more recently the synthesis of a smart plastic antibody material (SPAM) to target myoglobin in point-of-care. The design followed a bottom-up approach and the SPAM was tailored on top of disposable gold-screen printed electrodes (Au-SPE). The MIP cavities were visualized for the first time by AFM only in the SPAM network. Another outstanding feature was related to the rebinding of the target protein that only occurred in the SPAM materials, producing a linear electrical response against square wave voltammetry (SWV) assays. In contrast, the non-imprinted polymers showed a similar-to-random behaviour. The SPAM/Au-SPE device allowed the detection of myoglobin down to 1.5 µg/mL and 0.28 µg/mL in EIS and SWV, respectively. Using voltammetric assays, the SPAM materials showed negligible interference from troponin T, BSA and urea, displaying promising results for point-of-care applications [2] (Figure 4).

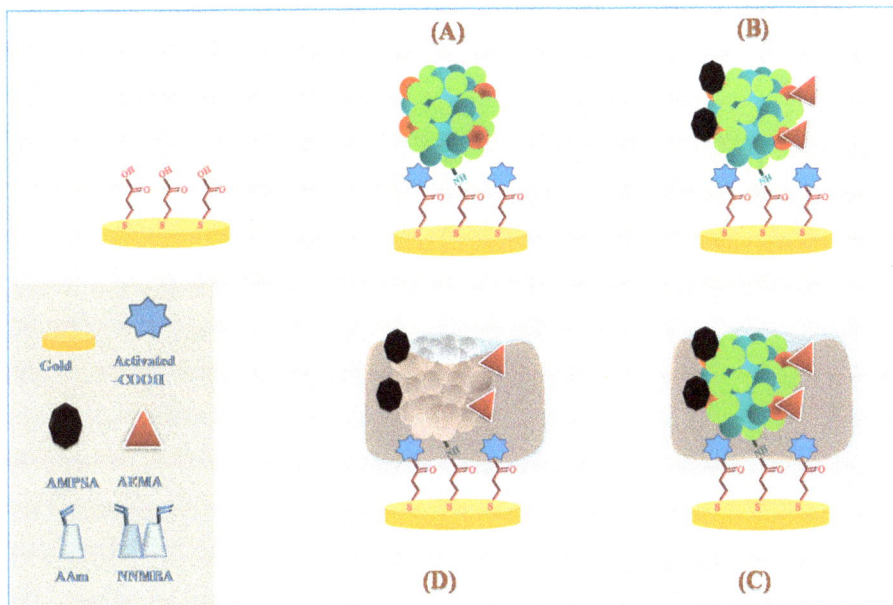

Figure 4. Synthesis of the SPAM material. (**A**) Protein bound; (**B**) Charged labels; (**C**) Polymerization; (**D**) Template removal.

3.2.4. Electropolymerization

Electropolymerization is typically conducted by applying a suitable potential or range of potentials to a solution containing the template with the monomer, originating a film formation on the surface of the electrode [95]. This simple approach is useful since by adjusting the electrochemical conditions (e.g., potential range, number of cycles and scan rate) and by using different conductive materials of various shape/size one can achieve a close control of the polymer thickness [18]. The electropolymerization around the template normally requires the use of a functional monomer, a porogenic solvent, and sometimes a cross-linking monomer in contact with the transducer surface. Also, no initiator is required, nor UV light or heat. Thus, electropolymerization is another method with plenty attractive features to overcome the difficulties of producing MIPs for macromolecules such as proteins, as reviewed elsewhere [96].

The electrosynthesis of MIPs involves conductive polymers (ECP) and insulators/non-conductive polymers (NCP). The charge transfer between the electrode substrate and the analyte that occupies the molecular cavities of the MIP is assured by incorporating ECP matrices in the film. Several research works in the literature report the use of electroactive monomers like 3,4-ethylenedioxythiophene (EDOT) [97–99], pyrrole [100–103], aniline [104,105], thiophene [106,107], and dopamine [108,109] for MIP electrosynthesis of organic compounds (Figure 5).

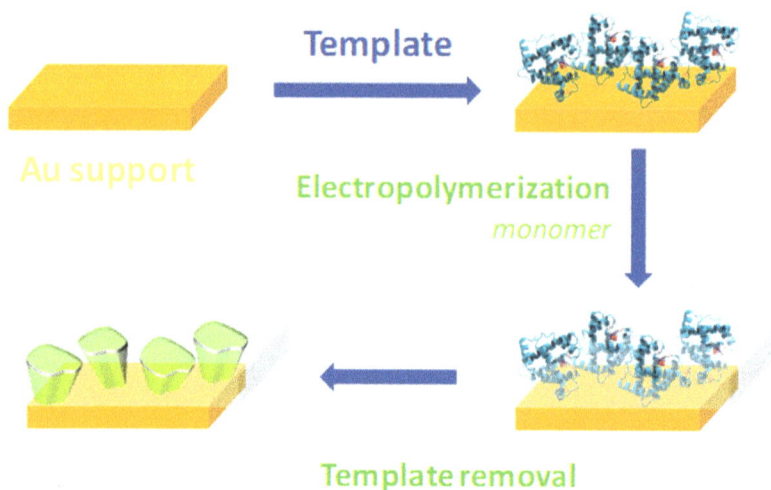

Figure 5. Electrosynthesis of MIPs.

Several non-conducting MIP films from the electro-synthesis of non-conductive monomers as phenol [32], aminophenol [110,111], phenylenediamine [112,113] among others have also been reported. Electropolymerization resulting in non-conducting MIP films are widely used for capacity chemosensors specially polyphenylenediamine and polyphenol. These materials form non-conductive, compact MIP films after electropolymerization. The prepared MIP biorecognition element can detect the analyte at nanoscale level with notable selectivity and low LOD. The selectivity of the electrosynthesized MIP can be improved by modifying the monomers with additional functional groups.

The resulting conducting or non-conducting MIPs show advantages and disadvantages. The deposition by electropolymerization of non-conducting MIP films has to be tightly controlled related to the polymer thickness. This self-limiting characteristic is due to the need of stopping the deposition upon reaching a thickness at which the polymer insulates the fundamental conducting electrode surface. On the other hand, deposition of ECPs by electropolymerization may occur indeterminately as the deposition conditions control the polymer thickness. The method of choice for signal transduction is related with the conductivity of the polymer. The impedance variations can more easily be detected with a NCP.

3.3. Epitope Imprinting

Another approach for macromolecular imprinting that relies on using a partial, short peptide as template, has been designed namely to improve the recognition of proteins and polypeptides. In the epitope strategy, that closely mimics the natural recognition antigen-antibody, suitable moieties have to be properly identified to represent the parent macromolecule [114]. Short fragments from exposed domains of the proteins have been generally selected as template. The use of epitope imprinting can thus be particularly valuable to target membrane proteins, since the recognition occurs through the part of the protein localized on the cell surface [115]. Interestingly, this approach is very sensitive to any mismatch in the amino acid sequence, as determined by imprinting native and single-point mutated template peptides for BSA. This experience demonstrated that the MIP film imprinted with the mismatched peptides presented lower selectivity of detection [116]. Similar results were obtained when imprinting target peptides relevant for cancer diagnosis, since the MIP was able to differentiate single amino acid mismatches [117].

Some advantages are attributed to this process, namely because it is a small template that is impressed, reducing non-specific interactions pertain to biomolecules complexity as well as conformational instability and improved binding kinetics [3,116,118]. Nonetheless, as MIPs are able to bind the whole analyte, diffusion constraints on rebinding studies have to be addressed and a combination with surface imprinting techniques facilitates protein access and defined orientation [117,119]. Such epitope-based imprinted materials show stronger affinity to the target protein with prominent selectivity when performing competitive binding assays [118,119]. Detection of cytochrome c is an example of the successful use of epitope imprinting together with electrochemical template removal and control of polymer film thickness [120]. Several studies show that the epitope approach combined with surface-confined imprinting using hybrid nanostructured materials can provide improved protein recognition [121,122].

4. Electrochemical Transduction

In biosensors with electrochemical transduction the target analyte binds the biorecognition element selectively. The detection of the analyte is based in the variations at electrode surface in terms of alteration in current and or voltage.

The first MIP-based electrochemical sensor was described in the early 1990s by Mosbach's group [123]. Three years later, Hedborg and co-authors reported the first thin polymer membrane-based molecular imprint. The membrane was composed by the imprinted sites of the L-phenylalanine anilide and was casted as a sensing material film in field-effect capacitors [124].

Electrochemical biosensors are widely synthesized by electropolymerization of such monomers as pyrrole [101], aniline [104], o-phenylenediamine (o-PD) [125,126], phenol [32], aminophenol [110,111,127] and EDOT [128,129], self-assembled monolayers (SAMs) [2,130,131] and sol-gel [93] materials.

The electrochemical transduction can be categorized according to the analytical signal output. In potentiometry the measurement is the potential (V), in amperometry and voltammetry is the current (A), in impedance is the resistance (Ohm) and in conductance, Siemens (S). Figure 6 illustrates the basic mechanisms of electrochemical MIP-based biomimetic sensors.

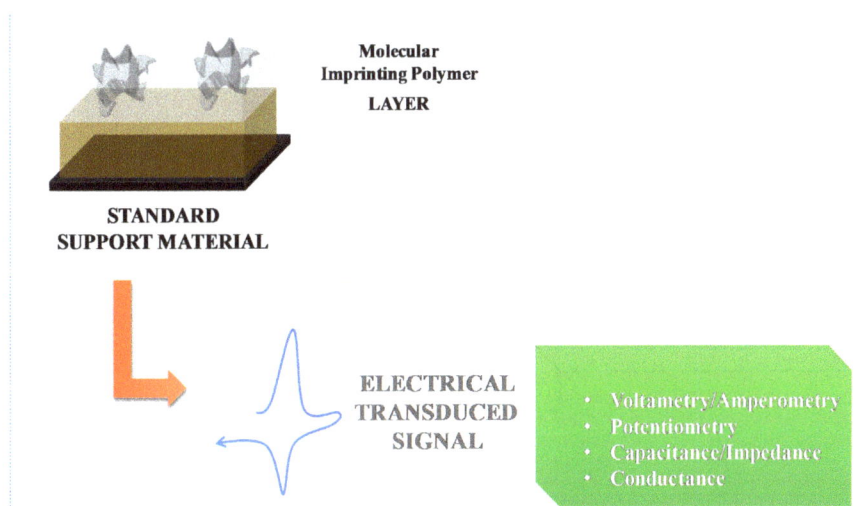

Figure 6. Basic mechanisms of electrochemical MIP-based biomimetic sensors.

4.1. Voltammetry/Amperometry

In voltammetric/amperometric MIP-based sensors the electroactive species are detected directly involving a linear or logarithmic correlation between the amount of species and the current measured [18,132]. Voltammetry is widely used due to inherent oxidation and reduction potential features of the electrodes, sensory materials and target analytes. The voltammetric techniques usually applied include: cyclic voltammetry (CV) linear sweep voltammetry (LSV), differential pulse voltammetry (DPV), square wave voltammetry (SWV) and cronoamperometry (CA). Table 1 depicts a summary of various voltammetric/amperometric MIPs and their analytical features for different organic compounds.

Some researchers incorporated the redox probe in the polymeric matrix. One example is the work developed by Udomsap et al. [133] that reported a versatile electrochemical MIP-based sensor with the introduction of a redox probe (vinylferrocene) inside the binding cavities of a crosslinked MIP. The biomimetic sensor detected easily the analyte benzo[a]pyrene with LOD of 0.09 μmol/L [133].

Some efforts have been developed by several research groups by using different nanomaterials and polymers in order to obtain a stable, sensitive and selective electrochemical MIP sensor. The high surface to volume ratio presented by these materials is a property that helps the molecule to access the multiplicity of the binding sites.

In order to increase the sensitivity of the biomimetic materials, some metallic nanoparticles or nanomaterials can be added to the polymeric matrix. These include iron oxide, gold nanoparticles (AuNPs), silver, and graphene oxide, among others. An innovative biomimetic material fabricated with magnetic graphene oxide and AuNPs was created and applied as bioreceptor to build a dibutyl phthalate electrochemical device. Under optimized conditions, the sensor showed a LOD of 8.0×10^{-10} mol/L [134].

Recently, Tan and co-authors described an electrochemical MIP-based sensor synthesized on glassy carbon (GC) electrode surface for carbofuran detection. Graphene oxide and AuNPs composites were used to improve the analytical signal upon binding of the target analyte on MIP receptor. The polymerization around the template molecule was carried out with the presence of functional monomer (methyl acrylic acid) and cross-linker (ethylene glycol maleic rosinate acrylate). The LOD was 2.0×10^{-8} mol/L [135].

Moreover, magnetic nanoparticles have been incorporated in conductive polymer and the combination of both properties resulted in an enhancement of the sensitivity of the biomimetic sensor. This device was described by Zamora-Gálvez and co-authors for sulfonamide detection with a LOD of 1×10^{-12} mol/L with similar results to the obtained by liquid chromatography and mass spectrometry [92]. Another technique to improve the sensitivity of MIP sensors is the polymerization of the electroactive materials concomitantly with substrate-guided dopant immobilization [136].

Table 1. Voltammetric transduction for MIP based electrochemical sensors (2014–2016).

Analyte Category	Template/Analyte	Monomer	Electrode	Detection Technique	LOD (M)	Linear Range (M)	Reference
Drugs	Ractopamine	Aminothiophenol	Screen printed electrode	DPV	4.23×10^{-11}	5.0×10^{-11}–1.0×10^{-9}	[137], 2016
	Famciclovir	Methacrylic acid and vinyl pyridine	Carbon paste electrode	CV	7.5×10^{-7}	2.5×10^{-6}–1.0×10^{-3}	[138], 2015
	Artemisinin	Acrylamide	Glassy carbon electrode	CV	2.0×10^{-9}	1.0×10^{-8}–4.0×10^{-5}	[139], 2015
	Dopamine	Aminophenol	Metallic microrod	CV	7.63×10^{-14}	2.0×10^{-13}–2.0×10^{-8}	[140], 2016
	Metronidazole	1,2-dimethylimidazole, dimetridazole, o-phenylenediamine	Nanoporous gold leaf	CV	1.8×10^{-11}	5.0×10^{-11}–1.0×10^{-9} and 1.0×10^{-9}–1.4×10^{-6}	[141], 2015
	Theophylline	4-amino-5-hydroxy-2,7-naphthalenedisulfonic acid	Glassy carbon electrode	CA	0.32×10^{-6}	0.4–17×10^{-6}	[142], 2016
	Epinephrine	Pyrrole	Indium tin oxide	DPV	–	1–10×10^{-6} and 10–800×10^{-6}	[143], 2016
	Carbofuran	Methyl acrylic acid	Glassy carbon electrode	DPV	2.0×10^{-8}	5.0×10^{-8}–2.0×10^{-5}	[135], 2015
	Eugenol	Aminobenzenethiol-co-p-aminobenzoic acid	Glass carbon electrode	LSV	1.0×10^{-7}	5.0×10^{-7}–2.0×10^{-5}	[144], 2016
	Cholesterol	Aminothiophenol	Glassy carbon electrode	DPV	3.3×10^{-14}	1.0×10^{-13}–1.0×10^{-9}	[145], 2015
Organic molecules	Melamine	Methacrylic acid	Diazonium-modified gold electrodes	SWV	1.75×10^{-12}	1.0×10^{-11}–1.0×10^{-4}	[146], 2015
	Glyphosate	p-aminothiophenol	Gold electrode	LSV	5×10^{-15}	5.9×10^{-15}–5.9×10^{-9}	[147], 2015
	Ascorbic acid	Polyvinylpyrrolidone	Glass carbon electrode	DPV	3.0×10^{-6}	10–1000×10^{-6}	[148], 2015
	Dibutyl phthalate	Methacrylic	Gold electrode	DPV	8.0×10^{-10}	2.5×10^{-9}–5.0×10^{-6}	[134], 2015
	Protein A	Aminophenol	SWCNT-Screen printed electrode	SWV	0.6×10^{-9}	23.8×10^{-9}–4.76×10^{-6}	[110], 2016
Biomacromolecules	Guanine-rich DNA (G-rich DNA)	Methacrylic acid and guanine	MWCNT electrode	DPV	7.52×10^{-9}	0.05–1×10^{-6} and 5–30×10^{-6}	[149], 2016
	Benzo[a]pyrene	Vinylferrocene	Carbon paste electrode	SWV	0.09×10^{-6}	0.08×10^{-6}–3.97×10^{-6}	[133], 2014
	Carcinoembryonic antigen	Pyrrole	Silver-Screen printed electrode	SWV and CV	2.8×10^{-16}	2.8×10^{-16}–6.9×10^{-15}	[101], 2016
	Human serum albumin	bis(2,2'-bithien-5-yl)methan	Gold electrode	DPV	0.25×10^{-12}	12×10^{-12}–300×10^{-12}	[150], 2015
	DNA	Pyridine	Carbon paste electrodes	SWV	1.38×10^{-6}	0–7.9×10^{-6}	[151], 2015
	Troponin T	Pyrrole	Screen printed electrode	SWV	1.64×10^{-13}	2.74×10^{-13}–2.74×10^{-12}	[152], 2016

4.2. Potentiometry

In potentiometry the potential is evaluated in flow or batch conditions. The potential obtained is used to evaluate or quantify the amount of an ion in solution [153]. Potentiometry shows some advantages in terms of simplicity, fast response time and low cost for detection of different compounds. Potentiometric sensors can be allocated in two groups: ion selective electrodes (ISEs) and field-effect transistors (FETs) [154]. In general, MIP potentiometric based sensors show outstanding features in terms of design, reusability, stability, low cost, response time and good sensitivity and selectivity.

4.2.1. ISE Systems

In potentiometry the species do not need to diffuse through the membrane, thus there is no size constrain on the template molecule. Due to these benefits, several MIP-based sensors have been described with potentiometric transduction [155] for detection of organic and inorganic compounds (Table 2). Murray et al. stated for the first time an ion-selective electrode based on a MIP as a biorecognition element with potentiometric transduction [156] for lead ions selective detection.

The main concern of the ISE includes the use of a PVC membrane, as it needs high volatile solvents and the membrane thickness is difficult to control, it may cause a reproducibility problem at the time of production.

Table 2. Potentiometric transduction for MIP based electrochemical sensors (2013–2016).

Analyte Category	Template/Analyte	Monomer	Electrode	Detection Technique	LOD (M)	Linear Range (M)	Reference
Drugs	Azithromycin	Acrylic acid and 2-vinyl pyridine	Graphite electrode	Potentiometry	1.0×10^{-7}	1.0×10^{-1}–1.0×10^{-6}	[157], 2015
	Losartan	Methacrylic acid	Graphene/carbon paste electrode	Potentiometry	1.82×10^{-9}	3.0×10^{-9}–1.0×10^{-2}	[158], 2015
	Clenbuterol	Chitosan	Carbon paste electrode	Potentiometry	0.91×10^{-11}	1.0×10^{-7}–1.0×10^{-12}	[159], 2016
	Taurine	3,4-Ethylenedioxythiophene	Glassy carbon disc electrodes	Potentiometry	-	1.0×10^{-2}–1.0×10^{-4}	[160], 2016
	Histamine	Methacrylic acid	Solid phase extraction	Potentiometry	1.12×10^{-6}	1.0×10^{-6}–1.0×10^{-2}	[161], 2014
	Carnitine	Vinylbenzyl trimethylammonium chloride and 4-styrenesulfonic acid	Graphite and ITO/FTO	Potentiometry	3.6×10^{-5}	1.0×10^{-6}–1.7×10^{-3}	[162], 2014
	Neopterin	Bis-bithiophene, bithiophene derivatized with boronic acid	Pt disk-working electrode	Potentiometry	22×10^{-6}	0.15×10^{-3}–2.5×10^{-3}	[163], 2016
	Dopamine	Acrylamide grafted MWCNTs	Cu electrode surface	Potentiometry	1.0×10^{-9}	1.0×10^{-9}–1.0×10^{-5}	[164], 2014
	Urea	Poly(methyl methacrylate)	ISFET	Potentiometry	1.0×10^{-4}	1.0×10^{-4}–1.0×10^{-1}	[165], 2016
	Memantine hydrochloride	Methacrylic acid	—	Potentiometry	6.0×10^{-6}	1.0×10^{-5}–1.0×10^{-1}	[155], 2013
	Chlorogenic acid	Pyrrole	Graphite electrode	Potentiometry	1.0×10^{-6}	1.0×10^{-2}–1×10^{-6}	[166], 2016
Biomacromolecules	Prostate specific antigen	Vinylbenzyl(trimethylammonium chloride), vinyl benzoate	Conductive carbon over a syringe	Potentiometry	5.8×10^{-11}	5.83×10^{-11}–2.62×10^{-9}	[167], 2016
	Carcinoembryonic antigen	11-Mercapto-1-undecanol	Gold electrode	Potentiometry	2.8×10^{-12}	2.8–8.3×10^{-11}	[168], 2016

Some imprinted PVC membranes have been recently reported for some drugs and metabolites as: azithromycin [157], losartan [158], clenbuterol [159], taurine [160] and carnitine [162,169], among others. In these, a sensitive MIP for azithromycin has been reported. The polymerization occurred in the presence of the monomers, cross-linker, initiator and target analyte, being 2-vinyl pyridine and acrylic acid, ethylene glycol dimethacrylate, benzoyl peroxide and azithromycin respectively. After the polymerization, the sensor was entrapped in a polymeric membrane with di-butyl phosphate (DBP) and di-octyl phthalate (DOP) as solvent plasticizers. The sensor shows good overall analytical performance in terms of slope and LOD (49.1 and 51.3 mV/decade and 10×10^{-7} mol/L) respectively.

Imprinting proteins have also been successful integrated with potentiometric transduction. However, imprinting proteins can be a challenging task, since they change their conformation easily. Also, proteins hold multiple charge locations that change according to the specific conformation. In order to overcome these limitations Rebelo and co-authors proposed recently an imprinted sensor for prostate specific antigen detection, were the imprinted sites were obtained by adding charged monomers as labels around the template protein in order to obtain specific binding sites. The polymerization was carried out with uncharged monomers in neutral conditions. The sensors were entrapped in PVC membranes with a solvent plasticizer and applied to conventional solid-contact carbon electrodes. The analytical performance of the sensor was evaluated by potentiometric assays by means of calibration curve, selectivity study and sample analysis [167].

4.2.2. FET Systems

The integration of MIP-based sensors with organic/inorganic compounds within an ion-selective field-effect transistor (ISFET) configuration is also a very attractive approach. The first report of ISFET was a short communication by Bergveld in 1970. This work described for the first time a miniaturized silicon-based chemical sensor, followed by a more extensive paper in 1972 [170,171]. FET biosensors provide huge advantages in terms of response time and dimension, enabling biosensor integration with on-chip arrays as a promising tool concerning the production of cheap and portable microanalysis devices. A work published in 2016 described the integration of MIP with ISFET for creatinine and urea. This biosensing device showed good reproducibility, repeatability and stability, as well as high selectivity [165]. Despite ISFET advantages, the classical FET systems may present compatibility limitations related to necessary chemical modifications when incorporating a recognition layer for sensing applications [172].

With the progress of research in technological applications, extended-gate field-effect transistor (EG-FET) has emerged, a practical and useful alternative as extension of ISFET that allows producing different sensors, including those based on MIP techniques. These devices based on the measurement of the current changes at constant gate voltage, allow the flexibility of shape of the extended gate structure, high sensitivity to distinguish homologous analytes, and good insensitivity to the light [172,173].

Several works have described EG-FET systems based on MIPs. In 2016, Dabrowski and co-authors reported a MIP film developed for D-arabitol detection in biological samples like urine using the EG-FET as transducer. The analyses were performed under stagnant-solution binding conditions, presenting a linear dynamic concentration range between 0.12 and 1.00 mmol/L with a LOD of 0.12 mmol/L. In general, these devices were considered good sensors able to detect very low concentrations of D-arabitol in urine, using a simple and inexpensive procedure [172]. Likewise, another work published by Iskierko and co-authors described two EG-FET based MIP sensors for a selective detection of specific phenylalanine enantiomers, D- and L-phenylalanine (Phe). Thus, the authors present successful sensors able to recognize levels of D- and L-Phe under 13 to 100 µmol/L with a LOD of 13 µmol/L and a good discrimination of interfering substances, such as D-proline, D-alanine, and D-tyrosine [174].

4.3. Capacitance/Impedance

The progress observed in electrochemical biosensors brought some promising research in capacitive and impedimetric approaches [175]. An impedimetric biosensor is defined as the application of impedance, as the transduction principle, in a biosensor system, which through measurements and/or monitoring of targeted analyte (antigen-antibody interactions, oligonucleotide-DNA interactions, among other biomolecules) on electrode surface, provides the output of an electrical impedance signal proportional to analyte activity [176].

Electrochemical Impedance Spectroscopy (EIS) allows monitoring of the electrochemical events occurring at the interface electrode and solution. This method allows the calculation of several parameters of the fitted model based on the ratio between the voltage and the current at a specific frequency response of the electrochemical system. The first impedimetric biosensor based on MIP technology with ultrathin insulating membranes was described by Panasyuk and co-authors in 1999 [177].

Special attention has been given to capacitance/impedance technique due to its outstanding properties in terms of high sensitivity, no need for labelling and possible real time monitoring (Table 3).

Recently, in 2016, Khan reported a bulk imprinting assay for Protein A (PA) detection in point of care. The MIP was developed on single walled carbon nanotubes (CNTs)-based screen printed electrodes [110]. The imprinted stage consisted in the mixture of the monomer (3-aminophenol) with PA following its electropolymerization by CV technique. The template was removed from polymeric matrix by proteolytic action of an enzyme. The control of the surface modification and PA detection was followed by EIS. Furthermore, more research work should be performed in order to get a reliable biosensing device with suitable thickness, homogeneous and stable polymeric surfaces.

In another recent work, Moreira and co-authors presented a MIP sensor synthesized on top of a silver SPE fabricated by PCB technology to detect a protein, carcinoembryonic antigen, in point-of-care. The protein was entrapped in the polypyrrole polymeric matrix and removed by enzymatic action of proteinase K. The analytical performance of the biomimetic sensor was evaluated by electrochemical techniques. The sensor shows linear response after 0.05 pg/mL against logarithm concentration [101].

EIS is considered a powerful tool to investigate dynamics of the bound or charge transfer in bulk or at interfacial region of such system. The non-destructive measurements make EIS a promising tool for studying MIP biosensing devices without perturbing its operation.

Table 3. Impedimetric/Capacitive transduction for MIP based electrochemical sensors (2013–2016).

Analyte Category	Template/Analyte	Monomer	Electrode	Detection Technique	LOD (M)	Linear Range (M)	Reference
Drug	Carnosine	Carboxy and 18-crown-6 ether and bis(2,20-bithien-5 yl)methane	Gold electrode	EIS	20×10^{-6}	0.1×10^{-3}–0.75×10^{-3}	[178], 2016
	Theophylline	Pyrrole	Silicon substrates	EIS	-	0.1×10^{-9}–1.0×10^{-6}	[179], 2015
	Aflatoxin B1	Ovalbumin and glutaraldheide	Gold electrode	EIS	6.3×10^{-12}	3.2×10^{-6}–3.2×10^{-9}	[180], 2016
Biomacromolecule	Prostate specific antigen	Dopamine	Gold electrode	EIS	2.94×10^{-14}	2.94×10^{-9}–2.94×10^{-12}	[181], 2016
	Protein A	Aminophenol	SWCNT-screen printed electrode	EIS	16.8×10^{-9}	23.8×10^{-9}–2.38×10^{-6}	[110], 2016
	Carcinoembryonic antigen	Pyrrole	Silver- screen printed electrode	EIS	2.8×10^{-16}	2.8×10^{-16}–6.9×10^{-15}	[101], 2016
	Carnitine	3,4-ethylenedioxythiophene (EDOT)	Carbon-cellulose paper	EIS	2.15×10^{-10}	1.0×10^{-8}–1.0×10^{-3}	[182], 2016

5. Labelling Methods

Commonly, electrochemical detection approaches have fallen into two main categories: labeled and label-free (direct) detection systems. According to the definition, the former rely on the attachment of a label to the molecule of interest to be detected either by its molecular presence or activity. This chemical (covalent bonding) or temporary intermolecular interaction can potentially promote changes on their intrinsic properties and, thus, produce an electrical signal. Instead, a label-free detection system is defined as an approach that do not require the use of biological or chemical receptors (colorimetric, fluorescent, luminescent or radiometric), in order to afford measurements [183–186].

5.1. Label-Free MIP-Based Sensors

Recent trends in the development of sensor structures focus on the use of inexpensive plastic materials, including sensor fabrication on long and continuous sheets of plastic film and in their incorporation into single-use disposable devices. As label-free detection instrumentation becomes more simple and compact, its integration in other biochemical analysis methods, as liquid chromatography, fluorescence assays, and mass spectrometry, will turn the systems capable of characterizing biomolecules with more complex information. The advances in sensors concept, inspire the continuous replacement of label-based assays, like devices based on fluorescence labelling, radiolabels, among others, with label-free detection methods [185]. Unlike what happens with label-based technologies, that simply allow to confirm the presence or absence of a detector molecule, the main advantage of label-free detection is to provide more detailed and direct information (such as selectivity, affinity, stoichiometry, kinetics or, thermodynamics of an interaction), since these methods only investigate native proteins and ligands that constitute a sample [176,185–187]. Furthermore, a major advantage of label-free detections is their high sensitivity enabling real-time detection and, thus, simplifying the time and effort of assay development [186,188].

Label-free biosensors have promoted a great advance in different fields, as in material sciences, computational design and nanofabrication. Due to physical properties of some target analytes in terms of dimension, electrical impedance or dielectric permittivity their determination is perhaps difficult to achieve, making it still necessary to rely on label detection.

5.2. Labeling MIP-Based Sensors

Several nanomaterials can be used as a label in order to improve the sensitivity of MIP-based sensors. These materials include AuNPs, quantum dots (QDs), CNTs, nanowires, magnetic nanoparticles, graphene, among others. Some research work reports the effect of the presence of CNTs and graphene in the overall performance of MIP sensors.

Bai et al. [189] described an ultrasensitive electrochemical sensor for determination of diethyl-stilbestrol. They conjugated the AuNPs and MWCNT-chitosan composite by drop casting on GC electrode in order to increase the surface area promoting an enhancement of the electron transfer rate, and finally amplifying the sensor signal. In addition, sol-gel MIP was then electrodeposited GC electrode surface. The sensor was characterized using CV and EIS. Electrochemical measurements were performed via DPV. Under optimal conditions, the device showed a detection range of 1.0×10^{-10}–1.0×10^{-6} mg/mL with a LOD of 24.3 fg/mL [189]. Wang and co-authors used a similar strategy, AuNPs and carboxylated MWCNT dropcasted on GC electrode, to produce a MIP sensor for olanquindox (OLA). The MIP was obtained by electropolymerization using OLA as a template and o-PD as monomer, as the novel sensor showed an LOD of 2.7 nmol/L [190].

A biomimetic sensor for sunset yellow detection was created with β-cyclodextrin as a monomer and functionalized with AuNPs and magnetic graphene oxide. The analytical performance of the hybrid nanomaterials was evaluated by CV and EIS. The linear range of the biomimetic sensor after optimization was from 5.0×10^{-9} to 2.0×10^{-6} mol/L and the LOD of 2.0×10^{-9} mol/L [191].

Overall, we can conclude that these nanomaterials show good biocompatibility, special catalytic activity and the convenience of a controlled fabrication process. They can improve the performance of the sensor, such as sensitivity, selectivity and so on.

6. Conclusions and Perspectives

This review describes different strategies to produce MIP materials that are combined with electrochemical transduction to fabricate advanced biomimetic sensors. These assays show several advantages including: (i) simple design and procedures; (ii) low price; (iii) portability; (iv) miniaturization and (v) point-of-care.

In general, this paper reviews several technical novelties to construct MIP electrochemical biomimetic sensors for screening different target compounds. In particular, electrochemical sensors show several advantages that have been addressed herein in detail. These devices provide the possibility of multi-analyte measurements, with a low response time, being a promising approach for screening different target analytes as proteins, antibiotics, pesticides, among many others. In addition, these sensors enable multi-target detection with minimal labour and small sample volumes.

One of the major challenges in MIP technology is the ongoing enhancement of MIP selectivity/ sensitivity for biological targets. Appropriate chemical conditions and polymer materials compatible with physiological requirements can preserve the natural structure and function (e.g., protein conformation, enzyme activity) of the analyte. MIP synthesis has proven its maturity but continuous research efforts remain, combining new nanomaterials and nanocomposites for preparing MIPs. This approach relies on the enormous advantages offered by nano-based imprinted materials (e.g., better assembly ability, fine-tunability, favorable mass transfer, better binding capacity and kinetics, superior removal of template). There is an enormous development in this field by using nanostructured MIPs (e.g., nanobeads, nanorods, nanowires) with simultaneous advances in porous materials, like hollow microspheres, which combined can provide competitive and versatile well-designed MIPs to empower the electrochemical response in terms of enhanced performance.

Overall, the sensitivity and LOD can be improved by incorporating conductive nanomaterials and electroactive complexes. Some nanomaterials as AuNPs, CNTs, graphene, nanowires and magnetic nanoparticles have been included in the polymeric matrix of MIP-based sensors providing a synergetic effect between catalytic activity, conductivity and biocompatibility, and improving the transduction signal. Increasingly, the integration of different fields as bioinformatics, nanotechnology, materials science, and synthetic biology will favor the achievement of reliable features in terms of analytical performance of the biomimetic sensors.

Moreover, it is possible to configure the MIP-based sensing devices towards multi-sensing array platforms for simultaneous detection of different target analytes. The results should be compared with standard clinical assays/methodologies.

Acknowledgments: The authors acknowledge the financial support of European Research Council though the Starting Grant, ERC-StG-3P's/2012, GA 311086 (to MGF Sales). One of the authors (Felismina Moreira) gratefully acknowledges Fundação para a Ciência e a Tecnologia for the financial support (Post-Doc grant reference SFRH/BPD/97891/2013 entitled "Biomedical devices for easier and quicker screening procedures of the Alzheimer disease".

Conflicts of Interest: The authors declare no conflict of interest.

References

1. Moreira, F.T.C.; Dutra, R.A.F.; Noronha, J.P.C.; Cunha, A.L.; Sales, M.G.F. Artificial antibodies for troponin T by its imprinting on the surface of multiwalled carbon nanotubes: Its use as sensory surfaces. *Biosens. Bioelectron.* **2011**, *28*, 243–250. [CrossRef] [PubMed]
2. Moreira, F.T.C.; Sharma, S.; Dutra, R.A.F.; Noronha, J.P.C.; Cass, A.E.G.; Sales, M.G.F. Smart plastic antibody material (SPAM) tailored on disposable screen printed electrodes for protein recognition: Application to myoglobin detection. *Biosens. Bioelectron.* **2013**, *45*, 237–244. [CrossRef] [PubMed]

3. Kryscio, D.R.; Peppas, N.A. Critical review and perspective of macromolecularly imprinted polymers. *Acta Biomater.* **2012**, *8*, 461–473. [CrossRef] [PubMed]
4. Whitcombe, M.J.; Chianella, I.; Larcombe, L.; Piletsky, S.A.; Noble, J.; Porter, R.; Horgan, A. The rational development of molecularly imprinted polymer-based sensors for protein detection. *Chem. Soc. Rev.* **2011**, *40*, 1547–1571. [CrossRef] [PubMed]
5. Scorrano, S.; Mergola, L.; Del Sole, R.; Vasapollo, G. Synthesis of molecularly imprinted polymers for amino acid derivates by using different functional monomers. *Int. J. Mol. Sci.* **2011**, *12*, 1735–1743. [CrossRef] [PubMed]
6. Sellergren, B.; Ekberg, B.; Mosbach, K. Molecular imprinting of amino acid derivatives in macroporous polymers—Demonstration of substrate- and enantio-selectivity by chromatographic resolution of racemic mixtures of amino acid derivatives. *J. Chromatogr. A* **1985**, *347*, 1–10. [CrossRef]
7. Wang, J.-F.; Zhou, L.-M.; Liu, X.-L.; Wang, Q.-H.; Zhu, D.-Q. Novel polymer system for molecular imprinting polymer against amino acid derivatives. *Chin. J. Chem.* **2000**, *18*, 621–625.
8. Singh, L.K.; Singh, M.; Singh, M. Biopolymeric receptor for peptide recognition by molecular imprinting approach-synthesis, characterization and application. *Mater. Sci. Eng. C* **2014**, *45*, 383–394. [CrossRef] [PubMed]
9. Janiak, D.S.; Kofinas, P. Molecular imprinting of peptides and proteins in aqueous media. *Anal. Bioanal. Chem.* **2007**, *389*, 399–404. [CrossRef] [PubMed]
10. Rachkov, A.; Minoura, N. Recognition of oxytocin and oxytocin-related peptides in aqueous media using a molecularly imprinted polymer synthesized by the epitope approach. *J. Chromatogr. A* **2000**, *889*, 111–118. [CrossRef]
11. Hu, J.; Mao, X.; Cao, S.; Yuan, X. Recognition of proteins and peptides: Rational development of molecular imprinting technology. *Polym. Sci. Ser. A* **2010**, *52*, 328–339. [CrossRef]
12. Tsunemori, H.; Araki, K.; Uezu, K.; Goto, M.; Furusaki, S. Surface imprinting polymers for the recognition of nucleotides. *Bioseparation* **2001**, *10*, 315–321. [CrossRef] [PubMed]
13. Shen, X.; Zhu, L.; Wang, N.; Ye, L.; Tang, H. Molecular imprinting for removing highly toxic organic pollutants. *Chem. Commun.* **2012**, *48*, 788–798. [CrossRef] [PubMed]
14. Uzuriaga-Sánchez, R.J.; Khan, S.; Wong, A.; Picasso, G.; Pividori, M.I.; Sotomayor, M.D.P.T. Magnetically separable polymer (Mag-MIP) for selective analysis of biotin in food samples. *Food Chem.* **2016**, *190*, 460–467. [CrossRef] [PubMed]
15. Kamel, A.H.; Almeida, S.A.A.; Sales, M.G.F.; Moreira, F.T.C. Sulfadiazine-potentiometric sensors for flow and batch determinations of sulfadiazine in drugs and biological fluids. *Anal. Sci.* **2009**, *25*, 365–371. [CrossRef] [PubMed]
16. Moreira, F.T.C.; Kamel, A.H.; Guerreiro, J.R.L.; Sales, M.G.F. Man-tailored biomimetic sensor of molecularly imprinted materials for the potentiometric measurement of oxytetracycline. *Biosens. Bioelectron.* **2010**, *26*, 566–574. [CrossRef] [PubMed]
17. Cheong, W.J.; Yang, S.H.; Ali, F. Molecular imprinted polymers for separation science: A review of reviews. *J. Sep. Sci.* **2013**, *36*, 609–628. [CrossRef] [PubMed]
18. Sharma, P.S.; Pietrzyk-Le, A.; D'Souza, F.; Kutner, W. Electrochemically synthesized polymers in molecular imprinting for chemical sensing. *Anal. Bioanal. Chem.* **2012**, *402*, 3177–3204. [CrossRef] [PubMed]
19. Tada, M.; Iwasawa, Y. Design of molecular-imprinting metal-complex catalysts. *J. Mol. Catal. A Chem.* **2003**, *199*, 115–137. [CrossRef]
20. Luliński, P. Molecularly imprinted polymers as the future drug delivery devices. *Acta Pol. Pharm.* **2013**, *70*, 601–609. [PubMed]
21. Uzun, L.; Turner, A.P.F. Molecularly-imprinted polymer sensors: Realizing their potential. *Biosens. Bioelectron.* **2016**, *76*, 131–144. [CrossRef] [PubMed]
22. Ye, L.; Mosbach, K. Molecular imprinting: Synthetic materials as substitutes for biological antibodies and receptors. *Chem. Mater.* **2008**, *20*, 859–868. [CrossRef]
23. Ravelet, C.; Peyrin, E. Recent developments in the HPLC enantiomeric separation using chiral selectors identified by a combinatorial strategy. *J. Sep. Sci.* **2006**, *29*, 1322–1331. [CrossRef] [PubMed]
24. Cabral-Miranda, G.; Gidlund, M.; Sales, M.G.F. Backside-surface imprinting as a new strategy to generate specific plastic antibody materials. *J. Mater. Chem. B* **2014**, *2*, 3087–3095. [CrossRef]
25. Henry, O.Y.F.; Piletsky, S.A.; Cullen, D.C. Fabrication of molecularly imprinted polymer microarray on a chip by mid-infrared laser pulse initiated polymerisation. *Biosens. Bioelectron.* **2008**, *23*, 1769–1775. [CrossRef] [PubMed]

26. Yang, K.; Li, S.; Liu, J.; Liu, L.; Zhang, L.; Zhang, Y. Multiepitope templates imprinted particles for the simultaneous capture of various target proteins. *Anal. Chem.* **2016**, *88*, 5621–5625. [CrossRef] [PubMed]

27. Piletsky, S.A.; Piletska, E.V.; Chen, B.; Karim, K.; Weston, D.; Barrett, G.; Lowe, P.; Turner, A.P.F. Chemical grafting of molecularly imprinted homopolymers to the surface of microplates. Application of artificial adrenergic receptor in enzyme-linked assay for β-agonists determination. *Anal. Chem.* **2000**, *72*, 4381–4385. [CrossRef] [PubMed]

28. Titirici, M.-M.; Sellergren, B. Thin molecularly imprinted polymer films via reversible addition-fragmentation chain transfer polymerization. *Chem. Mater.* **2006**, *18*, 1773–1779. [CrossRef]

29. Piletsky, S.A.; Turner, A.P.F. Electrochemical sensors based on molecularly imprinted polymers. *Electroanalysis* **2002**, *14*, 317–323. [CrossRef]

30. Ronkainen, N.J.; Halsall, H.B.; Heineman, W.R. Electrochemical biosensors. *Chem. Soc. Rev.* **2010**, *39*, 1747–1763. [CrossRef] [PubMed]

31. Guan, G.; Liu, B.; Wang, Z.; Zhang, Z. Imprinting of molecular recognition sites on nanostructures and its applications in chemosensors. *Sensors* **2008**, *8*, 8291–8320. [CrossRef] [PubMed]

32. Cai, D.; Ren, L.; Zhao, H.; Xu, C.; Zhang, L.; Yu, Y.; Wang, H.; Lan, Y.; Roberts, M.F.; Chuang, J.H.; et al. A molecular-imprint nanosensor for ultrasensitive detection of proteins. *Nat. Nanotechnol.* **2010**, *5*, 597–601. [CrossRef] [PubMed]

33. Mani, V.; Chikkaveeraiah, B.V.; Patel, V.; Gutkind, J.S.; Rusling, J.F. Ultrasensitive immunosensor for cancer biomarker proteins using gold nanoparticle film electrodes and multienzyme-particle amplification. *ACS Nano* **2009**, *3*, 585–594. [CrossRef] [PubMed]

34. Mayes, A.G.; Whitcombe, M.J. Synthetic strategies for the generation of molecularly imprinted organic polymers. *Adv. Drug Deliv. Rev.* **2005**, *57*, 1742–1778. [CrossRef] [PubMed]

35. Lofgreen, J.E.; Ozin, G.A. Controlling morphology and porosity to improve performance of molecularly imprinted sol-gel silica. *Chem. Soc. Rev.* **2014**, *43*, 911–933. [CrossRef] [PubMed]

36. Cieplak, M.; Kutner, W. Artificial biosensors: How can molecular imprinting mimic biorecognition? *Trends Biotechnol.* **2016**, *34*, 922–941. [CrossRef] [PubMed]

37. Wulff, G. Molecular imprinting in cross-linked materials with the aid of molecular templates—A way towards artificial antibodies. *Angew. Chem. Int. Ed. Engl.* **1995**, *34*, 1812–1832. [CrossRef]

38. Mosbach, K.; Ramström, O. The emerging technique of molecular imprinting and its future impact on biotechnology. *Nat. Biotechnol.* **1996**, *14*, 163–170. [CrossRef]

39. Ge, Y.; Turner, A.P.F. Too large to fit? Recent developments in macromolecular imprinting. *Trends Biotechnol.* **2008**, *26*, 218–224. [CrossRef] [PubMed]

40. Arifuzzaman, M.D.; Zhao, Y. Water-soluble molecularly imprinted nanoparticle receptors with hydrogen-bond-assisted hydrophobic binding. *J. Org. Chem.* **2016**, *81*, 7518–7526. [CrossRef] [PubMed]

41. Haupt, K.; Mosbach, K. Plastic antibodies: Developments and applications. *Trends Biotechnol.* **1998**, *16*, 468–475. [CrossRef]

42. Cormack, P.A.G.; Elorza, A.Z. Molecularly imprinted polymers: Synthesis and characterisation. *J. Chromatogr. B* **2004**, *804*, 173–182. [CrossRef] [PubMed]

43. Spivak, D.A. Optimization, evaluation, and characterization of molecularly imprinted polymers. *Adv. Drug Deliv. Rev.* **2005**, *57*, 1779–1794. [CrossRef] [PubMed]

44. Piletsky, S.A.; Andersson, H.S.; Nicholls, I.A. Combined hydrophobic and electrostatic interaction-based recognition in molecularly imprinted polymers. *Macromolecules* **1999**, *32*, 633–636. [CrossRef]

45. Byrne, M.E.; Park, K.; Peppas, N.A. Molecular imprinting within hydrogels. *Adv. Drug Deliv. Rev.* **2002**, *54*, 149–161. [CrossRef]

46. Liao, J.-L.; Wang, Y.; Hjertén, S. A novel support with artificially created recognition for the selective removal of proteins and for affinity chromatography. *Chromatographia* **1996**, *42*, 259–262. [CrossRef]

47. Hawkins, D.M.; Stevenson, D.; Reddy, S.M. Investigation of protein imprinting in hydrogel-based molecularly imprinted polymers (HydroMIPs). *Anal. Chim. Acta* **2005**, *542*, 61–65. [CrossRef]

48. Bueno, L.; El-Sharif, H.F.; Salles, M.O.; Boehm, R.D.; Narayan, R.J.; Paixão, T.R.L.C.; Reddy, S.M. MIP-based electrochemical protein profiling. *Sens. Actuators B Chem.* **2014**, *204*, 88–95. [CrossRef]

49. EL-Sharif, H.F.; Hawkins, D.M.; Stevenson, D.; Reddy, S.M. Determination of protein binding affinities within hydrogel-based molecularly imprinted polymers (HydroMIPs). *Phys. Chem. Chem. Phys.* **2014**, *16*, 15483–15489. [CrossRef] [PubMed]

50. EL-Sharif, H.F.; Yapati, H.; Kalluru, S.; Reddy, S.M. Highly selective BSA imprinted polyacrylamide hydrogels facilitated by a metal-coding MIP approach. *Acta Biomater.* **2015**, *28*, 121–127. [CrossRef] [PubMed]

51. Turner, N.W.; Jeans, C.W.; Brain, K.R.; Allender, C.J.; Hlady, V.; Britt, D.W. From 3D to 2D: A review of the molecular imprinting of proteins. *Biotechnol. Prog.* **2006**, *22*, 1474–1489. [CrossRef] [PubMed]

52. Bergmann, N.M.; Peppas, N.A. Molecularly imprinted polymers with specific recognition for macromolecules and proteins. *Prog. Polym. Sci.* **2008**, *33*, 271–288. [CrossRef]

53. Chen, W.; Ma, Y.; Pan, J.; Meng, Z.; Pan, G.; Sellergren, B. Molecularly imprinted polymers with stimuli-responsive affinity: Progress and perspectives. *Polymers* **2015**, *7*, 1689–1715. [CrossRef]

54. Mujahid, A.; Lieberzeit, P.A.; Dickert, F.L. Chemical sensors based on molecularly imprinted sol-gel materials. *Materials* **2010**, *3*, 2196–2217. [CrossRef]

55. Zhang, Z.; Liao, H.; Li, H.; Nie, L.; Yao, S. Stereoselective histidine sensor based on molecularly imprinted sol-gel films. *Anal. Biochem.* **2005**, *336*, 108–116. [CrossRef] [PubMed]

56. Díaz-García, M.E.; Laíño, R.B. Molecular imprinting in sol-gel materials: Recent developments and applications. *Microchim. Acta* **2005**, *149*, 19–36. [CrossRef]

57. Hansen, D.E. Recent developments in the molecular imprinting of proteins. *Biomaterials* **2007**, *28*, 4178–4191. [CrossRef] [PubMed]

58. Brown, M.E.; Puleo, D.A. Protein binding to peptide-imprinted porous silica scaffolds. *Chem. Eng. J.* **2008**, *137*, 97–101. [CrossRef] [PubMed]

59. Yang, D.-H.; Takahara, N.; Lee, S.-W.; Kunitake, T. Fabrication of glucose-sensitive TiO$_2$ ultrathin films by molecular imprinting and selective detection of monosaccharides. *Sens. Actuators B Chem.* **2008**, *130*, 379–385. [CrossRef]

60. Mizutani, N.; Yang, D.-H.; Selyanchyn, R.; Korposh, S.; Lee, S.-W.; Kunitake, T. Remarkable enantioselectivity of molecularly imprinted TiO$_2$ nano-thin films. *Anal. Chim. Acta* **2011**, *694*, 142–150. [CrossRef] [PubMed]

61. Zhang, Z.; Long, Y.; Nie, L.; Yao, S. Molecularly imprinted thin film self-assembled on piezoelectric quartz crystal surface by the sol-gel process for protein recognition. *Biosens. Bioelectron.* **2006**, *21*, 1244–1251. [CrossRef] [PubMed]

62. Ciriminna, R.; Fidalgo, A.; Pandarus, V.; Béland, F.; Ilharco, L.M.; Pagliaro, M. The sol-gel route to advanced silica-based materials and recent applications. *Chem. Rev.* **2013**, *113*, 6592–6620. [CrossRef] [PubMed]

63. Ang, Q.Y.; Zolkeflay, M.H.; Low, S.C. Configuration control on the shape memory stiffness of molecularly imprinted polymer for specific uptake of creatinine. *Appl. Surf. Sci.* **2016**, *369*, 326–333. [CrossRef]

64. Cohen, T.; Starosvetsky, J.; Cheruti, U.; Armon, R. Whole cell imprinting in sol-gel thin films for bacterial recognition in liquids: Macromolecular fingerprinting. *Int. J. Mol. Sci.* **2010**, *11*, 1236–1252. [CrossRef] [PubMed]

65. Gao, W.; Wan, F.; Ni, W.; Wang, S.; Zhang, M.; Yu, J. Electrochemical sensor for detection of trichlorfon based on molecularly imprinted sol-gel films modified glassy carbon electrode. *J. Inorg. Organomet. Polym.* **2012**, *22*, 37–41. [CrossRef]

66. Katz, A.; Davis, M.E. Molecular imprinting of bulk, microporous silica. *Nature* **2000**, *403*, 286–289. [PubMed]

67. Schirhagl, R.; Ren, K.N.; Zare, R.N. Surface-imprinted polymers in microfluidic devices. *Sci. China Chem.* **2012**, *55*, 469–483. [CrossRef]

68. Chen, L.; Wang, X.; Lu, W.; Wu, X.; Li, J. Molecular imprinting: Perspectives and applications. *Chem. Soc. Rev.* **2016**, *45*, 2137–2211. [CrossRef] [PubMed]

69. Hayden, O.; Bindeus, R.; Haderspöck, C.; Mann, K.-J.; Wirl, B.; Dickert, F.L. Mass-sensitive detection of cells, viruses and enzymes with artificial receptors. *Sens. Actuators B Chem.* **2003**, *91*, 316–319. [CrossRef]

70. Truskett, V.N.; Watts, M.P.C. Trends in imprint lithography for biological applications. *Trends Biotechnol.* **2006**, *24*, 312–317. [CrossRef] [PubMed]

71. Kane, R.S.; Takayama, S.; Ostuni, E.; Ingber, D.E.; Whitesides, G.M. Patterning proteins and cells using soft lithography. *Biomaterials* **1999**, *20*, 2363–2376. [CrossRef]

72. Shi, H.; Tsai, W.-B.; Garrison, M.D.; Ferrari, S.; Ratner, B.D. Template-imprinted nanostructured surfaces for protein recognition. *Nature* **1999**, *398*, 593–597. [PubMed]

73. Chou, P.-C.; Rick, J.; Chou, T.-C. C-reactive protein thin-film molecularly imprinted polymers formed using a micro-contact approach. *Anal. Chim. Acta* **2005**, *542*, 20–25. [CrossRef]

74. Su, W.-X.; Rick, J.; Chou, T.-C. Selective recognition of ovalbumin using a molecularly imprinted polymer. *Microchem. J.* **2009**, *92*, 123–128. [CrossRef]

75. Lin, H.-Y.; Hsu, C.-Y.; Thomas, J.L.; Wang, S.-E.; Chen, H.-C.; Chou, T.-C. The microcontact imprinting of proteins: The effect of cross-linking monomers for lysozyme, ribonuclease A and myoglobin. *Biosens. Bioelectron.* **2006**, *22*, 534–543. [CrossRef] [PubMed]

76. Hsu, C.-Y.; Lin, H.-Y.; Thomas, J.L.; Wu, B.-T.; Chou, T.-C. Incorporation of styrene enhances recognition of ribonuclease A by molecularly imprinted polymers. *Biosens. Bioelectron.* **2006**, *22*, 355–363. [CrossRef] [PubMed]

77. Ertürk, G.; Berillo, D.; Hedström, M.; Mattiasson, B. Microcontact-BSA imprinted capacitive biosensor for real-time, sensitive and selective detection of BSA. *Biotechnol. Rep.* **2014**, *3*, 65–72. [CrossRef]

78. Wulff, G.; Knorr, K. Stoichiometric noncovalent interaction in molecular imprinting. *Bioseparation* **2001**, *10*, 257–276. [CrossRef] [PubMed]

79. Xu, H.; Schönhoff, M.; Zhang, X. Unconventional layer-by-layer assembly: Surface molecular imprinting and its applications. *Small* **2012**, *8*, 517–523. [CrossRef] [PubMed]

80. Ertürk, G.; Mattiasson, B. From imprinting to microcontact imprinting—A new tool to increase selectivity in analytical devices. *J. Chromatogr. B* **2016**, *1021*, 30–44. [CrossRef] [PubMed]

81. Idil, N.; Hedström, M.; Denizli, A.; Mattiasson, B. Whole cell based microcontact imprinted capacitive biosensor for the detection of *Escherichia coli*. *Biosens. Bioelectron.* **2017**, *87*, 807–815. [CrossRef] [PubMed]

82. Welch, M.; Rastogi, A.; Ober, C. Polymer brushes for electrochemical biosensors. *Soft Matter* **2011**, *7*, 297–302. [CrossRef]

83. Prime, K.L.; Whitesides, G.M. Adsorption of proteins onto surfaces containing end-attached oligo(ethylene oxide): A model system using self-assembled monolayers. *J. Am. Chem. Soc.* **1993**, *115*, 10714–10721. [CrossRef]

84. Herrwerth, S.; Eck, W.; Reinhardt, S.; Grunze, M. Factors that determine the protein resistance of oligoether self-assembled monolayers—Internal hydrophilicity, terminal hydrophilicity, and lateral packing density. *J. Am. Chem. Soc.* **2003**, *125*, 9359–9366. [CrossRef] [PubMed]

85. Ostuni, E.; Chapman, R.G.; Holmlin, R.E.; Takayama, S.; Whitesides, G.M. A survey of structure-property relationships of surfaces that resist the adsorption of protein. *Langmuir* **2001**, *17*, 5605–5620. [CrossRef]

86. Zdyrko, B.; Hoy, O.; Luzinov, I. Toward protein imprinting with polymer brushes. *Biointerphases* **2009**, *4*, FA17–FA21. [CrossRef] [PubMed]

87. Wang, Y.; Zhang, Z.; Jain, V.; Yi, J.; Mueller, S.; Sokolov, J.; Liu, Z.; Levon, K.; Rigas, B.; Rafailovich, M.H. Potentiometric sensors based on surface molecular imprinting: Detection of cancer biomarkers and viruses. *Sens. Actuators B Chem.* **2010**, *146*, 381–387. [CrossRef]

88. Yang, X.; Chen, D.; Zhao, H. Silica particles with immobilized protein molecules and polymer brushes. *Acta Biomater.* **2016**, *29*, 446–454. [CrossRef] [PubMed]

89. Patel, A.K.; Sharma, P.S.; Prasad, B.B. Electrochemical sensor for uric acid based on a molecularly imprinted polymer brush grafted to tetraethoxysilane derived sol-gel thin film graphite electrode. *Mater. Sci. Eng.* **2009**, *29*, 1545–1553. [CrossRef]

90. Prasad, B.B.; Tiwari, K.; Singh, M.; Sharma, P.S.; Patel, A.K.; Srivastava, S. Ultratrace analysis of dopamine using a combination of imprinted polymer-brush-coated SPME and imprinted polymer sensor techniques. *Chromatographia* **2009**, *69*, 949–957. [CrossRef]

91. Iskierko, Z.; Sharma, P.S.; Bartold, K.; Pietrzyk-Le, A.; Noworyta, K.; Kutner, W. Molecularly imprinted polymers for separating and sensing of macromolecular compounds and microorganisms. *Biotechnol. Adv.* **2016**, *34*, 30–46. [CrossRef] [PubMed]

92. Zamora-Gálvez, A.; Ait-Lahcen, A.; Mercante, L.A.; Morales-Narváez, E.; Amine, A.; Merkoçi, A. Molecularly imprinted polymer-decorated magnetite nanoparticles for selective sulfonamide detection. *Anal. Chem.* **2016**, *88*, 3578–3584. [CrossRef] [PubMed]

93. Moreira, F.T.C.; Dutra, R.A.F.; Noronha, J.P.C.; Sales, M.G.F. Myoglobin-biomimetic electroactive materials made by surface molecular imprinting on silica beads and their use as ionophores in polymeric membranes for potentiometric transduction. *Biosens. Bioelectron.* **2011**, *26*, 4760–4766. [CrossRef] [PubMed]

94. Lv, Y.; Tan, T.; Svec, F. Molecular imprinting of proteins in polymers attached to the surface of nanomaterials for selective recognition of biomacromolecules. *Biotechnol. Adv.* **2013**, *31*, 1172–1186. [CrossRef] [PubMed]

95. Syritski, V.; Reut, J.; Menaker, A.; Gyurcsányi, R.E.; Öpik, A. Electrosynthesized molecularly imprinted polypyrrole films for enantioselective recognition of L-aspartic acid. *Electrochim. Acta* **2008**, *53*, 2729–2736. [CrossRef]

96. Erdőssy, J.; Horváth, V.; Yarman, A.; Scheller, F.W.; Gyurcsányi, R.E. Electrosynthesized molecularly imprinted polymers for protein recognition. *Trends Anal. Chem.* **2016**, *79*, 179–190. [CrossRef]

97. Hamedi, M.; Herland, A.; Karlsson, R.H.; Inganäs, O. Electrochemical devices made from conducting nanowire networks self-assembled from amyloid fibrils and alkoxysulfonate PEDOT. *Nano Lett.* **2008**, *8*, 1736–1740. [CrossRef] [PubMed]

98. Li, Y.; Hsieh, C.-H.; Lai, C.-W.; Chang, Y.-F.; Chan, H.-Y.; Tsai, C.-F.; Ho, J.-A.A.; Wu, L.-C. Tyramine detection using PEDOT:PSS/AuNPs/1-methyl-4-mercaptopyridine modified screen-printed carbon electrode with molecularly imprinted polymer solid phase extraction. *Biosens. Bioelectron.* **2017**, *87*, 142–149. [CrossRef] [PubMed]

99. Yeh, W.-M.; Ho, K.-C. Amperometric morphine sensing using a molecularly imprinted polymer-modified electrode. *Anal. Chim. Acta* **2005**, *542*, 76–82. [CrossRef]

100. Özcan, L.; Şahin, Y. Determination of paracetamol based on electropolymerized-molecularly imprinted polypyrrole modified pencil graphite electrode. *Sens. Actuators B Chem.* **2007**, *127*, 362–369. [CrossRef]

101. Moreira, F.T.C.; Ferreira, M.J.M.S.; Puga, J.R.T.; Sales, M.G.F. Screen-printed electrode produced by printed-circuit board technology. Application to cancer biomarker detection by means of plastic antibody as sensing material. *Sens. Actuators B Chem.* **2016**, *223*, 927–935. [CrossRef]

102. Mazzotta, E.; Picca, R.A.; Malitesta, C.; Piletsky, S.A.; Piletska, E.V. Development of a sensor prepared by entrapment of MIP particles in electrosynthesised polymer films for electrochemical detection of ephedrine. *Biosens. Bioelectron.* **2008**, *23*, 1152–1156. [CrossRef] [PubMed]

103. Xia, J.; Cao, X.; Wang, Z.; Yang, M.; Zhang, F.; Lu, B.; Li, F.; Xia, L.; Li, Y.; Xia, Y. Molecularly imprinted electrochemical biosensor based on chitosan/ionic liquid-graphene composites modified electrode for determination of bovine serum albumin. *Sens. Actuators B Chem.* **2016**, *225*, 305–311. [CrossRef]

104. Roy, A.K.; Nisha, V.S.; Dhand, C.; Malhotra, B.D. Molecularly imprinted polyaniline film for ascorbic acid detection. *J. Mol. Recognit.* **2011**, *24*, 700–706. [CrossRef] [PubMed]

105. Wang, P.; Sun, G.; Ge, L.; Ge, S.; Yu, J.; Yan, M. Photoelectrochemical lab-on-paper device based on molecularly imprinted polymer and porous Au-paper electrode. *Analyst* **2013**, *138*, 4802–4811. [CrossRef] [PubMed]

106. Huynha, T.P.; Sharma, P.S.; Sosnowska, M.; D'Souza, F.; Kutner, W. Functionalized polythiophenes: Recognition materials for chemosensors and biosensors of superior sensitivity, selectivity, and detectability. *Prog. Polym. Sci.* **2015**, *47*, 1–25. [CrossRef]

107. Tiu, B.D.B.; Pernites, R.B.; Tiu, S.B.; Advincula, R.C. Detection of aspartame via microsphere-patterned and molecularly imprinted polymer arrays. *Colloids Surf. A* **2016**, *495*, 149–158. [CrossRef]

108. Yin, Y.; Yan, L.; Zhang, Z.; Wang, J. Magnetic molecularly imprinted polydopamine nanolayer on multi-walled carbon nanotubes surface for protein capture. *Talanta* **2015**, *144*, 671–679. [CrossRef] [PubMed]

109. Lynge, M.E.; Westen, R.; Postma, A.; Städler, B. Polydopamine—A nature-inspired polymer coating for biomedical science. *Nanoscale* **2011**, *3*, 4916–4928. [CrossRef] [PubMed]

110. Khan, M.A.R.; Moreira, F.T.C.; Riu, J.; Sales, M.G.F. Plastic antibody for the electrochemical detection of bacterial surface proteins. *Sens. Actuators B Chem.* **2016**, *233*, 697–704. [CrossRef]

111. Moreira, F.T.C.; Sharma, S.; Dutra, R.A.F.; Noronha, J.P.C.; Cass, A.E.G.; Sales, M.G.F. Protein-responsive polymers for point-of-care detection of cardiac biomarker. *Sens. Actuators B Chem.* **2014**, *196*, 123–132. [CrossRef]

112. Malitesta, C.; Losito, I.; Zambonin, P.G. Molecularly imprinted electrosynthesized polymers: New materials for biomimetic sensors. *Anal. Chem.* **1999**, *71*, 1366–1370. [CrossRef] [PubMed]

113. Bai, H.; Wang, C.; Peng, J.; Wu, Y.; Yang, Y.; Cao, Q. A novel sensitive electrochemical sensor for podophyllotoxin assay based on the molecularly imprinted poly-o-phenylenediamine film. *J. Nanosci. Nanotechnol.* **2015**, *15*, 2456–2463. [CrossRef] [PubMed]

114. Bossi, A.M.; Sharma, P.S.; Montana, L.; Zoccatelli, G.; Laub, O.; Levi, R. Fingerprint-imprinted polymer: Rational selection of peptide epitope templates for the determination of proteins by molecularly imprinted polymers. *Anal. Chem.* **2012**, *84*, 4036–4041. [CrossRef] [PubMed]

115. Zhang, Y.; Deng, C.; Liu, S.; Wu, J.; Chen, Z.; Li, C.; Lu, W. Active targeting of tumors through conformational epitope imprinting. *Angew. Chem. Int. Ed.* **2015**, *54*, 5157–5160. [CrossRef] [PubMed]

116. Nishino, H.; Huang, C.-S.; Shea, K.J. Selective protein capture by epitope imprinting. *Angew. Chem. Int. Ed.* **2006**, *45*, 2392–2396. [CrossRef] [PubMed]

117. Tang, A.-N.; Duan, L.; Liu, M.; Dong, X. An epitope imprinted polymer with affinity for kininogen fragments prepared by metal coordination interaction for cancer biomarker analysis. *J. Mater. Chem. B* **2016**, *4*, 7464–7471. [CrossRef]

118. Schwark, S.; Sun, W.; Stute, J.; Lütkemeyer, D.; Ulbricht, M.; Sellergren, B. Monoclonal antibody capture from cell culture supernatants using epitope imprinted macroporous membranes. *RSC Adv.* **2016**, *6*, 53162–53169. [CrossRef]

119. Çorman, M.E.; Armutcu, C.; Uzun, L.; Say, E.; Denizli, A. Self-oriented nanoparticles for site-selective immunoglobulin G recognition via epitope imprinting approach. *Colloids Surf. B* **2014**, *123*, 831–837. [CrossRef] [PubMed]

120. Dechtrirat, D.; Jetzschmann, K.J.; Stöcklein, W.F.M.; Scheller, F.W.; Gajovic-Eichelmann, N. Protein rebinding to a surface-confined imprint. *Adv. Funct. Mater.* **2012**, *22*, 5231–5237. [CrossRef]

121. Li, D.-Y.; Zhang, X.-M.; Yan, Y.-J.; He, X.-W.; Li, W.-Y.; Zhang, Y.-K. Thermo-sensitive imprinted polymer embedded carbon dots using epitope approach. *Biosens. Bioelectron.* **2016**, *79*, 187–192. [CrossRef] [PubMed]

122. Yan, Y.-J.; He, X.-W.; Li, W.-Y.; Zhang, Y.-K. Nitrogen-doped graphene quantum dots-labeled epitope imprinted polymer with double templates via the metal chelation for specific recognition of cytochrome c. *Biosens. Bioelectron.* **2017**, *91*, 253–261. [CrossRef] [PubMed]

123. Andersson, L.I.; Miyabayashi, A.; O'Shannessy, D.J.; Mosbach, K. Enantiomeric resolution of amino acid derivatives on molecularly imprinted polymers as monitored by potentiometric measurements. *J. Chromatogr. A* **1990**, *516*, 323–331. [CrossRef]

124. Hedborg, E.; Winquist, F.; Lundström, I.; Andersson, L.I.; Mosbach, K. Some studies of molecularly-imprinted polymer membranes in combination with field-effect devices. *Sens. Actuators A Phys.* **1993**, *37–38*, 796–799. [CrossRef]

125. Peng, Y.; Wu, Z.; Liu, Z. An electrochemical sensor for paracetamol based on an electropolymerized molecularly imprinted o-phenylenediamine film on a multi-walled carbon nanotube modified glassy carbon electrode. *Anal. Methods* **2014**, *6*, 5673–5681. [CrossRef]

126. Karimian, N.; Turner, A.P.F.; Tiwari, A. Electrochemical evaluation of troponin T imprinted polymer receptor. *Biosens. Bioelectron.* **2014**, *59*, 160–165. [CrossRef] [PubMed]

127. Neto, J.D.R.M.; Santos, W.D.J.R.; Lima, P.R.; Tanaka, S.M.C.N.; Tanaka, A.A.; Kubota, L.T. A hemin-based molecularly imprinted polymer (MIP) grafted onto a glassy carbon electrode as a selective sensor for 4-aminophenol amperometric. *Sens. Actuators B Chem.* **2011**, *152*, 220–225. [CrossRef]

128. Pan, Y.; Zhao, F.; Zeng, B. Electrochemical sensors of octylphenol based on molecularly imprinted poly(3,4-ethylenedioxythiophene) and poly(3,4-ethylenedioxythiophene-gold nanoparticles). *RSC Adv.* **2015**, *5*, 57671–57677. [CrossRef]

129. Pardieu, E.; Cheap, H.; Vedrine, C.; Lazerges, M.; Lattach, Y.; Garnier, F.; Remita, S.; Pernelle, C. Molecularly imprinted conducting polymer based electrochemical sensor for detection of atrazine. *Anal. Chim. Acta* **2009**, *649*, 236–245. [CrossRef] [PubMed]

130. Mathur, A.; Blais, S.; Goparaju, C.M.V.; Neubert, T.; Pass, H.; Levon, K. Development of a biosensor for detection of pleural mesothelioma cancer biomarker using surface imprinting. *PLoS ONE* **2013**, *8*, e57681. [CrossRef] [PubMed]

131. Wang, Y.; Zhou, Y.; Sokolov, J.; Rigas, B.; Levon, K.; Rafailovich, M. A potentiometric protein sensor built with surface molecular imprinting method. *Biosens. Bioelectron.* **2008**, *24*, 162–166. [CrossRef] [PubMed]

132. Zhang, J.; Xu, L.; Wang, Y.-Q.; Lue, R.-H. Electrochemical sensor for bisphenol A based on molecular imprinting technique and electropolymerization membrane. *Chin. J. Anal. Chem.* **2009**, *37*, 1041–1044.

133. Udomsap, D.; Branger, C.; Culioli, G.; Dollet, P.; Brisset, H. A versatile electrochemical sensing receptor based on a molecularly imprinted polymer. *Chem. Commun.* **2014**, *50*, 7488–7491. [CrossRef] [PubMed]

134. Li, X.; Wang, X.; Li, L.; Duan, H.; Luo, C. Electrochemical sensor based on magnetic graphene oxide@gold nanoparticles-molecular imprinted polymers for determination of dibutyl phthalate. *Talanta* **2015**, *131*, 354–360. [CrossRef] [PubMed]

135. Tan, X.; Hu, Q.; Wu, J.; Li, X.; Li, P.; Yu, H.; Lei, F. Electrochemical sensor based on molecularly imprinted polymer reduced graphene oxide and gold nanoparticles modified electrode for detection of carbofuran. *Sens. Actuators B Chem.* **2015**, *220*, 216–221. [CrossRef]

136. Komarova, E.; Aldissi, M.; Bogomolova, A. Design of molecularly imprinted conducting polymer protein-sensing films via substrate-dopant binding. *Analyst* **2015**, *140*, 1099–1106. [CrossRef] [PubMed]

137. Ma, M.; Zhu, P.; Pi, F.; Ji, J.; Sun, X. A disposable molecularly imprinted electrochemical sensor based on screen-printed electrode modified with ordered mesoporous carbon and gold nanoparticles for determination of ractopamine. *J. Electroanal. Chem.* **2016**, *775*, 171–178. [CrossRef]

138. El Gohary, N.A.; Madbouly, A.; El Nashar, R.M.; Mizaikoff, B. Synthesis and application of a molecularly imprinted polymer for the voltammetric determination of famciclovir. *Biosens. Bioelectron.* **2015**, *65*, 108–114. [CrossRef] [PubMed]

139. Bai, H.; Wang, C.; Chen, J.; Peng, J.; Cao, Q. A novel sensitive electrochemical sensor based on in-situ polymerized molecularly imprinted membranes at graphene modified electrode for artemisinin determination. *Biosens. Bioelectron.* **2015**, *64*, 352–358. [CrossRef] [PubMed]

140. Li, Y.; Song, H.; Zhang, L.; Zuo, P.; Ye, B.C.; Yao, J.; Chen, W. Supportless electrochemical sensor based on molecularly imprinted polymer modified nanoporous microrod for determination of dopamine at trace level. *Biosens. Bioelectron.* **2016**, *78*, 308–314. [CrossRef] [PubMed]

141. Li, Y.; Liu, Y.; Liu, J.; Tang, H.; Cao, C.; Zhao, D.; Ding, Y. Molecularly imprinted polymer decorated nanoporous gold for highly selective and sensitive electrochemical sensors. *Sci. Rep.* **2015**, *5*, 7699. [CrossRef] [PubMed]

142. Aswini, K.K.; Vinu Mohan, A.M.; Biju, V.M. Molecularly imprinted poly(4-amino-5-hydroxy-2,7-naphthalenedisulfonic acid) modified glassy carbon electrode as an electrochemical theophylline sensor. *Mater. Sci. Eng. C* **2016**, *65*, 116–125. [CrossRef] [PubMed]

143. Li, H.-H.; Wang, H.-H.; Li, W.-T.; Fang, X.-X.; Guo, X.-C.; Zhou, W.-H.; Cao, X.; Kou, D.-X.; Zhou, Z.-J.; Wu, S.-X. A novel electrochemical sensor for epinephrine based on three dimensional molecularly imprinted polymer arrays. *Sens. Actuators B Chem.* **2016**, *222*, 1127–1133. [CrossRef]

144. Yang, L.; Zhao, F.; Zeng, B. Electrochemical determination of eugenol using a three-dimensional molecularly imprinted poly (p-aminothiophenol-co-p-aminobenzoic acids) film modified electrode. *Electrochim. Acta* **2016**, *210*, 293–300. [CrossRef]

145. Ji, J.; Zhou, Z.; Zhao, X.; Sun, J.; Sun, X. Electrochemical sensor based on molecularly imprinted film at Au nanoparticles-carbon nanotubes modified electrode for determination of cholesterol. *Biosens. Bioelectron.* **2015**, *66*, 590–595. [CrossRef] [PubMed]

146. Bakas, I.; Salmi, Z.; Jouini, M.; Geneste, F.; Mazerie, I.; Floner, D.; Carbonnier, B.; Yagci, Y.; Chehimi, M.M. Picomolar detection of melamine using molecularly imprinted polymer-based electrochemical sensors prepared by UV-graft photopolymerization. *Electroanalysis* **2015**, *27*, 429–439. [CrossRef]

147. Do, M.H.; Florea, A.; Farre, C.; Bonhomme, A.; Bessueille, F.; Vocanson, F.; Tran-Thi, N.-T.; Jaffrezic-Renault, N. Molecularly imprinted polymer-based electrochemical sensor for the sensitive detection of glyphosate herbicide. *Int. J. Environ. Anal. Chem.* **2015**, *95*, 1489–1501. [CrossRef]

148. Zhai, Y.; Wang, D.; Liu, H.; Zeng, Y.; Yin, Z.; Li, L. Electrochemical molecular imprinted sensors based on electrospun nanofiber and determination of ascorbic acid. *Anal. Sci.* **2015**, *31*, 793–798. [CrossRef] [PubMed]

149. You, M.; Yang, S.; Jiao, F.; Yang, L.-Z.; Zhang, F.; He, P.-G. Label-free electrochemical multi-sites recognition of G-rich DNA using multi-walled carbon nanotubes-supported molecularly imprinted polymer with guanine sites of DNA. *Electrochim. Acta* **2016**, *199*, 133–141. [CrossRef]

150. Cieplak, M.; Szwabinska, K.; Sosnowska, M.; Chandra, B.K.C.; Borowicz, P.; Noworyta, K.; D'Souza, F.; Kutner, W. Selective electrochemical sensing of human serum albumin by semi-covalent molecular imprinting. *Biosens. Bioelectron.* **2015**, *74*, 960–966. [CrossRef] [PubMed]

151. Muti, M.; Soysal, M.; Nacak, F.M.; Gençdağ, K.; Karagözler, A.E. A novel DNA probe based on molecularly imprinted polymer modified electrode for the electrochemical monitoring of DNA. *Electroanalysis* **2015**, *27*, 1368–1377. [CrossRef]

152. Silva, B.V.M.; Rodríguez, B.A.G.; Sales, G.F.; Sotomayor, M.D.P.T.; Dutra, R.F. An ultrasensitive human cardiac troponin T graphene screen-printed electrode based on electropolymerized-molecularly imprinted conducting polymer. *Biosens. Bioelectron.* **2016**, *77*, 978–985. [CrossRef] [PubMed]

153. Christopher, M.A.; Brett, A.M.O.B. *Electrochemistry: Principles, Methods, and Applications*; Oxford University Press, Incorporated: Oxford, UK, 1993; p. 427.

154. Arnold, B.R.; Euler, A.C.; Jenkins, A.L.; Uy, O.M.; Murray, G.M. Progress in the development of molecularly imprinted polymer sensors. *Johns Hopkins APL Tech. Dig.* **1999**, *20*, 190–198.

155. Arvand, M.; Samie, H.A. A biomimetic potentiometric sensor based on molecularly imprinted polymer for the determination of memantine in tablets. *Drug Test. Anal.* **2013**, *5*, 461–467. [CrossRef] [PubMed]

156. Murray, G.M.; Jenkins, A.L.; Bzhelyansky, A.; Uy, O.M. Molecularly imprinted polymers for the selective sequestering and sensing of ions. *Johns Hopkins APL Tech. Dig.* **1997**, *18*, 464–472.

157. Abu-Dalo, M.A.; Nassory, N.S.; Abdulla, N.I.; Al-Mheidat, I.R. Azithromycin-molecularly imprinted polymer based on PVC membrane for Azithromycin determination in drugs using coated graphite electrode. *J. Electroanal. Chem.* **2015**, *751*, 75–79. [CrossRef]

158. Bagheri, H.; Shirzadmehr, A.; Rezaei, M. Designing and fabrication of new molecularly imprinted polymer-based potentiometric nano-graphene/ionic liquid/carbon paste electrode for the determination of losartan. *J. Mol. Liq.* **2015**, *212*, 96–102. [CrossRef]

159. Özkütük, E.B.; Uğurağ, D.; Ersöz, A.; Say, R. Determination of clenbuterol by multiwalled carbon nanotube potentiometric sensors. *Anal. Lett.* **2016**, *49*, 778–789. [CrossRef]

160. Kupis-Rozmysłowicz, J.; Wagner, M.; Bobacka, J.; Lewenstam, A.; Migdalski, J. Biomimetic membranes based on molecularly imprinted conducting polymers as a sensing element for determination of taurine. *Electrochim. Acta* **2016**, *188*, 537–544. [CrossRef]

161. Basozabal, I.; Guerreiro, A.; Gomez-Caballero, A.; Goicolea, M.A.; Barrio, R.J. Direct potentiometric quantification of histamine using solid-phase imprinted nanoparticles as recognition elements. *Biosens. Bioelectron.* **2014**, *58*, 138–144. [CrossRef] [PubMed]

162. Truta, L.A.A.N.A.; Ferreira, N.S.; Sales, M.G.F. Graphene-based biomimetic materials targeting urine metabolite as potential cancer biomarker: Application over different conductive materials for potentiometric transduction. *Electrochim. Acta* **2014**, *150*, 99–107. [CrossRef] [PubMed]

163. Sharma, P.S.; Wojnarowicz, A.; Sosnowska, M.; Benincori, T.; Noworyta, K.; D'Souza, F.; Kutner, W. Potentiometric chemosensor for neopterin, a cancer biomarker, using an electrochemically synthesized molecularly imprinted polymer as the recognition unit. *Biosens. Bioelectron.* **2016**, *77*, 565–572. [CrossRef] [PubMed]

164. Anirudhan, T.S.; Alexander, S.; Lilly, A. Surface modified multiwalled carbon nanotube based molecularly imprinted polymer for the sensing of dopamine in real samples using potentiometric method. *Polymer* **2014**, *55*, 4820–4831. [CrossRef]

165. Rayanasukha, Y.; Pratontep, S.; Porntheeraphat, S.; Bunjongpru, W.; Nukeaw, J. Non-enzymatic urea sensor using molecularly imprinted polymers surface modified based-on ion-sensitive field effect transistor (ISFET). *Surf. Coat. Technol.* **2016**, *306*, 147–150. [CrossRef]

166. Koirala, K.; Sevilla, F.B., III; Santos, J.H. Biomimetic potentiometric sensor for chlorogenic acid based on electrosynthesized polypyrrole. *Sens. Actuators B Chem.* **2016**, *222*, 391–396. [CrossRef]

167. Rebelo, T.S.C.R.; Santos, C.; Costa-Rodrigues, J.; Fernandes, M.H.; Noronha, J.P.; Sales, M.G.F. Novel Prostate Specific Antigen plastic antibody designed with charged binding sites for an improved protein binding and its application in a biosensor of potentiometric transduction. *Electrochim. Acta* **2014**, *132*, 142–150. [CrossRef]

168. Yu, Y.; Zhang, Q.; Buscaglia, J.; Chang, C.-C.; Liu, Y.; Yang, Z.; Guo, Y.; Wang, Y.; Levon, K.; Rafailovich, M. Quantitative real-time detection of carcinoembryonic antigen (CEA) from pancreatic cyst fluid using 3-D surface molecular imprinting. *Analyst* **2016**, *141*, 4424–4431. [CrossRef] [PubMed]

169. Moret, J.; Moreira, F.T.C.; Almeida, S.A.A.; Sales, M.G.F. New molecularly-imprinted polymer for carnitine and its application as ionophore in potentiometric selective membranes. *Mater. Sci. Eng. C* **2014**, *43*, 481–487. [CrossRef] [PubMed]

170. Bergveld, P. Development of an ion-sensitive solid-state device for neurophysiological measurements. *IEEE Trans. Biomed. Eng.* **1970**, *17*, 70–71. [CrossRef] [PubMed]

171. Schöning, M.J.; Poghossian, A. Recent advances in biologically sensitive field-effect transistors (BioFETs). *Analyst* **2002**, *127*, 1137–1151. [CrossRef] [PubMed]

172. Dabrowski, M.; Sharma, P.S.; Iskierko, Z.; Noworyta, K.; Cieplak, M.; Lisowski, W.; Oborska, S.; Kuhn, A.; Kutner, W. Early diagnosis of fungal infections using piezomicrogravimetric and electric chemosensors based on polymers molecularly imprinted with D-arabitol. *Biosens. Bioelectron.* **2016**, *79*, 627–635. [CrossRef] [PubMed]

173. Iskierko, Z.; Sharma, P.S.; Prochowicz, D.; Fronc, K.; D'Souza, F.; Toczydłowska, D.; Stefaniak, F.; Noworyta, K. Molecularly Imprinted Polymer (MIP) Film with Improved Surface Area Developed by Using Metal–Organic Framework (MOF) for Sensitive Lipocalin (NGAL) Determination. *ACS Appl. Mater. Interfaces* **2016**, *8*, 19860–19865. [CrossRef] [PubMed]

174. Iskierko, Z.; Checinska, A.; Sharma, P.S.; Golebiewska, K.; Noworyta, K.; Borowicz, P.; Fronc, K.; Bandi, V.; D'Souza, F.; Kutner, W. Molecularly imprinted polymer based extended-gate field-effect transistor chemosensors for phenylalanine enantioselective sensing. *J. Mater. Chem. C* **2017**, *5*, 969–977. [CrossRef]

175. Pohanka, M.; Skládal, P. Electrochemical Biosensors—Principles and applications. *J. Appl. Biomed.* **2008**, *6*, 57–64.

176. Berggren, C.; Bjarnason, B.; Johansson, G. Capacitive biosensors. *Electroanalysis* **2001**, *13*, 173–180. [CrossRef]

177. Panasyuk, T.L.; Mirsky, V.M.; Piletsky, S.A.; Wolfbeis, O.S. Electropolymerized molecularly imprinted polymers as receptor layers in a capacitive chemical sensors. *Anal. Chem.* **1999**, *71*, 4609–4613. [CrossRef]

178. Wojnarowicz, A.; Sharma, P.S.; Sosnowska, M.; Lisowski, W.; Huynh, T.-P.; Pszona, M.; Borowicz, P.; D'Souza, F.; Kutner, W. An electropolymerized molecularly imprinted polymer for selective carnosine sensing with impedimetric capacity. *J. Mater. Chem. B* **2016**, *4*, 1156–1165. [CrossRef]

179. Baleviciute, I.; Ratautaite, V.; Ramanaviciene, A.; Balevicius, Z.; Broeders, J.; Croux, D.; McDonald, M.; Vahidpour, F.; Thoelen, R.; De Ceuninck, W.; et al. Evaluation of theophylline imprinted polypyrrole film. *Synth. Met.* **2015**, *209*, 206–211. [CrossRef]

180. Gutierrez, R.A.V.; Hedström, M.; Mattiasson, B. Bioimprinting as a tool for the detection of aflatoxin B1 using a capacitive biosensor. *Biotechnol. Rep.* **2016**, *11*, 12–17. [CrossRef]

181. Jolly, P.; Tamboli, V.; Harniman, R.L.; Estrela, P.; Allender, C.J.; Bowen, J.L. Aptamer-MIP hybrid receptor for highly sensitive electrochemical detection of prostate specific antigen. *Biosens. Bioelectron.* **2016**, *75*, 188–195. [CrossRef] [PubMed]

182. Tavares, A.P.M.; Ferreira, N.S.; Truta, L.A.A.N.A.; Sales, M.G.F. Conductive paper with antibody-like film for electrical readings of biomolecules. *Sci. Rep.* **2016**, *6*, 26132. [CrossRef] [PubMed]

183. Fowler, J.M.; Wong, D.K.Y.; Halsall, H.B.; Heineman, W.R. Recent developments in electrochemical immunoassays and immunosensors. In *Electrochemical Sensors, Biosensors and Their Biomedical Applications*, 1st ed.; Zhang, X., Ju, H., Wang, J., Eds.; Elsevier Inc.: FL/AZ, USA; Nanjing, China, 2008; pp. 115–140.

184. Suri, C.R. Immunosensors for pesticides monitoring. In *Advances in Biosensors: Perspectives in Biosensors*, 1st ed.; Malhotra, A.P.F., Turner, B.D., Eds.; Elsevier Science B.V.: Amsterdam, The Netherlands, 2003; Volume 5, pp. 161–176.

185. Cooper, M.A. *Label-Free Biosensors Techniques and Applications*, 1st ed.; Cambridge University Press: New York, NY, USA, 2009; pp. 1–279.

186. Syahir, A.; Usui, K.; Tomizaki, K.-Y.; Kajikawa, K.; Mihara, H. Label and Label-Free Detection Techniques for Protein Microarrays. *Microarrays* **2015**, *4*, 228–244. [CrossRef] [PubMed]

187. Vestergaard, M.; Kerman, K.; Tamiya, E. An overview of label-free electrochemical protein sensors. *Sensors* **2007**, *7*, 3442–3458. [CrossRef]

188. Hunt, H.K.; Armani, A.M. Label-free biological and chemical sensors. *Nanoscale* **2010**, *2*, 1544–1559. [CrossRef] [PubMed]

189. Bai, J.; Zhang, X.; Peng, Y.; Hong, X.; Liu, Y.; Jiang, S.; Ning, B.; Gao, Z. Ultrasensitive sensing of diethylstilbestrol based on AuNPs/MWCNTs-CS composites coupling with sol-gel molecularly imprinted polymer as a recognition element of an electrochemical sensor. *Sens. Actuators B Chem.* **2017**, *238*, 420–426. [CrossRef]

190. Wang, H.; Yao, S.; Liu, Y.; Wei, S.; Su, J.; Hu, G. Molecularly imprinted electrochemical sensor based on Au nanoparticles in carboxylated multi-walled carbon nanotubes for sensitive determination of olaquindox in food and feedstuffs. *Biosens. Bioelectron.* **2017**, *87*, 417–421. [CrossRef] [PubMed]

191. Li, J.; Wang, X.; Duan, H.; Wang, Y.; Bu, Y.; Luo, C. Based on magnetic graphene oxide highly sensitive and selective imprinted sensor for determination of sunset yellow. *Talanta* **2016**, *147*, 169–176. [CrossRef] [PubMed]

sensors

MDPI

Article

Ultratrace Detection of Histamine Using a Molecularly-Imprinted Polymer-Based Voltammetric Sensor

Maedeh Akhoundian, Axel Rüter and Sudhirkumar Shinde *

Department of Biomedical Sciences, Faculty of Health and Society, Malmö University, Malmö SE-20506, Sweden; m.akhoundian@yahoo.com (M.A.); axel.ruter@fkem1.lu.se (A.R.)
* Correspondence: sudhirkumar.shinde@mah.se; Tel.: +46-703-463-963

Academic Editor: Alexander Star
Received: 6 February 2017; Accepted: 17 March 2017; Published: 21 March 2017

Abstract: Rapid and cost-effective analysis of histamine, in food, environmental, and diagnostics research has been of interest recently. However, for certain applications, the already-existing biological receptor-based sensing methods have usage limits in terms of stability and costs. As a result, robust and cost-effective imprinted polymeric receptors can be the best alternative. In the present work, molecularly-imprinted polymers (MIPs) for histamine were synthesized using methacrylic acid in chloroform and acetonitrile as two different porogens. The binding affinity of the MIPs with histamine was evaluated in aqueous media. MIPs synthesized in chloroform displayed better imprinting properties for histamine. We demonstrate here histamine MIPs incorporated into a carbon paste (CP) electrode as a MIP-CP electrode sensor platforms for detection of histamine. This simple sensor format allows accurate determination of histamine in the sub-nanomolar range using an electrochemical method. The sensor exhibited two distinct linear response ranges of 1×10^{-10}–7×10^{-9} M and 7×10^{-9}–4×10^{-7} M. The detection limit of the sensor was calculated equal to 7.4×10^{-11} M. The specificity of the proposed electrode for histamine is demonstrated by using the analogous molecules and other neurotransmitters such as serotonin, dopamine, etc. The MIP sensor was investigated with success on spiked serum samples. The easy preparation, simple procedure, and low production cost make the MIP sensor attractive for selective and sensitive detection of analytes, even in less-equipped laboratories with minimal training.

Keywords: histamine imprinted polymers; Ultratraces; sensor

1. Introduction

Histamine (β-imidazolylethylamine) is a biogenic amine. It is an important mediator involved in various physiological and pathological processes, including neurotransmission and numerous brain functions, secretion of some hormones, regulation of gastrointestinal, circulatory functions, and inflammatory reactions [1]. However, a high level of histamine in the human body causes an allergy-like syndrome called histamine intolerance [2] or histamine poisoning (toxic level 50 mg per 100 g of product) [3,4]. The source of elevated histamine level is due to fermented foods, such as fish, cheese, sauerkraut, beer, wine, processed meat [2], and spoilage of the foodstuff, in particular seafood, due to uncontrolled microbial growth [5]. Therefore, monitoring of histamine is a critical task for the food industry and food safety.

A number of methods for the determination of histamine have been reported including thin-layer chromatography [6,7], gas chromatography [8,9], capillary zone electrophoresis [10], and high-performance liquid chromatography [11–14], as well as fluorimetric [15] and colorimetric assays [16]. These methods require extensive sample processing such as pretreatment of samples.

In addition, it involves qualified analysts. These tedious and time-consuming methods result in expensive and slow sample throughput. In this regard, enzyme-based methods [e.g., enzyme-linked immunosorbent assay (ELISA)] [17,18] offer rapid means of detection, but necessitate the use of unstable enzymes, expensive test kits, and tend to overestimate histamine [19].

Different types of electrochemical sensors with chemical modifications [20–22] or with immobilized amine oxidases and dehydrogenases [23–28] have been described in several reports of histamine determination.

The development of stable histamine receptor with a capacity to detect low histamine concentrations (nM range) is an urgent need in the biomedical and diagnostics research [29–34]. In this direction, MIP-based sensors have attracted much interest due to easy preparation, good stability, and robustness. In the literature, histamine-imprinted polymers and MIP-based sensors have already been reported for histamine recognition [35–37] in surface enhanced Raman spectroscopy (SER) [38], thermal [39], quartz crystal microbalance (QCM) [40], amperometric [41], and impedimetric [34,42] sensors.

In the current work, a histamine MIP has been developed and used for fabrication of voltammetric sensor. The developed sensor has been successfully applied for histamine determination in serum samples. The developed methodology offers advantages such as simplicity, precision, short analysis time, low cost of analysis and instrumentation, with comparable selectivity and sensitivity with advanced instruments.

2. Materials and Methods

Dopamine hydrochloride, histamine, and serotonin hydrochloride was received from Sigma, (Steinheim, Germany). H-His-OH (99%) was received from Bachem Biochemica GmBH (Heidelberg, Germany) and Boc-His-OH was received from Calbiochem-Novabiochem AG (Läufelfingen, Switzerland). Methacrylic acid (MAA) and ethylene glycol dimethacrylate (EGDMA) were purchased from Sigma-Aldrich Chemie GmbH (Taufkirchen, Germany). EGDMA was washed consecutively with 10% NaOH, water, and brine and then dried over $MgSO_4$, and filtered prior to distillation under reduced pressure. MAA was also distilled under reduced pressure. The initiator azo-N,N'-bisdivaleronitrile (ABDV) was purchased from Wako Chemicals and used without further purification. Chloroform ($CHCl_3$), extra dry, and acetonitrile (MeCN), extra dry, were received from Acros Organics (Geel, Belgium). The porogens were kept under nitrogen atmosphere over molecular sieves and were used without further purification. Graphite flake powder (325 mesh) was received from Alfa Aesar and carbon (mesoporous nanopowder, <500 nm particle size) was received from Aldrich (Steinheim, Germany). Paraffin oil was received from Kebo AB (Stockholm, Sweden).

2.1. Polymer Synthesis

Imprinted polymers were prepared using bulk polymerization (Table 1) in the following manner: Template histamine (0.24 mmol), functional monomer MAA (1.2 mmol), and EDGMA (6 mmol) were dissolved in 1.5 mL dry chloroform (MIP1) and 1.5 mL dry acetonitrile (MIP2). The initiator ABDV (1% *w/w* of total monomers) was added to the solution which was transferred to screw-capped glass vials, cooled to 0 °C, and purged with a flow of dry nitrogen for 10 min. The glass vials were sealed with silicone tape while still under cooling and the polymerization was initiated by placing the vials at 50 °C for 48 h. The polymers were lightly crushed and the template was extracted with methanol: acetic acid/90:10 for 24 h. This was followed by further lightly crushing the particles without fractionation to evaluate their binding properties. Corresponding non-imprinted polymers (NIP1/NIP2) were prepared in the same manner described above, but with the omission of the template molecule from the pre-polymerization solution.

Table 1. Composition of histamine imprinted polymer by bulk polymerization.

Polymer	Template (mmol)	MAA (mmol)	EGDMA (mmol)	Solvent
MIP1	0.24	1.2	6	$CHCl_3$
NIP1	-	1.2	6	$CHCl_3$
MIP2	0.24	1.2	6	MeCN
NIP2	-	1.2	6	MeCN

2.2. Binding Analysis

Increasing amount of histamine (0 to 1 mM) 1 mL solution was suspended in the 10 mg MIP and NIP particles. Rebinding tests of the polymeric materials were all performed in a 50 mM PBS solution (pH 7.4) with an equilibration time of 4 h followed by analysis with UV-VIS spectroscopy measuring absorbance at 208 nm. Measurements were performed on a BIOTEK POW-ERWAVE XS plate reader in a 96-well quartz plate.

2.3. Preparation of the Sensors

In order to fabricate the sensor, 11 mg of nano-carbon (NC) and 0.2 g of graphite was dispersed in 2 mL of dimethylformamide (DMF) and sonicated for 30 min, and allowed to dry in an oven at 80 °C overnight. This mixture was then added to the 11 mg of MIP1/NIP1 and homogenized in a mortar. Subsequently, 57.6 mg of paraffin oil was added to the mixture to form a homogeneous paste. This paste was used to fill a hole (4.00 mm in diameter, 3 mm in depth) at the end of the electrode body. The excess material from the surface of the electrode was removed by polishing on a paper sheet, thus devising the MIP-CP electrode. For the normal mixing electrode, all of the steps were repeated except the dispersion and sonication steps otherwise described.

2.4. General Method for Electrochemical Measurements

Electrochemical data was obtained with a system using an Ivium-stat potentiostat/galvanostat. The measurements were performed in a three-electrode system: working electrode (MIP/NIP-based CP electrodes), a counter electrode (platinum), and a reference electrode (Ag/AgCl). Electrochemical measurement of histamine concentration was performed according to the following procedure: The solution containing 20 mL of potassium hexacyanoferrate (0.1 M $K_3[Fe(CN)_6]$) and potassium chloride (0.1 M KCl) was added to the cell as a blank. Before each determination, histamine stock solution was prepared in PBS (0.1 M, pH = 7), spiked to the blank solution and stirred for 15 s. The MIP-CP electrode was immediately placed into the electrochemical cell and the cyclic voltammetry technique was applied using scan rate = 50 mV/s, Estep = 10 mV, and equilibration time = 5 s. The modified electrode was rinsed with milliQ water and polished on the paper sheet after each measurement.

2.5. The Measurement of Histamine in Real Samples

In order to determine the histamine in real sample, a certain amount of histamine stock solution was spiked in to the serum sample and it was diluted with milliQ water in a 1:20 ratio. The prepared sensor was then immersed in to the spiked serum sample and solution was stirred for 15 s. Cyclic voltammetry responses were recorded immediately after equilibration time of the program, which was 5 s.

3. Results

3.1. Molecularly- Imprinted Polymers for Histamine

3.1.1. Polymer Synthesis

The preparation of histamine MIPs was done following established procedure with slight modifications (Table 1) starting from prepolymerization mixtures containing histamine as the template, methacrylic acid (MAA) as functional monomers, and ethylenglycol dimethacrylate (EGDMA) as a crosslinking monomer. In order to study the effect of porogen on histamine imprinting, we chose $CHCl_3$ and MeCN over highly-hygroscopic polar dimethyl sulfoxide (DMSO) among mostly used porogens reported earlier. After free radical polymerization, the resulting polymers were freed from the template by washing with acidic methanol, leaving binding sites complementary to a relatively narrow group of histamine.

3.1.2. Optical Batch Rebinding Experiment

At this stage, we were interested in measuring histamine in biological fluids at neutral pH and aqueous environment. The functional monomer MAA [43] (the pKa 6.5), which was used by Bongaers et al. [42], Horemans et al. [40], and Trikka et al. [35], is a deprotonated form (COO^-) at neutral pH, which binds to histamine. The MIP synthesized from MAA is suitable for rebinding of histamine at pH 7.4 [34,35]. Therefore, we performed rebinding experiments for synthesized MIPs at pH 7.4 using PBS buffer followed by analysis with UV spectrophotometer. For the rebinding experiment, 10 mg of MIP and NIP powder were added to 1 mL of histamine between 0 to 1 mM concentration ranges. The resulting suspensions were shaken for 4 h at room temperature. The supernatant was collected after centrifugation and the free concentration of histamine was determined by UV-VIS spectroscopy. Hereby, the amount of bound histamine per gram of MIP or NIP was calculated and the binding isotherms were constructed. The binding isotherms for the MIP and NIP were prepared in $CHCl_3$ and MeCN are demonstrated in Figure 1a,b. The polymer prepared in $CHCl_3$ displayed better imprinting performance compared to those in MeCN. This is due to the difference in polarity (chloroform is less polar than acetonitrile) a large concentration of prepolymer complex would be expected using chloroform as a porogen [44]. Further, Trikka et al. [35] has confirmed by ^1H NMR spectroscopy that interaction between histamine and MAA are mainly hydrogen bonding when chloroform is used as a solvent. The best performing MIP and NIP prepared in $CHCl_3$ was further used for MIP-modified carbon paste (CP) electrodes.

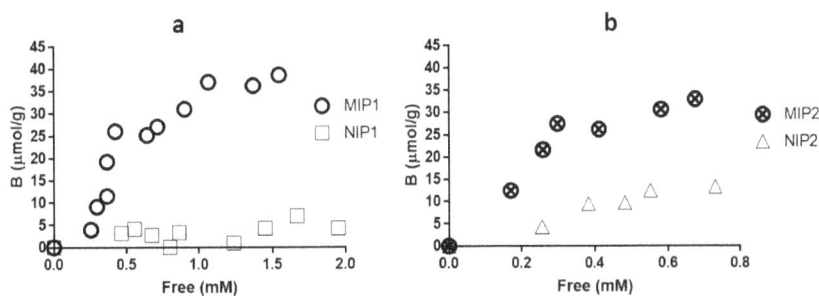

Figure 1. Rebinding isotherm of histamine (**a**) MIP1, NIP1; (**b**) MIP2, NIP2 in 50 mM PBS buffer (pH 7.4). MIP1 and MIP2 (corresponding NIPs) were prepared in $CHCl_3$ and MeCN, respectively.

3.2. MIP-CP Electrode

3.2.1. Fabrication of MIP-CP Electrode

In the first steps of the fabrication process for optimization of composite, we have tested different sequences of mixing of nano carbon (NC), graphite and MIP/NIP as follows: (1) graphite + NC + MIP/NIP normal mix, these powders were physically mixed together as a dry powder; (2) graphite + NC + MIP/NIP (dispersed) powders were mixed together and dispersed in DMF, sonicated, and dried; (3) graphite + NC (dispersed) + MIP/NIP, graphite, and NC was dispersed in DMF, sonicated, and dried followed by mixing of MIP/NIP powder as described in material and methods. Paraffin oil was added to prepare the paste in all the composites. Subsequently, electrochemical responses of hexacyanoferrate solution with different concentrations of histamine were investigated for all electrodes (Figure 2). The MIP-CP electrode prepared by the graphite + NC (dispersed) + MIP/NIP method gave a linear electrochemical signal in cyclic voltammetric measurement (Figure 3). This is most likely related to accessibility of MIP for recognition of histamine in the composite materials.

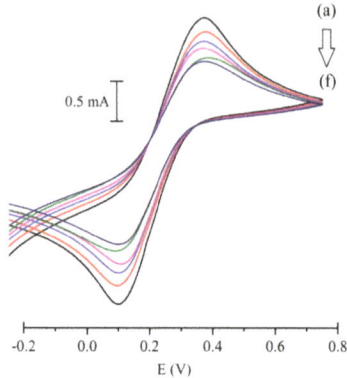

Figure 2. MIP-CP electrode responses to histamine. Cyclic voltammograms in different molar concentrations of histamine: 1×10^{-10} (**a**); 2×10^{-9} (**b**); 4×10^{-9} (**c**); 7×10^{-9} (**d**); 2×10^{-7} (**e**); and 4×10^{-7} (**f**). All of the histamine solutions were made in 0.1 M solution of hexacyanoferrate (III) and KCl as blank.

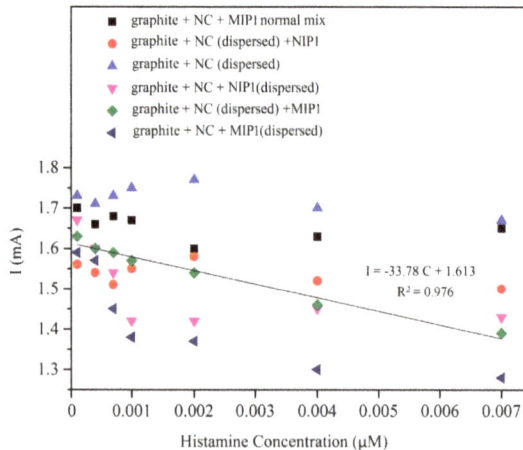

Figure 3. Comparison of voltammetric responses to histamine for different electrode composites. All of the histamine solutions were made in the 0.1 M solution of hexacyanoferrate (III) and KCl as a blank.

3.2.2. Selectivity of the Modified CP Electrode

The MIP-CP electrode was evaluated using structurally-similar compounds and neurotransmitters. In order to accomplish these experiments, the MIP-CP electrode was immersed in individual solution (concentration $= 7 \times 10^{-9}$ M) and the electrochemical detection process was carried out according to the above mentioned procedure. Figure 4 illustrates the responses of histamine and other histamine-like compounds, including the neurotransmitters serotonin and dopamine, in the histamine MIP-CP electrode under the exact same conditions. As can be seen, the responses of histamine to the MIP-CP electrode is the highest compared to other analytes, followed by H-His-OH amino acid having a similar structure. It is interesting to note that when the MIP-CP electrode was tested for Boc-His-OH, a lower response was observed. This may suggest that primary amine of histamine and MAA is important for interaction involved in molecular recognition.

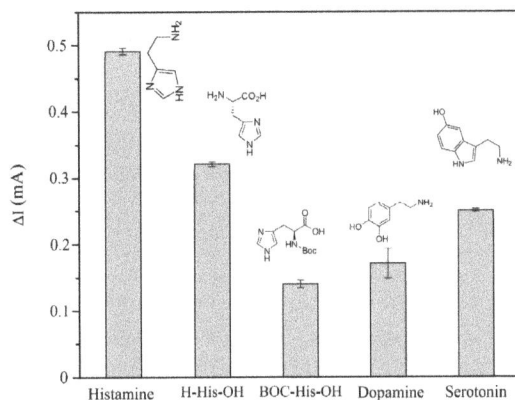

Figure 4. Selectivity investigation by cyclic voltammetry; electrochemical response of MIP-CP electrode for histamine and other similar structure compounds. [Histamine and all analytes] = 0.004 μM, scan rate = 50 mV/s, E_{step} = 10 mV and equilibration time = 5 s.

3.3. Analytical Characterization

3.3.1. Calibration of the MIP-CP Electrode

The developed sensor was used for calibration curve plotting. It is worth noting that the values of the current response used for the calibration curve are actually the absolute values of the oxidative peak current observed for blank solution and after spiking of different concentrations of histamine solutions. The calibration graph obtained for histamine determination of the prepared sensor is shown in Figure 5 that exhibited two distinct linear response ranges of 1×10^{-10}–7×10^{-9} and 7×10^{-9}–4×10^{-7} M with the detection limit of 7.4×10^{-11} M (S/N = 3).

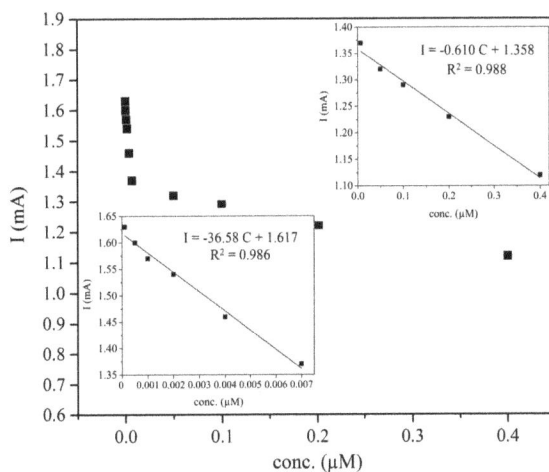

Figure 5. Calibration curve obtained for the developed sensor; (insets are showing two linear ranges of histamine, 1×10^{-10}–7×10^{-9} M and 7×10^{-9}–4×10^{-7} M).

3.3.2. Histamine Determination in Human Plasma

The analytical utility of the method was assessed by applying it to the determination of histamine in human serum samples. Percent recovery of histamine was obtained demonstrating that the proposed sensor is a promising approach in sensor preparation and histamine analysis. The results for the determination of histamine in different samples are summarized in Table 2.

Table 2. Determination of histamine in human serum.

Sample	Spiked (mol·L^{-1})	Found (mol·L^{-1})	Recovery (%)	RSD (%)
	5.0×10^{-10}	5.2×10^{-10}	104	2.02
Human serum	4.0×10^{-9}	4.2×10^{-9}	105	3.58
	2.0×10^{-7}	1.9×10^{-7}	95	3.42

3.4. Comparison of the Developed Method and Other Previously Reported Electrochemical Methods

In Table 3, the analytical parameter of reported electrochemical methods for quantification of histamine are summarized. Compared to most of other methods, the approach presented here shows a wide dynamic range and has a lower detection limit. The proposed procedures for making MIP-CP electrode and analytical method developed using bulk imprinted polymer is very simple and inexpensive. In this proposed sensor, the low detection limit, wide linear working range, and comparable sensitivity and selectivity to the advanced instrumentation is noteworthy.

Table 3. Comparison of analytical parameters of the proposed sensor and some other previously-reported histamine electrochemical sensors.

Method	Electrode	Linear Range (mol·L^{-1})	Detection Limit (mol·L^{-1})	Reference
Impedimetry	Polymer-coated Al	1.2×10^{-8}–2.0×10^{-9}	2.0×10^{-9}	[42]
Voltammetry	Glassy carbon	2.0×10^{-4}–5.0×10^{-6}	0.3×10^{-6}	[21]
Amperometry	Screen-printed	6.0×10^{-5}–8.0×10^{-6}	8.1×10^{-6}	[45]
Amperometry	Heterogeneous carbon	8.9×10^{-5}–4.5×10^{-6}	1.8×10^{-6}	[46]
Voltammetry	SWCNT-modified carbon paste	7.2×10^{-4}–4.5×10^{-6}	1.3×10^{-6}	[20]
Chronopotentiometry	Gold	8.9×10^{-4}–1.8×10^{-5}	2.4×10^{-6}	[28]
Chronopotentiometry	Glassy carbon	8.1×10^{-4}–1.8×10^{-5}	1.2×10^{-5}	[47]
Amperometry	Boron-doped diamond	8.1×10^{-3}–4.5×10^{-5}	4.0×10^{-5}	[48]
Voltammetry	Gold micro electrode	4.9×10^{-8}–9.9×10^{-12}	3.1×10^{-12}	[49]
Voltammetry	NC/MIP/CPE	4×10^{-7}–7×10^{-9} and 7×10^{-9}–10×10^{-10}	7.4×10^{-11}	This work

4. Conclusions

Histamine, a biogenic amine, is an indicator in pathophysiology, microbial infection and food safety. Routinely, faster detection of histamine is performed by time-consuming chemical methods or ELISA. Therefore, MIPs, as synthetic receptors, are a cost-effective alternative. In this study, histamine-imprinted polymers were prepared by bulk imprinted polymerization in chloroform and acetonitrile. The imprinted polymer prepared in chloroform displayed better imprinting performance in a neutral buffer. Modification of carbon paste (CP) electrode with histamine MIPs as recognition elements lead to an excellent sensor for histamine. As a proof of principle, the measurement of histamine in the human serum is demonstrated using MIP-based voltammetric sensor.

Acknowledgments: We acknowledge Börje Sellergren and Tautgirdas Ruzgas for discussion. We thank Bo Mattiasson and Gizem Ertürk for the invitation to submit in the special issue entitled "Biosensors and Molecular Imprinting".

Author Contributions: S.S. designed the experiments; A.R. synthesized MIP and performed histamine rebinding study; M.A. performed the sensor construction, testing, and validation study. S.S. and M.A. analyzed data and wrote the manuscript.

Conflicts of Interest: The authors declare no conflict of interest.

References

1. Jutel, M.; Watanabe, T.; Akdis, M.; Blaser, K.; Akdis, C.A. Immune regulation by histamine. *Curr. Opin. Immunol.* **2002**, *14*, 735–740. [CrossRef]
2. Maintz, L.; Novak, N. Histamine and histamine intolerance. *Am. J. Clin. Nutr.* **2007**, *85*, 1185–1196. [PubMed]
3. Taylor, S.L.; Eitenmiller, R.R. Histamine food poisoning: Toxicology and clinical aspects. *CRC Crit. Rev. Toxicol.* **1986**, *17*, 91–128. [CrossRef] [PubMed]
4. Lehane, L.; Olley, J. Histamine fish poisoning revisited. *Int. J. Food Microbiol.* **2000**, *58*, 1–37. [CrossRef]
5. Fernandes, J.O.; Judas, I.C.; Oliveira, M.B.; Ferreira, I.M.P.L.V.O.; Ferreira, M.A. A GC-MS method for quantitation of histamine and other biogenic amines in beer. *Chromatographia* **2001**, *53*, S327–S331. [CrossRef]
6. Tao, Z.; Sato, M.; Han, Y.; Tan, Z.; Yamaguchi, T.; Nakano, T. A simple and rapid method for histamine analysis in fish and fishery products by tlc determination. *Food Control* **2011**, *22*, 1154–1157. [CrossRef]
7. Lieber, E.R.; Taylor, S.L. Thin-layer chromatographic screening methods for histamine in tuna fish. *J. Chromatogr. A* **1978**, *153*, 143–152. [CrossRef]
8. Antoine, F.R.; Wei, C.I.; Otwell, W.S.; Sims, C.A.; Littell, R.C.; Hogle, A.D.; Marshall, M.R. Gas chromatographic analysis of histamine in mahi-mahi (coryphaena hippurus). *J. Agric. Food Chem.* **2002**, *50*, 4754–4759. [CrossRef] [PubMed]
9. Hwang, B.-S.; Wang, J.-T.; Choong, Y.-M. A rapid gas chromatographic method for the determination of histamine in fish and fish products. *Food Chem.* **2003**, *82*, 329–334. [CrossRef]
10. Zhang, L.-Y.; Sun, M.-X. Determination of histamine and histidine by capillary zone electrophoresis with pre-column naphthalene-2,3-dicarboxaldehyde derivatization and fluorescence detection. *J. Chromatogr. A* **2004**, *1040*, 133–140. [CrossRef] [PubMed]
11. Bauza, T.; Blaise, A.; Daumas, F.; Cabanis, J.C. Determination of biogenic amines and their precursor amino acids in wines of the vallée du rhône by high-performance liquid chromatography with precolumn derivatization and fluorimetric detection. *J. Chromatogr. A* **1995**, *707*, 373–379. [CrossRef]
12. Jensen, T.B.; Marley, P.D. Development of an assay for histamine using automated high-performance liquid chromatography with electrochemical detection. *J. Chromatogr. B: Biomed. Sci. Appl.* **1995**, *670*, 199–207. [CrossRef]
13. Önal, A. A review: Current analytical methods for the determination of biogenic amines in foods. *Food Chem.* **2007**, *103*, 1475–1486. [CrossRef]
14. Proestos, C.; Loukatos, P.; Komaitis, M. Determination of biogenic amines in wines by HPLC with precolumn dansylation and fluorimetric detection. *Food Chem.* **2008**, *106*, 1218–1224. [CrossRef]
15. Bjornsdottir-Butler, K.; Bencsath, F.A.; Benner, J.R.A. Modification and single-laboratory validation of aoac official method 977.13 for histamine in seafood to improve sample throughput. *J. AOAC Int.* **2015**, *98*, 622–627. [PubMed]
16. Patange, S.B.; Mukundan, M.K.; Ashok Kumar, K. A simple and rapid method for colorimetric determination of histamine in fish flesh. *Food Control* **2005**, *16*, 465–472. [CrossRef]
17. Pessatti, T.L.; Fontana, J.D.; Pessatti, M.L. Spectrophotometric determination of histamine in fisheries using an enzyme immunoassay method. *Meth. Mol. Biol.* **2004**, *268*, 311–316.
18. Cinquina, A.L.; Longo, F.; Calí, A.; De Santis, L.; Baccelliere, R.; Cozzani, R. Validation and comparison of analytical methods for the determination of histamine in tuna fish samples. *J. Chromatogr. A* **2004**, *1032*, 79–85. [CrossRef] [PubMed]
19. Ben-Gigirey, B.; Craven, C.; An, H. Histamine formation in albacore muscle analyzed by aoac and enzymatic methods. *J. Food Sci.* **1998**, *63*, 210–214. [CrossRef]
20. Stojanović, Z.S.; Mehmeti, E.; Kalcher, K.; Guzsvány, V.; Stanković, D.M. Swcnt-modified carbon paste electrode as an electrochemical sensor for histamine determination in alcoholic beverages. *Food Anal. Methods* **2016**, *9*, 2701–2710. [CrossRef]
21. Degefu, H.; Amare, M.; Tessema, M.; Admassie, S. Lignin modified glassy carbon electrode for the electrochemical determination of histamine in human urine and wine samples. *Electrochim. Acta* **2014**, *121*, 307–314. [CrossRef]

22. Geto, A.; Tessema, M.; Admassie, S. Determination of histamine in fish muscle at multi-walled carbon nanotubes coated conducting polymer modified glassy carbon electrode. *Synth. Met.* **2014**, *191*, 135–140. [CrossRef]

23. Bao, L.; Sun, D.; Tachikawa, H.; Davidson, V.L. Improved sensitivity of a histamine sensor using an engineered methylamine dehydrogenase. *Anal. Chem.* **2002**, *74*, 1144–1148. [CrossRef] [PubMed]

24. Keow, C.M.; Abu Bakar, F.; Salleh, A.B.; Heng, L.Y.; Wagiran, R.; Bean, L.S. An amperometric biosensor for the rapid assessment of histamine level in tiger prawn (penaeus monodon) spoilage. *Food Chem.* **2007**, *105*, 1636–1641. [CrossRef]

25. Niculescu, M.; Frébort, I.; Peč, P.; Galuszka, P.; Mattiasson, B.; Csöregi, E. Amine oxidase based amperometric biosensors forhistamine detection. *Electroanalysis* **2000**, *12*, 369–375. [CrossRef]

26. Yamamoto, K.; Takagi, K.; Kano, K.; Ikeda, T. Bioelectrocatalytic detection of histamine using quinohemoprotein amine dehydrogenase and the native electron acceptor cytochrome c-550. *Electroanalysis* **2001**, *13*, 375–379. [CrossRef]

27. Zeng, K.; Tachikawa, H.; Zhu, Z.; Davidson, V.L. Amperometric detection of histamine with a methylamine dehydrogenase polypyrrole-based sensor. *Anal. Chem.* **2000**, *72*, 2211–2215. [CrossRef] [PubMed]

28. Sarada, B.V.; Rao, T.N.; Tryk, D.A.; Fujishima, A. Electrochemical oxidation of histamine and serotonin at highly boron-doped diamond electrodes. *Anal. Chem.* **2000**, *72*, 1632–1638. [CrossRef] [PubMed]

29. Barbara, G.; Stanghellini, V.; De Giorgio, R.; Cremon, C.; Cottrell, G.S.; Santini, D.; Pasquinelli, G.; Morselli-Labate, A.M.; Grady, E.F.; Bunnett, N.W.; et al. Activated mast cells in proximity to colonic nerves correlate with abdominal pain in irritable bowel syndrome. *Gastroenterology* **2004**, *126*, 693–702. [CrossRef] [PubMed]

30. Wood, J.D. Histamine, mast cells, and the enteric nervous system in the irritable bowel syndrome, enteritis, and food allergies. *Gut* **2006**, *55*, 445–447. [CrossRef] [PubMed]

31. Lin, R.Y.; Schwartz, L.B.; Curry, A.; Pesola, G.R.; Knight, R.J.; Lee, H.-S.; Bakalchuk, L.; Tenenbaum, C.; Westfal, R.E. Histamine and tryptase levels in patients with acute allergic reactions: An emergency department–based study. *J. Allergy Clin. Immunol.* **2000**, *106*, 65–71. [CrossRef] [PubMed]

32. Chhabra, J.; Li, Y.Z.; Alkhouri, H.; Blake, A.E.; Ge, Q.; Armour, C.L.; Hughes, J.M. Histamine and tryptase modulate asthmatic airway smooth muscle gm-csf and rantes release. *Eur. Respir. J.* **2007**, *29*, 861. [CrossRef] [PubMed]

33. Adlesic, M.; Verdrengh, M.; Bokarewa, M.; Dahlberg, L.; Foster, S.J.; Tarkowski, A. Histamine in rheumatoid arthritis. *Scand. J. Immunol.* **2007**, *65*, 530–537. [CrossRef] [PubMed]

34. Peeters, M.; Troost, F.J.; Mingels, R.H.G.; Welsch, T.; van Grinsven, B.; Vranken, T.; Ingebrandt, S.; Thoelen, R.; Cleij, T.J.; Wagner, P. Impedimetric detection of histamine in bowel fluids using synthetic receptors with ph-optimized binding characteristics. *Anal. Chem.* **2013**, *85*, 1475–1483. [CrossRef] [PubMed]

35. Trikka, F.A.; Yoshimatsu, K.; Ye, L.; Kyriakidis, D.A. Molecularly imprinted polymers for histamine recognition in aqueous environment. *Amino Acids* **2012**, *43*, 2113–2124. [CrossRef] [PubMed]

36. Allender, C.J.; Richardson, C.; Woodhouse, B.; Heard, C.M.; Brain, K.R. Pharmaceutical applications for molecularly imprinted polymers. *Int. J. Pharm.* **2000**, *195*, 39–43. [CrossRef]

37. Tong, A.; Dong, H.; Li, L. Molecular imprinting-based fluorescent chemosensor for histamine using zinc(ii)–protoporphyrin as a functional monomer. *Anal. Chim. Acta* **2002**, *466*, 31–37. [CrossRef]

38. Gao, F.; Grant, E.; Lu, X. Determination of histamine in canned tuna by molecularly imprinted polymers-surface enhanced raman spectroscopy. *Anal. Chim. Acta* **2015**, *901*, 68–75. [CrossRef] [PubMed]

39. Peeters, M.; Kobben, S.; Jiménez-Monroy, K.L.; Modesto, L.; Kraus, M.; Vandenryt, T.; Gaulke, A.; van Grinsven, B.; Ingebrandt, S.; Junkers, T.; et al. Thermal detection of histamine with a graphene oxide based molecularly imprinted polymer platform prepared by reversible addition–fragmentation chain transfer polymerization. *Sens. Actuators B Chem.* **2014**, *203*, 527–535. [CrossRef]

40. Horemans, F.; Alenus, J.; Bongaers, E.; Weustenraed, A.; Thoelen, R.; Duchateau, J.; Lutsen, L.; Vanderzande, D.; Wagner, P.; Cleij, T.J. MIP-based sensor platforms for the detection of histamine in the nano- and micromolar range in aqueous media. *Sens. Actuators B Chem.* **2010**, *148*, 392–398. [CrossRef]

41. Pietrzyk, A.; Suriyanarayanan, S.; Kutner, W.; Chitta, R.; D'Souza, F. Selective histamine piezoelectric chemosensor using a recognition film of the molecularly imprinted polymer of bis(bithiophene) derivatives. *Anal. Chem.* **2009**, *81*, 2633–2643. [CrossRef] [PubMed]

42. Bongaers, E.; Alenus, J.; Horemans, F.; Weustenraed, A.; Lutsen, L.; Vanderzande, D.; Cleij, T.J.; Troost, F.J.; Brummer, R.J.; Wagner, P. A MIP-based biomimetic sensor for the impedimetric detection of histamine in different ph environments. *Phys. Status Solidi (A)* **2010**, *207*, 837–843. [CrossRef]

43. Sellergren, B. Molecular imprinting by noncovalent interactions. Enantioselectivity and binding capacity of polymers prepared under conditions favoring the formation of template complexes. *Die Makromol. Chem.* **1989**, *190*, 2703–2711. [CrossRef]

44. Yan, M. *Molecularly Imprinted Materials: Science and Technology*; CRC Press: Boca Raton, FL, USA, 2004.

45. Henao-Escobar, W.; del Torno-de Román, L.; Domínguez-Renedo, O.; Alonso-Lomillo, M.A.; Arcos-Martínez, M.J. Dual enzymatic biosensor for simultaneous amperometric determination of histamine and putrescine. *Food Chem.* **2016**, *190*, 818–823. [CrossRef] [PubMed]

46. Veseli, A.; Vasjari, M.; Arbneshi, T.; Hajrizi, A.; Švorc, Ľ.; Samphao, A.; Kalcher, K. Electrochemical determination of histamine in fish sauce using heterogeneous carbon electrodes modified with rhenium(IV) oxide. *Sens. Actuators B Chem.* **2016**, *228*, 774–781. [CrossRef]

47. Stojanović, Z.S.; Švarc-Gajić, J.V. A simple and rapid method for histamine determination in fermented sausages by mediated chronopotentiometry. *Food Control* **2011**, *22*, 2013–2019. [CrossRef]

48. Švarc-Gajić, J.; Stojanović, Z. Determination of histamine in cheese by chronopotentiometry on a thin film mercury electrode. *Food Chem.* **2011**, *124*, 1172–1176. [CrossRef]

49. Akbari-adergani, B.; Norouzi, P.; Ganjali, M.R.; Dinarvand, R. Ultrasensitive flow-injection electrochemical method for determination of histamine in tuna fish samples. *Food Res. Int.* **2010**, *43*, 1116–1122. [CrossRef]

Review

Molecular Imprinting Applications in Forensic Science

Erkut Yılmaz [1], Bora Garipcan [2], Hirak K. Patra [3] and Lokman Uzun [4,*]

[1] Department of Biotechnology and Molecular Biology, Aksaray University, 68100 Aksaray, Turkey; yilmazerkut@yandex.com
[2] Institute of Biomedical Engineering, Bogazici University, 34684 Istanbul, Turkey; bgaripcan@gmail.com
[3] Department of Clinical and Experimental Medicine, Linkoping University, 58225 Linköping, Sweden; hirak.kumar.patra@liu.se
[4] Department of Chemistry, Hacettepe University, 06381 Ankara, Turkey
* Correspondence: lokman@hacettepe.edu.tr or lokmanuzun@gmail.com;
Tel.: +90-312-780-7337; Fax: +90-312-299-2163

Academic Editors: Bo Mattiasson and Gizem Ertürk
Received: 16 February 2017; Accepted: 23 March 2017; Published: 28 March 2017

Abstract: Producing molecular imprinting-based materials has received increasing attention due to recognition selectivity, stability, cast effectiveness, and ease of production in various forms for a wide range of applications. The molecular imprinting technique has a variety of applications in the areas of the food industry, environmental monitoring, and medicine for diverse purposes like sample pretreatment, sensing, and separation/purification. A versatile usage, stability and recognition capabilities also make them perfect candidates for use in forensic sciences. Forensic science is a demanding area and there is a growing interest in molecularly imprinted polymers (MIPs) in this field. In this review, recent molecular imprinting applications in the related areas of forensic sciences are discussed while considering the literature of last two decades. Not only direct forensic applications but also studies of possible forensic value were taken into account like illicit drugs, banned sport drugs, effective toxins and chemical warfare agents in a review of over 100 articles. The literature was classified according to targets, material shapes, production strategies, detection method, and instrumentation. We aimed to summarize the current applications of MIPs in forensic science and put forth a projection of their potential uses as promising alternatives for benchmark competitors.

Keywords: molecular imprinting; forensic science; toxicology; analytical methods; pre-concentration

1. Introduction

The area of forensic science emerged due to the need for scientific techniques for investigating and proving crimes. Forensic science is quite a multidisciplinary area of study with ten and more subdivisions like chemistry, biology, toxicology, geology, archeology, anthropology, astronomy, engineering, etc. All these subdivisions have different methods for problem solving and use a series of specialized tools. In general, "problem solving" in the area of forensic analysis have two meanings: First, identifying the physical evidence or a questioned sample, and the second one is comparing the results with a known material to find the origin of the questioned sample. Molecular imprinting techniques present solutions for both of these requirements as well. Due to their versatility, molecular imprinted polymers (MIPs) have gained many applications in a variety of areas [1]. Versatile usage, stability and recognition capabilities make them a perfect candidate for the use in forensic sciences [2]. MIPs can be prepared in different physical shape and sizes while conferring them with some multi-functional smart material capabilities, like magnetic, stimuli-responsive, fluorescence labelling, etc. These functions support many possible application areas in the field of forensic sciences.

2. A Brief Theory of Molecular Imprinting

Molecular imprinting is the method of producing tailor-made complementary cavities against a targeted structure called the template. By means of these smart and complementary cavities, the resulting molecules have both chemical and physical recognition capabilities; therefore, they are also classified as biomimetic receptors or plastic antibodies [3]. Molecular imprinting is a method for producing selective binding sites in highly cross-linked synthetic polymeric matrices [4]. It is generally achieved via the self-assembly of functional monomers around a "template" and then polymerization of these pre-polymer complexes in the presence of extensive crosslinkers, which is only one of the generally followed synthetic routes and is called non-covalent imprinting. These monomers and ratio in the pre-polymer complex are chosen according to their affinity towards the template. After the template removal from the polymeric matrix, the imprinted cavities come out with both chemical and physical recognition capabilities (Figure 1) [5].

Figure 1. The molecular imprinting principle. a: functional monomers; b: cross-linker; c: template molecule; 1: assembly of the prepolymerisation complex; 2: polymerization; 3: extraction; 4: rebinding. Reprinted with permission from [5].

In imprinting history, researchers have integrated several polymerization techniques into the imprinting process which has resulted in five main types of molecular imprinting techniques. These can be summarized as covalent, non-covalent, semi-covalent, ionic, and metal coordination methods [2]. A schematic representation of these molecular imprinting techniques is shown in Figure 2.

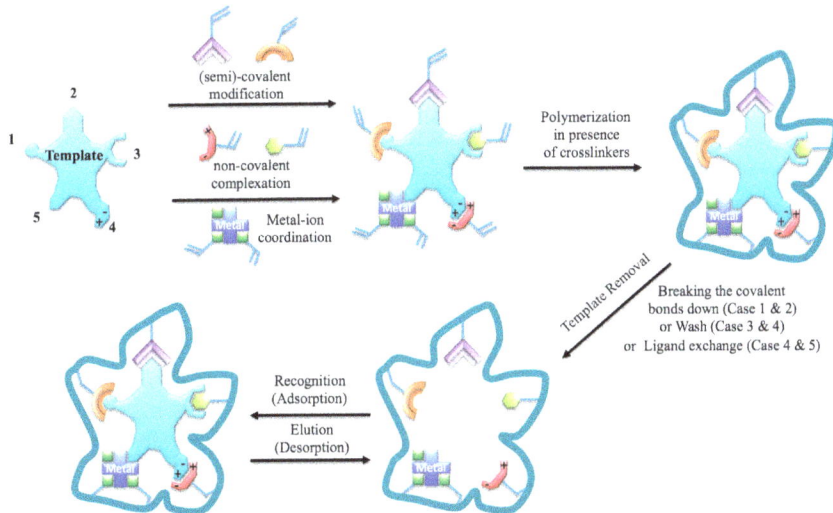

Figure 2. Schematic representation of five main types of interactions used for molecular imprinting purpose. 1: covalent; 2: semi-covalent; 3: covalent; 4: ionic; 5: metal ion coordination.

Each of these techniques has certain advantages with respect to the affinity, selectivity, kinetics and reproducibility of the final polymers.

3. Imprinting Approaches for Forensic Science

In this review, we have focused our attention on compiling the imprinting literature for forensic science applications after briefly summarizing the history of imprinted polymers. In this context, we have clustered the studies into groups related to target structures, material shapes, production strategy, application method and detection platforms. In each subsection, we summarize the related studies while mentioning its novelty, contribution and importance for forensic science applications.

3.1. Target Structures

Availability of molecular imprinting as an analytical tool for variety of target structures is advantageous on forensic chemistry and forensic toxicology. Some of target molecules related to forensic sciences were summarized below:

Legal/illicit Drugs: There is a wide range of abused legal or illicit drugs that come across in forensic cases. Some of the most common drugs are given below with the examples from the molecular imprinting literature. Pain relievers, cough/cold medicines, prescription sedatives like benzodiazepines, and barbiturate sleeping aid medicines are the most studied drugs [6–9]. Ariffin et al. reported the extraction of diazepam and other benzodiazepines from hair samples. They reported a high recovery up to 93% with a good precision (RSD = 1.5%) and a limit of detection (LOD) and quantification (LOQ) of 0.09 and 0.14 ng/mg, respectively [6]. Anderson et al. compared the benzodiazepine extraction performance of an imprinted solid-phase extraction (SPE) system for 10-post-mortem scalp hair samples. The samples were chosen as blood samples of drug-related deaths with a positive benzodiazepine result. They simultaneously analyzed the samples through parallel experiments with classical and molecularly imprinted solid-phase extraction systems while detection was performed by liquid chromatography-tandem mass spectrometry (LC-MS-MS) measurements. They concluded that molecularly imprinted cartridges have a higher selectivity than the classical ones and might be used as complementary method for chronic users [7]. Figueiredo et al. reported a direct extraction and quantitation of benzodiazepines in human plasma by using MIPs. They utilized the electrospray ionization mass spectrometry (ESI-MS) as a detection platform with a high and selective extraction capability, ionic suppression and a short analysis time and a high analytical speed. They reported a linear calibration curve in the range of 10–250 µg/L ($r > 0.98$) with a low LOQ quantification < 10 µg/L that also had an acceptable precision and accuracy for day-to-day and in-day measurements [8]. Rezaei et al. utilized MIPs with ESI-ion mobility spectrometry (IMS) system for detection of primidone (an antiepileptic drug) from complex matrices such as pharmaceutical and human samples. They concluded that the combination of MIPs with ESI-IMS was a very sensitive analytical tool for selective extraction and detection of the target molecule due to its wide linear dynamic range, good recovery, and low relative standard deviation (RSD) of 0.02–2.00 µg/mL, above 90%, and below 3%, respectively [9].

Cannabinoids like marijuana and hashish are the most commonly used illicit drug [10–12]. Nestic et al. reported a combination of MIP integrated with gas chromatography (GC-MS) for simultaneous determination of tetrahydrocannabinol and its main metabolite in urine samples. They reported that the performance of the method completely meets the requirements of toxicological analysis although the extraction recovery, LOD and lower limit of quantification (LLOQ) only suggested performance comparable with the described method in the light of results they achieved [10]. Sanchez-Gonzalez et al. also reported a micro-solid extractor for cannabinoids for assessing plasma and urine analysis of marijuana abusers by the combination of MIPS with a HPLC-MS/MS system. They reported LOQ values for plasma and urine samples in the ranges of 0.36–0.49 ng/L and 0.47–0.57 ng/L, respectively, with an accurate method for inter-day and intra-day analytical recovery performances [11]. Cela-Perez et al. also reported water-compatible imprinted pills for a combined cannabinoids extraction/detection method in urine and oral fluid. They optimized the extraction performance by tuning the MIP composition with respect to screening results of a non-imprinted polymer library. They developed a linear method for urine and oral fluid in the ranges of 1–500 ng/mL and 0.75–500 ng/mL, respectively. They finally applied the developed method to four urine and five oral fluid samples in which low imprecision (lower than 15%) and varied recovery (50%–111%) and good process efficiency (15.4%–54.5%) were determined [12].

Opioids like heroin and opium are substances that act on the opioid receptors to produce morphine-like effects [13–15]. Andersson et al. reported one of first studies including morphine and endogeneous neuropeptide-imprinted polymers. They demonstrated a high binding affinity and selectivity in aqueous buffers which allowed study with biological materials. They also observed high binding constants (as low as 10^{-7} M) at levels of selectivity similar to those of antibodies. They concluded that the high binding affinities and selectivity could lead to the use to MIPS in enzyme-based assays like ELISA and immune-affinity techniques for isolation/separation of water-soluble biologically related compounds [13]. Piletska et al. also developed a multisensory method for drugs of abuse while optimizing the MIP composition by computational techniques. They reported an imprinting factor of 3 for morphine by comparing the recognition capabilities of imprinted and non-imprinted polymers [14]. Devanathan et al. reported a covalently imprinted polymer having subpicomolar binding affinity in an aqueous environment and a well-defined and homogeneously distributed cavity. They utilized plasmon-waveguide resonance spectroscopy as a sensitive optical detection of the target molecules, which allowed achieving the tightest binding ability (up to 10^3-folds) in comparison to the competitor molecules [15].

Another class of the drugs often studied are stimulants like cocaine, amphetamine, and methamphetamine [14,16,17]. As mentioned before, Piletska et al. used a set of MIPs for detecting some drugs in combination with HPLC analysis. They reported the imprinting factors for cocaine, deoxyephedrine and methadone as 1.8-, 4.2-, 6-folds, respectively, under optimized conditions [14]. Li et al. reported a novel stimulant assay by combining colloidal crystals with MIPs for theophylline and ephedrine as template molecules. This approach led to a rapid, handy, sensitive and specific detection system due to its structural features such as highly ordered and interconnected macropores with thin hydrogel walls. Homogenously distributed nanocavities on the walls enabled a rapid, easy, sensitive and direct response during the molecular recognition process without any need forf transducers and analyte treatments. As concluded in the article, a synergetic contribution of structural features and MIPs results in an extremely high sensitivity at such an analyte concentration as 0.1 fM and specificity even in spiked urine samples [16]. Club drugs like methylenedioxymethamphetamine (MDMA), flunitrazepam, γ-hydroxybutyrate and dissociative drugs like ketamine, phencyclidine (PCP) and its analogs, *Salvia divinorum*, and dextromethorphan (DXM, found in some cough and cold medications) were extensively examined targets [14]. Djozan et al. reported a solvent-free and sensitive method for analyzing the methamphetamine, amphetamine and ecstasy levels in human urine. They combined inside-needle trap and MIPs while coating the internal surface of a hollow stainless steel needle with a MIP layer. Due to the fact there was no requirement for an extraction solvent, the method developed was quite fast and simple. They reported LOD, LOQ and relative recovery values of 12 ng/mL, 40 ng/mL and 81%–93%, respectively, with a low relative standard deviation of 4.9% for six repeated experiments [17].

Hallucinogens like lysergic acid diethylamide (LSD), mescaline, psilocybin (magic mushrooms) are another class of targets [18]. Chapuis-Hugon et al. followed a non-covalent imprinting approach to develop a selective extractor for LSD from hair and urine samples. For this aim, they performed offline extraction before HPLC-MS analysis and reported a successful detection of LSD at a low concentration of 0.1 ng/mg in hair samples with an extraction recovery of 82%. For urine samples, easy detection at only 0.5 ng/mL with extraction recovery of 83% was also reported [18].

The compounds including anabolic steroids, inhalants (solvents and gases), nicotine, and alcohol are other intensively examined targets [19–25]. Zhou et al. applied Pickering emulsion polymerization to synthesize MIPs for steroid recognition. They reported that the combination of hydrophobic and hydrogen-bond interactions that were located in imprinted cavities resulted in a high selectivity for similar steroid structures [19]. Tan et al. used MIP as a recognition coating on a quartz crystal thickness-shear mode sensor for nicotine detection in human serum and urine. They achieved a highly selective and sensitive response that was linear over a wide nicotine concentration range of 5.0×10^{-8}–1.0×10^{-4} M with a detection limit of 2.5×10^{-8} M [20]. Krupadam et al. synthesized nicotine-imprinted nanocavities of 24.0 ± 5.0 nm size which were homogeneously distributed through the polymeric structure. The MIPs developed showed a high selectivity with a dissociation constant (Kd) around 10^{-5} M, similar to those of the natural analog (acetylcholine esterase). They concluded that MIP-based artificial receptors are very useful for isolating and separating water-soluble biologically related compounds and detecting nicotine levels for addicted patients [21]. Zhou et al. also synthesized nicotine-imprinted polymers via a reversible addition-fragmentation chain transfer polymerization. They were able to form monodispersed beads with an average size of 1.55 µm. They concluded that these beads could be utilized as building blocks for developing chemical sensors and polymer-enzyme conjugates for analytical applications [22]. Matsuguchi and Uno developed a quartz crystal microbalance (QCM)-based sensor for volatile organic compound for detection of xylene and toluene in the vapor phase. They concluded that the simplicity and reliability of the developed sensor had promising potential in forming sensors combined with a MIP although some further improvements were needed in light of selectivity and response time performance [23]. Yang et al. developed a MIP-based SPE for assaying hair nicotine levels in smokers and non-smokers to investigate exposure to environmental tobacco smoke. They reported LOD and LOQ values of 0.2 ng/mL and 0.5 ng/mL, respectively, while determining a wide linear concentration range of 0.5–80 ng/mL with a regression coefficient greater than 0.987. The nicotine levels determined in smokers and non-smokers varied between 5.1–69.5 ng/mg hair and 0.50–9.3 ng/mg hair, respectively [24]. Wu et al. developed an amperometric sensor for nicotine by combining titanium dioxide, a conductive polymer [poly(2,3-ethylenedioxythiophene), PEDOT] and an imprinting approach. They reported a linear detection range, LOD and imprinting factors of 0–5 mM, 4.9 µM, and 1.24, respectively. They also evaluated the sensitivity enhancement with respect to the electroactive surface area and at-rest stability over 3 days, in which the current response remained at around 85% of its initial value at the end of the second day [25].

Forensic analysis approaches are also applicable for anti-doping purposes to determine doping with performance-enhancing drugs, stimulants, steroids, and corticosteroids [26–30]. Ozgur et al. developed a mass sensitive sensor for real time estradiol detection. They synthesized MIP nanoparticles and used them as recognition elements on quartz crystals. They reported that the resulting nanosensor had a high selectivity and sensitivity against target molecules in the concentration range of 3.67 nM–3.67 pM and the LOD and LOQ values they calculated were quite low, at 613 fM and 2.04 pM, respectively [27]. Zulfiqar et al. developed a MIP-SPE system for extracting and screening multiple steroids in urine. They analyzed a series of twelve structurally similar and commercially available compounds while imprinting only testosterone as a template. They reported effective LODs between 11.7 and 27.0 pg for individual steroids when investigating concentrations (equal for each steroid) between 0.234 and 0.540 ng/mL in urine. They also demonstrated multiple screening applications using a 10 ng/mL mixed sample [28]. Kellens et al. performed bulk and miniemulsion polymerization to produce colloidal particles, which were used for testosterone recognition. They compared the performances of bulk and colloidal MIPs and determined that the imprinting factor increased from 2.2 to 6.8 due to the smaller size, homogeneity and increased surface area of colloidal particles. They concluded that water-based stable MIP dispersions might be useful to construct sensing platforms via spin-coating or dropcasting methods [29]. Tu et al. developed a MIP-based plasmonic immunosandwich assay for erythropoietin recognition in human urine through surface-enhanced Raman scattering (SERS) measurements. They reported a specific detection level as low as 29 fM in a short analysis time of only 30 min in total. They also determined that the cross-reactivity of the assay varied in the range of 1.9%–9.6% for a 1000-fold higher concentration for interfering glycoproteins and non-glycoproteins and additionally it was only 0.8% for a 10,000-fold higher concentration of glucose [30] (Figure 3).

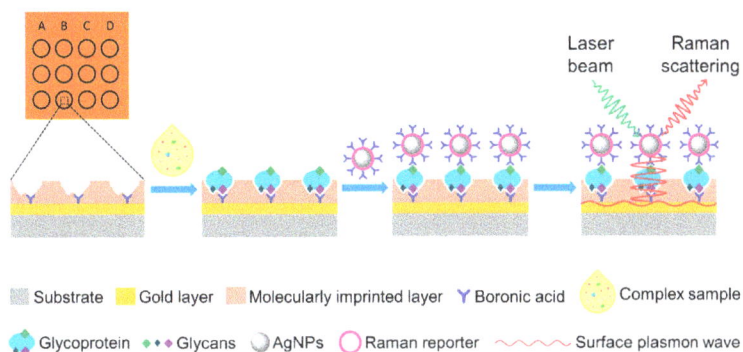

Figure 3. Schematic illustration of the MIP-based immunosandwich assay for recognizing the target (erythropoietin) glycoprotein. Reprinted with permission from [30]. Copyright (2016) American Chemical Society.

Poisons including cyanide, arsenic, nightshade, hemlock, curare, nicotine, caffeine, quinine, atropine, strychnine, and brucine were also intensively evaluated targets in the MIP literature [31–34]. Jackson et al. imprinted 2-aminothiazoline-4-carboxylic acid (ATCA), a chemically stable metabolite of cyanide, on the surface of a silica stir bar and used it for determining the endogenous level of ATCA in cases of cyanide poisoning. Without any derivatization requirements, the strategy enhanced the selectivity and sensitivity of ATCA detection in urine samples at a low concentration of around 400 ng/mL [31]. Liu et al. integrated electropolymerized MIPs with single-wall carbon nanotubes (SWNTs) for brucine detection in human serum. Linear concentration range, detection limit, and recoveries were determined as 6.2×10^{-7}–1.2×10^{-5} M, 2.1×10^{-7} M, and 99.5%–103.2%, respectively [32]. Alizadeh et al. developed arsenic-imprinted nanoparticles for electrochemical ion detection. They inserted a hydrophobic chains (dodecanol) on the nanoparticles to improve durability, lifetime, and analytical characteristic of the polymeric membrane electrodes. They used the sensor for arsenic determination in different water samples and reported a wide concentration range of 5.0×10^{-8}–1.0×10^{-1} M with a LOD value as 30 nM [33]. Nakamura et al. followed precipitation polymerization and multistep swelling and polymerization techniques to form monodisperse MIPs for strychnine recognition. They utilized liquid chromatography for separation and quantification purposes and evaluated the retention and molecular recognition performances of MIPs against not only the template (strychnine) but also some structurally relevant molecules, including brucine, quinine, quinidine, and indole. The retention factors and imprinting factors of strychnine were reported as 220 and 58 for the particles synthesized via precipitation polymerization, respectively, and 73 and 4.5 for the particles synthesized via multistep swelling and polymerization [34]. Xu et al. combined stimuli-responsive polymers with MIPs to synthesize dual (photonic and magnetic) responsive polymers for caffeine detection. They demonstrated that the recoveries ranged from 89.5% to 117.6% from real water and beverages after optimizing the adsorption/elution conditions while performing the experiment under UV (365 nm, adsorption) and visible light (release) [35].

DNA: Not only small molecules but also biomacromolecules, especially DNA, are supplying important information for forensic investigations. Although DNA sequences are 99.9% the same in every person, the remaining 0.01% is enough to distinguish one person from another, if you have the enough amount of DNA molecule to test [2]. Therefore, MIPs are very useful tools for the enrichment of DNA fragments from biological fluids [36–39]. Ogiso et al. used a MIP layer for electrophoretic DNA analysis and achieved the detection of target double-stranded DNA sequences in the presence of different sizes of interfering DNA fragments [36]. Diltemiz et al. developed a biomimicking sensor by creating thymine-imprinted cavities in MIPs with a synthesized adenine-based polymerizable monomer. They reported that the imprinted cavities were homogeneously distributed for thymine recognition with an affinity constant of 10 µM, whereas these sites interacted heterogeneously with uracil as a competitor nucleotide [37]. Ersöz et al. combined MIP-based SPE with a mass sensitive sensor for pre-concentration/detection of 8-hydroxy-2'-deoxyguanosine. They concluded that analytical performance of the proposed system was a promising alternative in comparison to electrophoresis [38]. Uzek et al. synthesized MIP-based monolithic cryogels for rapid plasmid DNA purification. They utilized hydrophobic interactions to recognize DNA molecules and directly integrated the developed column into a fast-protein liquid chromatography system [39].

Explosives: Nitrobenzene, dinitrotoluene (DNT), trinitrotoluene (TNT), and cyclotrimethylene-trinitramine (RDX) were extensively studied explosives for criminal investigations [40–49]. Gao et al. synthesized core-shell imprinted particles via asurface functional monomer-directing strategy for selective 2,4,6-trinitrotoluene (TNT) detection. They used silica nanoparticles as a core material while acrylamide and ethylene glycol dimethacrylate were used as monomer and crosslinker, respectively. They also compared the performances of core-shell and traditional imprinted particles and concluded that core-shell MIPs have five times higher capacity and fast kinetics [40] (Figure 4).

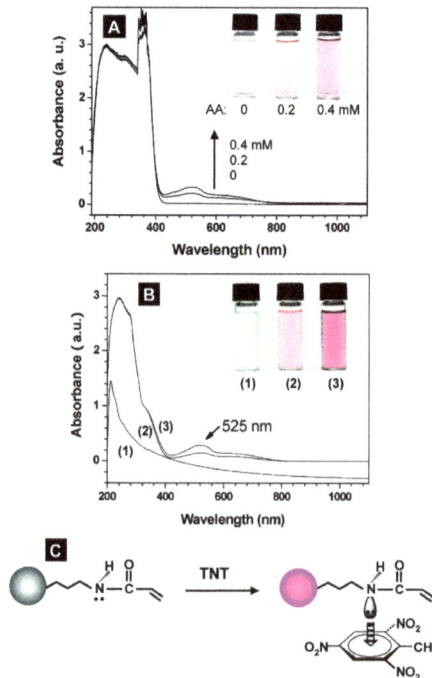

Figure 4. The interactions of TNT molecules with acrylamide and AA-APTS-silica nanoparticles. (**A**) The evolution of UV-visible spectra of TNT solution with increasing acrylamide amount (Inset colorful image shows the corresponding colors of TNT solutions); (**B**) The evolution of UV-visible spectra of AA-APTS-silica nanoparticles solution with increasing TNT amount: (1) without, (2) 0.25 and (3) 0.5 mM TNT (Inset colorful image shows the corresponding colors of nanoparticle solutions); (**C**) The schematic illustration for the charge-transfer complexing interactions between AA-APTS monolayer and TNT molecules. Reprinted with permission from [40]. Copyright (2007) American Chemical Society.

Guan et al. developed hollow polymer with holes in the shells and core-shell microspheres for TNT detection. They reported the maximum TNT binding capacities of hollow and core-shell microspheres as 6.2 and 1.5 µmol of TNT, respectively, for single-runs of 20 mg microspheres [41] (Figure 5).

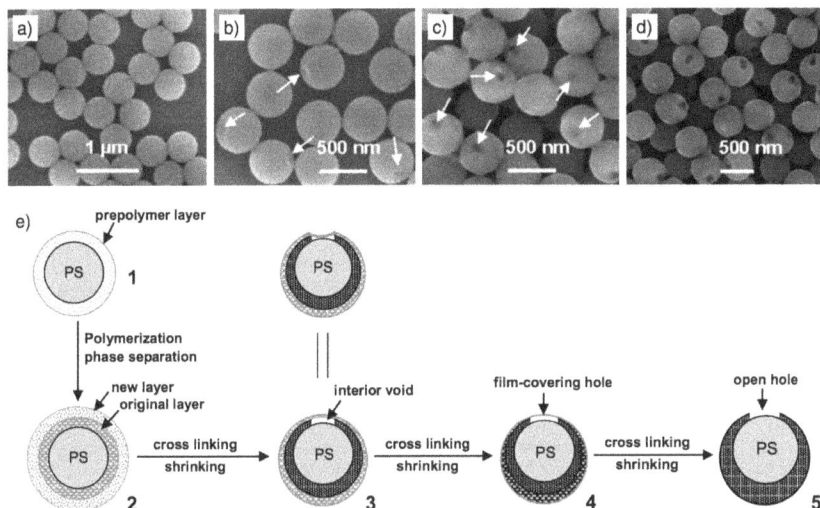

Figure 5. The formation mechanism of holes. (**a–d**) SEM images of four intermediate products with different polymerizing/cross-linking periods at a temperature of 60 °C: (**a**) Monodispersive core–shell microspheres (after 3 h); (**b**) core-shell microspheres with a dimple-like concave (after 6 h); (**c**) core-shell microspheres with a single hole covered with a layer of ultrathin film (after 12 h); and (**d**) core-shell microspheres with open holes (after 20 h); (**e**) Schematic illustration for the formation process of holes in the shells. The microphase separation and progressive volume shrinkage of shell materials leads to the formation of holes via five typical stages: (1) the formation of a prepolymer layer; (2) the microphase separation of two polymer shell layers; (3) the development of a small interior void at the shell; (4) formation of a film-covered hole; and (5) formation of an open hole in the polymer shells. Reprinted with permission from [41]. Copyright © 2007 WILEY-VCH Verlag GmbH & Co. KGaA, Weinheim, Germany.

Li et al. evaluated the chemosensing performance of fluorescent conjugated MIPs for the detection of TNT and related nitroaromatic compounds in vapor phase. They demonstrated that vapor exposures for ten minutes caused a substantial decrease in fluorescence intensity and this behavior was repeated without any noticeable irreversible quenching [42]. Holthoff et al. reported a new and interesting strategy for integrating TNT-imprinted polymers into a surface enhanced Raman scattering platform. They deposited sol-gel derived xerogels on a SERS-active surface and achieved an apparent dissociation constant and detection limit for TNT of 23 µM and 3 µM, respectively [43]. Riskin et al. developed an ultrasensitive surface plasmon resonance (SPR) detection method for hexahydro-1,3,5-trinitro-1,3,5-triazine (RDX) by using gold nanoparticles crosslinked with bisaniline. They electrochemically polymerized gold nanoparticles in the presence of Kemp's acid that allowed the selective and sensitive RDX detection with a quite low detection limit around 12 fM [44]. Riskin et al. followed similar strategy to develop a SPR sensor for pentaerythritol tetranitrate, nitroglycerin, and ethylene glycol dinitrate as well. They reported LOD values for pentaerythritol tetranitrate, nitroglycerin, and ethylene glycol dinitrate of 200 fM, 20 pM, and 400 fM, respectively [45]. Lordel et al. utilized MIPs as a selective extractor for the analysis of nitroaromatic explosives. They concluded that MIP-based extractors indicate a promising potential for miniaturized system due to a very large capacity value, higher than 3.2 mg/g [46,47]. Furthermore, they developed an online microextractor as well. By this way, they achieved a simultaneous extraction and determination of different nitroaromatic explosives with recovery values higher than 90% [48]. Mamo and Gonzalez-Rodriguez developed an electrochemical sensor for the detection of triacetone triperoxide (TATP) over a wide linear range of 82–44,300 µg/mL with a correlation coefficient of 0.996. They determined the LOD and LOQ values as 26.9 µg/L and 81.6 µg/L, respectively, with a quite good repeatability [49].

Gunshot residues: Components of gunpowder are not found in the general population, so the residues of gunshot on clothing or hands of a suspect are good indicator of a fired gun. Gunshot residues consist of inorganic (lead, antimony, barium, calcium, and silicon) and organic (diphenylamine, ethyl centralite and nitrodiphenylamine, dinitrotoluene, nitrobenzene) residues [50] (Figure 6). Studies show that there are at least 136 organic compounds like additives coolants, plasticizers, anti-wear additives that may contribute to gunshot residue [51]. Generally molecular imprinting methods are used for the sample preparation step before a chromatographic or mass spectroscopy analysis of gunshot residue analysis [52]. Pereira et al. developed MIPs for the retention of diphenylamine, one of the organic residues of gunshots, and demonstrated their capacity by HPLC and UV-visible spectroscopy measurements. They concluded that recognition of target molecules was really fast and reached a maximum retention in only the first five minutes [52].

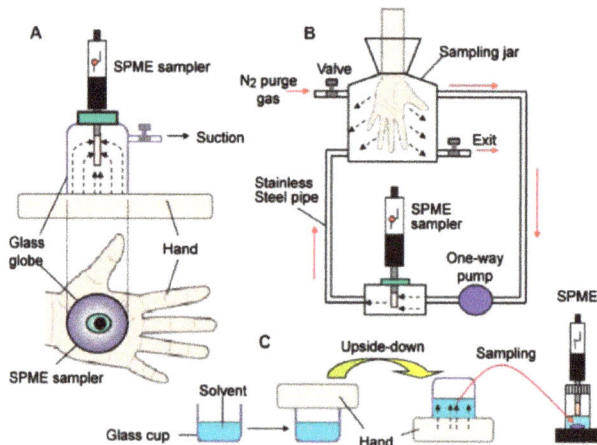

Figure 6. Systems for sampling volatile compounds emitted from the skin. (**A**) Direct SPME in sealed glass globes; (**B**) direct SPME in flow sampling chambers; and (**C**) liquid sampling in glass cup. Reprinted with permission from [53].

Fire accelerants: Generally a molecular imprinting method is used for the sample preparation step before a chromatographic or mass spectroscopy analysis, in fire debris analysis for accelerant compounds like gasoline, kerosene and alcohol [53,54]. Kabir et al. summarized the recent advances in micro-sample preparation for forensic science while considering fire debris analysis and toxicology [53]. Alizadeh et al. developed an ethanol sensor by combining multi-walled carbon nanotubes, nano-sized MIPs, and poly(methyl methacrylate) as conducting element, recognition element, and adhesive substance, respectively. They achieved a reversible sensor response with a low relative standard error around 2.6%. They also reported that the sensor has a linear response in the concentration range of 0.65–45.0 ppm with a LOD value as 0.5 ppm. They concluded that the sensor response did not vary significantly within 4 months (confidence level = 95%) that indicated good durability and a long shelf-life [54].

Chemical warfare agents: In terms of national security and defense, the detection of chemical warfare agents is a very important target for forensic studies [55–57]. Boyd et al. developed a waveguide sensor for pinacolyl methylphosphonate (PMP, a hydrolysis product of the chemical warfare agent soman). They utilized a fluoropolymer with a refractive index of 1.29 that is slightly less than water (1.33), to develop lanthanide-based fluorescent detection of PMP. They concluded that the synergetic effect of MIP (selective and sensitive recognition) and fluoropolymer (inherent sensitivity and fast response time) allowed detecting the target molecules within seconds and had potential to use for warfare agent release at or below the time-weighted average/airborne exposure limit that is as low as parts-per-trillion range for the nerve agents [55]. Prathish et al. developed a potentiometric biosensor for the specific recognition of methylphosphonic acid (MPA), which is the degradation product of nerve agents such as sarin, soman, VX, etc. They plasticized MIPs with 2-nitrophenyloctyl ether on a polyvinyl chloride matrix. The sensor gave a linear response in the concentration range of 5×10^{-5}–1×10^{-1} M with a LOD value as 5×10^{-8} M. They also reported that the sensor had a rapid response whereby 75% of the response was realized in 2 min and reached equilibrium in 5 min. They concluded that the sensor was stable, reusable, portable, and ready-to-use for in situ detection for not only for MPA, but also for actual chemical warfare agents and their simulants [56]. Lu et al. mentioned the importance and emergency of a real-time and on-site detection of chemical warfare agents due to the terrorist threats in their review article. They also summarized that MIPs serves as a promising potential to complete the requirements with their features including strong mechanical strength, flexibility, long-time storability, designing in required geometry and structure, and, of course, low cost [57].

Environmental forensics: Environmental forensic is the area of interest in which finding the source and age of an environmental contaminant like an oil spill, and heavy metal pollutants is investigated [58–60]. In terms of the legislative framework for environmental forensics, Mudge comprehensively summarized environmental forensics and the importance of source identification. Starting from national, regional and US legislations, the author mentioned the source identification methods and environmental forensic problems including illegal discharge, fugitive emissions or discharge, deliberate fly-tipping, historical discharges, and altered environmental processes. The author also compiled tools for source apportionments on the basis of chemical and biological approaches [58]. Davis et al. focused on determining release periods of petrochemicals to groundwater (in Whitehorse, Yukon) for use in geochemical forensics. In this article, they studied several forensic methods using a Geographic Information System (GIS)/Access©-based data visualization tool to investigate the source and timing of hydrocarbon releases at a site in Whitehorse, Yukon Territory [59]. Alizadeh et al. also applied a molecular imprinting approach to electrochemically detect TNT in different water and soil samples. The sensors worked with a dynamic linear range of 5×10^{-9}–1×10^{-6} M and a low LOD value of 1.5×10^{-9} M [60].

3.2. Uses of MIPs for Pre-Concentration/Sample Preparation/Extraction

Molecular imprinted materials have a variety usage purposes like drug delivery, biomimetic enzyme catalysis, separation, extraction, and sensing [61–67]. In the area of forensic sciences, the most common usages are extraction, pre-concentration, and detection of target molecules from a variety of body samples and fluids like hair, saliva, urine, pericardial fluid, blood and questioned samples like chewing gums, cigarette butts [68].

Even though the advanced analytical technique is the main step for quantification of target molecules, a pre-concentration step is a necessity in some cases like trace analysis. Extraction has widely been used in different fields including for environmental, food, natural products, pharmaceuticals, toxic and forensic samples [69]. Kabir et al. comprehensively summarized the major classes of extraction methods used for forensic analysis [53] (Figure 7). The main purpose of extraction is generally pre-concentration of target molecules before the detection step without further separation steps. There are two main approaches followed for extraction step: off-line and on-line extraction [46–48]. In general, solid-phase extraction is applied as off-line pre-concentration just before the quantification of target molecules (Figure 8).

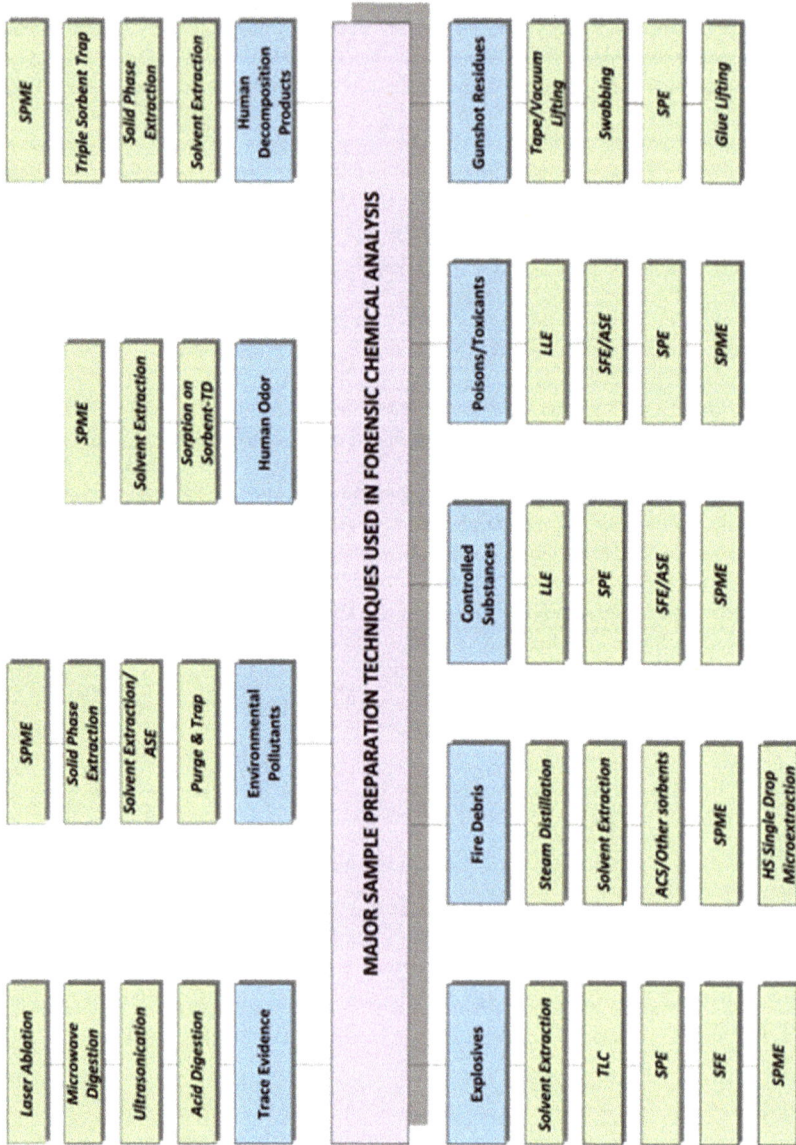

Figure 7. Major classes of forensic samples and their sample preparation techniques. Reprinted with permission from Ref. [53]. ACS: Activated carbon sorbents; ASE: accelerated solvent extraction; HS: Head space; LLE: Liquid-liquid extraction; SFE: supercritical fluid extraction; Sorbent-TD: Sorbent-Thermal desorption; SPE: Solid phase extraction; SPME: solid phase micro-extraction; TLC: Thin layer chromatography.

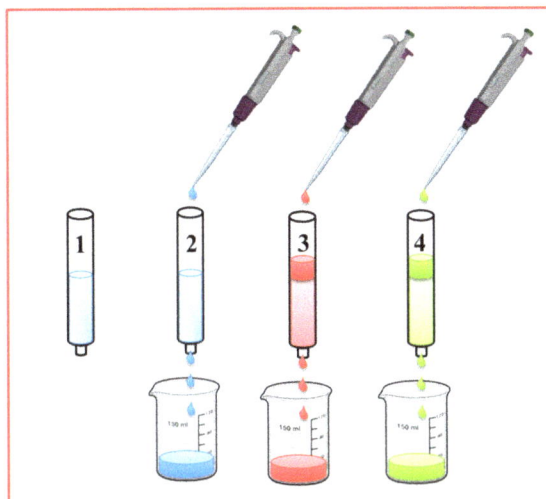

Figure 8. Schematic representation of off-line solid-phase extraction. 1. Equilibration of the column packed with MIP particle or monolith synthesized via in-situ polymerization; 2. Sample loading; 3. Washing out of interfering substances; 4. Elution of the analyte pre-concentrated.

In this approach, the extraction column is filled with MIP particles or synthesized monolithic structures via in-situ polymerization and equilibrated with solvent. Then, the sample is flowed through the column. After washing out the interfering substances, the target molecules are eluted by using a desorbing agent at higher concentration. Although this approach is simple and intensively used, it is really time-consuming and does not allow the automation of the detection and it is hard to handle samples in higher amounts with this approach. Therefore, on-line extraction/pre-concentration has attracted researchers' interests (Figure 9). In this approach, it is possible to attach a column to chromatographic system directly via some modifications of sampling step by using different and complicated valve-systems.

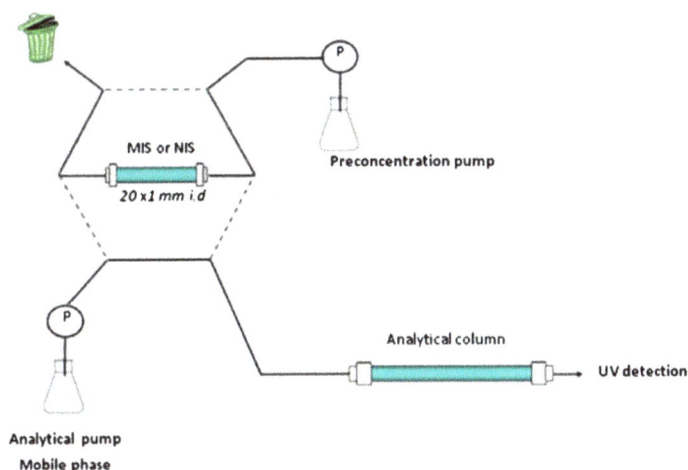

Figure 9. Scheme of the online coupling of the MIP-based extractor with HPLC. Reprinted with permission from [48]. Copyright © 2013 Springer International Publishing AG.

4. Instrumentation and Detection

Sensitivity enhancement is one of the advantages of molecular imprinting technology. Basically, MIPs are used as recognition elements on the transducers instead of biorecognition element such as antibodies, enzymes, DNAs, and aptamers etc. (Figure 10). The integration of MIPs into commercial systems offers several advantages due to the different possible MIP-based designs while keeping all instrumentation and detection methods constant. In this part of the review, we summarize some of the interesting methods utilizing MIPs for forensic analysis.

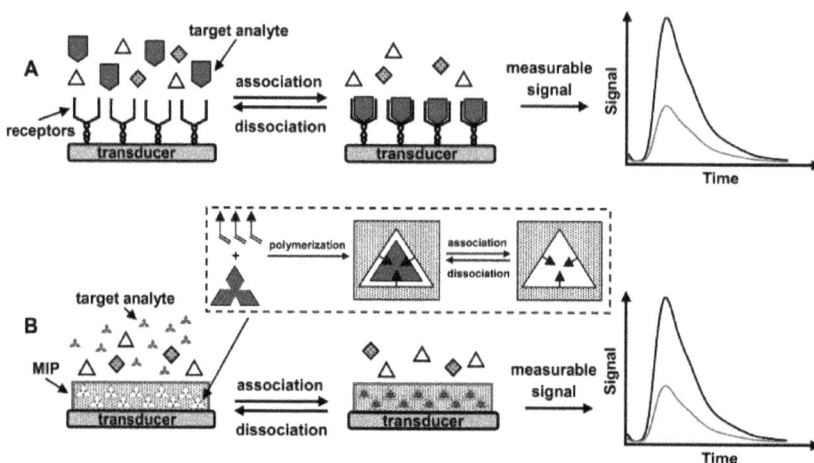

Figure 10. Schematic representations of (**A**) antibody-based chemosensor and (**B**) MIP-based biomimetic sensor. Inset in (**B**) shows the concept of molecular imprinting. Reprinted with permission from [70].

4.1. Spectroscopy

Surface-enhanced Raman scattering (SERS) is an extremely sensitive spectroscopic technique based on molecular vibrations but it has some drawbacks, as any molecule with a similar chemical structure may interfere with the SERS and cause doubtful results. Thus, molecular imprinting provides a good solution to increase the selectivity. Holthoff et al. presented a novel strategy to combine molecular imprinting technology with SERS to improve the detection of explosives via deposing imprinted xerogels which provide a high porosity along with large surface areas on SERS active surface [43]. By using the attractive structural features of xerogels in combination of MIPs, they were able to develop a quite sensitive SERS substrate for TNT detection (Figure 11).

Ion mobility spectrometry (IMS) is a widely used analytical method used for detection and identification of trace amounts of vapors based on the mobility of gas phase ions in a weak electric field. IMS has many uses in the area of forensic sciences, including personal markers, and detection of explosives, toxic substances and narcotics [71]. There have been many studies for implementing the molecular imprinting technology to IMS for the detection of explosives and drugs in the last decade [72–74].

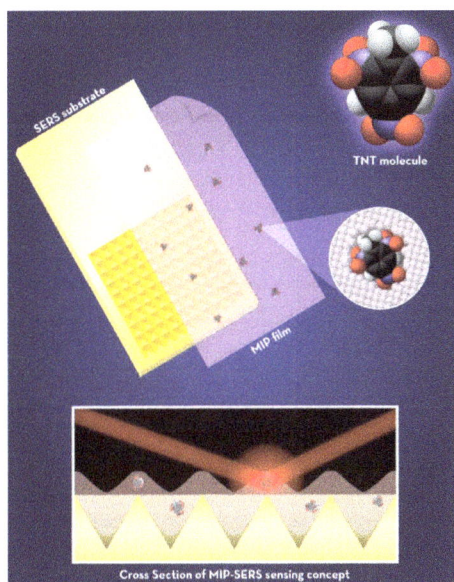

Figure 11. Xerogels based MIP-SERS detection concept. Reprinted with permission from Ref. [43].

4.2. Optical Detection

Surface plasmon resonance (SPR) is an optical sensor based on optical excitation of surface electrons of metal on a metal-dielectric interface. SPR is a widely used method due to its availability for mobility, ease of modification and production of sensor surfaces available for multi-analyte detection via spotting and sensitivity [75].

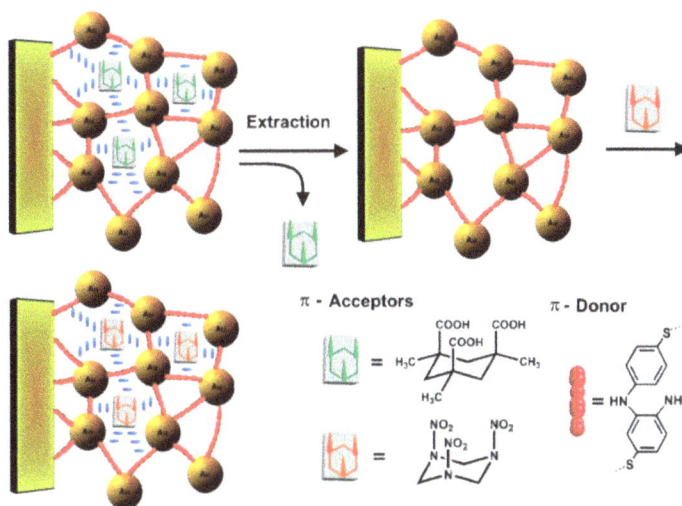

Figure 12. Imprinting of Kemp's acid molecular recognition sites into the composite of bisaniline-crosslinked Au NPs associated with a Au electrode, for RDX analysis. Reprinted with permission from [44]. Copyright © 2010 WILEY-VCH Verlag GmbH & Co. KGaA, Weinheim, Germany.

114

Molecular imprinting-based SPR sensors are used for a variety of biological structures like proteins, antibodies and antibody fragments, hormones, cells, viruses, aptamers [76,77] as well as drugs, explosives, and toxins. Riskin et al. reported molecularly imprinted gold nanoparticles for the detection of various explosives. The produced imprinted nanoparticles had high affinities and selectivity toward the imprinted explosives with detection levels of 200 fM for pentaerythritol tetranitrate and 20 pM for nitroglycerin [45] (Figure 12).

4.3. Colorimetric Sensing

Photonic crystals are periodic nanostructures that affect the motion of photons that can also be used as label-free detecting platforms due to their structural coloration ability. Hu et al. demonstrated a method for combining photonic crystals with molecular imprinting technology via hierarchical porous structured molecularly-imprinted hydrogel for sensor applications [78]. They achieved an enhanced response by implementing the molecular imprinting strategies into photonic crystals which found applications in forensic sciences. Meng et al. showed a semi-quantitative method for atropine, an important alkaloid in forensics, via a molecularly imprinting-based photonic hydrogel with a very low LOD value of 1 pg/mL [79]. There are some other studies showing sensitive colorimetric detection for organophosphorus-based nerve agents like sarin, soman, VX and R-VX, explosives, ketamine and morphine [16,79–82].

Chemiluminescence sensing is the one of the more sensitive colorimetric sensing strategies. Han et al. developed a molecular imprinting-based electroluminescence sensor which has the possibility of becoming an alternative to a trained dog for the detection of hidden drugs in luggage, mail, vehicles, and aircraft, and in the human body [83]. They achieved detection of methamphetamine hydrochloride (MA) and morphine via the combination of molecularly imprinted sol-gel polymers with a light emitting material and a multi-walled carbon nanotube composite. A detection limit of 4.0×10^{-15} M was achieved. This method is quite promising for detection of hidden drugs or explosives from their odor. There are also molecular imprinted chemiluminescence sensors for poisons in the literature. Liu et al. developed a molecular imprinted chemiluminescence sensor for the determination of brucine, which is a dangerous poison. The detection limit was reported as 2.0×10^{-9} g/mL [84].

Fluorescence sensing is another colorimetric sensing approach for ultrasensitive target detection in combination with MIPs. Fluorescence emission may be obtained by a variety of fluorophore dyes or different-sized quantum dots. Fluorescent-labeled MIP materials could be used for visualizing the recognition between the imprinted material and the template molecules. Use of fluorescent probes in combination with MIPs has mainly focused on the detection of poisons and vapors of explosives and fire debris analysis [42,85–87].

4.4. Mass Detection

Piezoelectric (PZ) transduction sensors have many sensing applications for a variety of target structures. Their advantages in gas phase detection are quite important for on-site analysis. Quartz crystal microbalances, surface acoustic wave-guides, microcantilevers, and micro-electromechanical systems are among intensively studied mass sensitive sensor systems.

Quartz crystal microbalance (QCM): Chianella et al. reported a molecular complete imprinting solution to detect microcystin-LR, a cyanobacteria-based toxin [88]. They prepared both MIP-based solid phase extraction cartridges for pre-concentration, and a MIP-based QCM sensor. They achieved a limit of detection of 0.35 nM of toxin.

Surface/bulk acoustic wave: Surface acoustic wave (SAW)- and/or bulk acoustic wave (BAW)-based sensors work according to the variation of velocity of the acoustic waves based on the increased mass on the transducer. Percival et al. studied a molecular imprinting-based surface acoustic wave sensor for the detection of an anabolic steroid, nandrolone. They synthesized MIP layers via a covalent imprinting approach and reported a frequency shift of up to 0.2 ppm with the sensor preferring the target, nandrolone, to analogous compounds. They also concluded that such acoustic wave devices could be integrated into lightweight and low cost oscillator circuits allowing inexpensive screening technique to be developed [89]. Pan et al. developed a novel molecular imprinting-based SAW sensor to the detect warfare agent VX. The detection limit of the produced sensor was reported as $0.15 \, \text{mg/m}^3$ and after 18 months, its detection signal decreased by about only 4.4% [90].

Even though caffeine is a widely used legal natural stimulant, it has also importance for forensic purposes. Tanada et al. reported caffeine analysis can be used for forensic hair discrimination [91]. Liang et al. prepared a biomimic BAW sensor by coating the surface with caffeine-imprinted polymers. The sensor was highly selective and gave a sensitive response in the linear concentration range of 5.0×10^{-9}–1.0×10^{-4} M at pH 8.0 with a LOD value of 5.0×10^{-9} M and high recoveries between 96.1% and 105.6% [92].

Tan et al. produced a molecular imprinting-based BAW sensor for the detection of paracetamol, a frequently encountered drug in suicide cases [20]. They reported the produced sensor was successfully used for the determination of paracetamol in human serum and urine with a limit of detection of 5.0×10^{-3} µM.

Microcantilevers: Use of atomic force microscopes (AFMs) for chemical and biological sensing is a trending area. A modified AFM microcantilever AFM was used as mechanical transducer via the increased mass on the modified cantilever with binding of complementary species. This method may be seen as an artificial nose due to the availability of odor detection at ultra-trace levels. Implementing microcantilever-based detection with molecular imprinting is a relatively new area of study [93].

Micro-electromechanical systems (MEMS): MEMS are made up of components between 1 and 100 micrometers in size for miniaturized sensors, actuators, and structures. The use of molecular imprinting technology is a relatively new area for this approach as well. Holthoff et al. developed a microbeam-based MEMS gas sensor with molecular imprinting for the detection of TNT and dimethyl methylphosphonate (DMMP), a simulant for the nerve gas sarin [43]. To form xerogel- based TNT and DMMP imprinted films on MEMS devices, complex mixtures containing TNT and DMMP were spin casted on MEMS devices.

4.5. Electrochemical Detection

Electrochemical sensors are a main sensor platform due to their simplicity, cost-efficiency and widely usability. Triacetone triperoxide (TATP) is one of the most common components of explosives and therefore Mamo et al. produced a highly sensitive electrochemical sensor for TATP via a differential pulse voltammetry-based molecular imprinted sensor. They produced a MIP-based glassy carbon electrode which demonstrated good performance at low concentrations for a linear and wide concentration range of 82–44,300 parts per billion (ppb) with a high correlation coefficient of $R^2 = 0.996$. LOD and LOQ values were reported as 26.9 and 81.6 ppb. They also showed very good repeatability with precision values ($n = 6$, expressed as relative standard deviation (RSD)%) of 1.098% and 0.55% for 1108 and 2216 ppb, respectively. They concluded that the sensor can selectively detect TATP in presence of other explosives including pentaerythritol tetranitrate, 1,3,5-trinitroperhydro-1,3,5-triazine, octahydro-1,3,5,7-tetranitro-1,3,5,7-tetrazocine, and 2,4,6-trinitrotoluene [49].

4.6. Chromatography

There are many studies on MIP-based chromatographic methods in the area of forensic sciences like determining gunshot residues from hand and clothes. These residues are obtained by application of adhesive tapes to suspected person hands and clothes which may interfere with the results due to the adhesives on the tapes. Pereira et al. produced molecularly imprinted polymers for diphenylamine removal from organic gunshot residues, which is one of the most common components of gunshot residue [52].

Capillary electrophoresis (CE) is a widely used method for trace analysis and it is a valuable tool in the area of forensic chemistry for analyzing inks, dyes, gunshot/explosive residues and drugs. Deng et al. produced a fiber-based SPME method by using molecular imprinting in capillary electrophoresis application for ephedrine and pseudoephedrine. Even though ephedrine and pseudoephedrine are therapeutic drugs, their amphetamine-like effects at high doses makes them important for forensic purposes. Limits of detection values were increased from 0.20 to 0.00096 µg/mL for ephedrine and 0.12 to 0.0011 µg/mL for pseudoephedrine via successful molecular imprinting-based SPME [94]. As mentioned before, Ogiso et al. utilized a CE system for selective DNA sequencing (Figure 13) [36]. They applied a mixture of target double-stranded (ds) DNA, non-target ds-DNA and standard DNA marker to capillary columns, imprinted and non-imprinted. The relative migration through the column was based on the size of the DNA as well as the selective interaction ability with polymer fillers (imprinted or non-imprinted). The place and length of the band indicated the sequence of the DNA fragments analyzed.

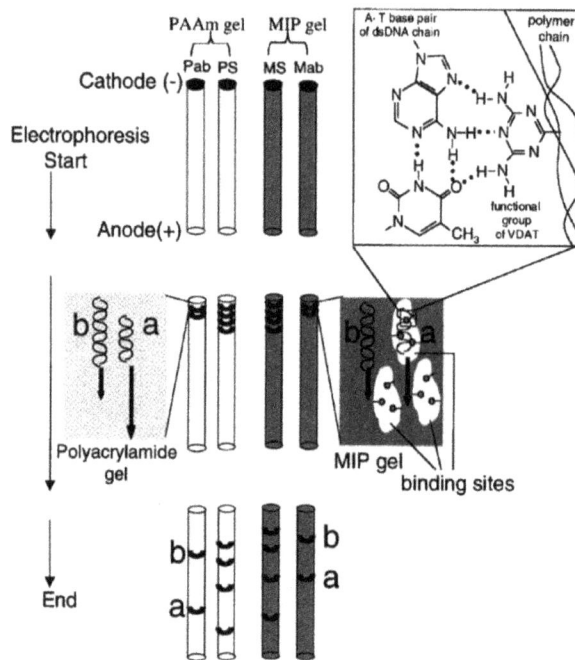

Figure 13. Principle of detection of the target dsDNA sequence by the dsDNA-imprinted polymer (MIP) gel in gel electrophoresis. A mixture of target dsDNA (**a**) and nontarget dsDNA (**b**) and the DNA standard size marker (S) is subjected to tube-gel electrophoresis with polyacrylamide gel (Pab and PS, respectively) and MIP gel (Mab and MS, respectively). During electrophoresis, migration of both nontarget (**b**) and target (**a**) dsDNA in polyacrylamide gel is dependent on the dsDNA fragment size. In the case of MIP gel, the migration of the target dsDNA (**a**) is dependent on both the fragment size and the capture effect of binding sites in the MIP gel. Consequently, the fragment size of the target dsDNA in the MIP gel should be larger than that in the polyacrylamide gel. While, the migration of the nontarget dsDNA (**b**) in MIP gel is dependent on the dsDNA fragment size. The fragment size of the nontarget dsDNA in the MIP gel should be the same as that in the polyacrylamide gel. Lengths of bold arrows indicate the velocity of dsDNA under the gradient of the electric field. The right top portion of the figure illustrates the binding site (imprinted cavity) of the target dsDNA in the MIP gel. Reprinted with permission from [36].

Heat transfer methods are one of the recent approaches to develop selective biosensors in combination of MIPs. Peters et al. presented a novel approach for molecular imprinting-based nicotine detection from saliva via differential heat transfer resistance [95]. With the use of this method temperature differences via adsorption-desorption processes were measured precisely with the use of thermistor devices and they applied impedance spectroscopy to validate the results (Figure 14). They reported quite good dose-response results in the low concentration range of 0.2–0.75 µM with a high correlation coefficient (R^2) of 0.97. LOD value reported as 125 nM was proper for biological samples in which nicotine levels vary in the range of 0–500 µM. Due to the importance of detecting small molecules for forensic purposes, this method offers a promising alternative.

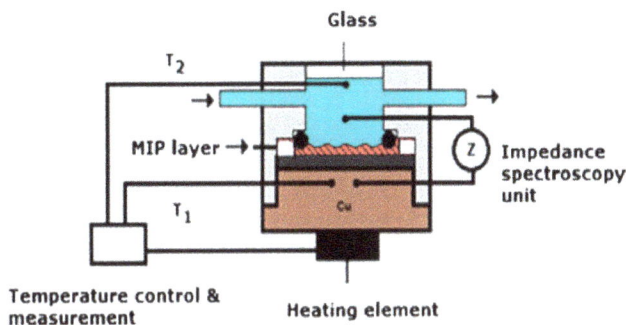

Figure 14. Schematic illustration of the general concept of the measuring setup. The temperature of the copper block, T1, is strictly controlled at 37.00 ± 0.02 °C. The heat flows from the copper block through the MIP layer to the liquid, where T2 is measured. Simultaneously with the temperature, the impedance is monitored. Reprinted with permission from [95].

5. Conclusions and Future Aspects

In this review, we aimed to draw a projection on the use of molecular imprinting in the related areas of forensic science. There are two main challenges with forensic science applications: (i) the complexity of the analytes and (ii) a quite low concentration of the analyte in this complex medium. Therefore, MIPs-based materials are intensively utilized as pre-concentrators before quantification through commercial techniques. Meanwhile, the integration of MIPs into traditional setups has recently attracted the efforts of the researchers (Table 1). Online pre-concentration, enhanced detection, improved selectivity and specificity, and excellent quantification limits as well as ease-of-production, robustness, cost-efficiency, and a variety of production strategies of MIP-based platform make them a promising alternative for achieving these aims and overcoming the mentioned drawbacks. Also, their other advantages such as chemical/physical stability, excellent compatibility with both of organic and aqueous media, reusability, long shelf-life and recognition capabilities make MIP-based design quite suitable for both sensing and sample preparation/pre-concentration of forensic targets including drugs, fire debris residues, explosives and gunshot residues, and chemical warfare agents. According to the reviewed literature reviewed, the attempts on MIP-based sample preparation/pre-concentration methods are much more closer to commercial products in respect to the sensory applications. Even though molecular imprinting-based materials have a variety of applications in many different areas, the evaluation of molecular imprinting-based sensor and sample preparation platforms should be expanded to the demands of the forensic sciences. Some of the recent studies have utilized computational approaches for determining the composition of imprinting materials, which would make possible to implement computationally well-designed systems to be used as effective, time saving and useful tools in the area of molecular imprinting. As a conclusion, MIP-based systems are rapidly and continuously growing platforms and their application in forensic science is at the beginning stages yet, demanding many novel designs.

Sensors **2017**, *17*, 691

Table 1. An overview of the literature mentioned in this review in respect to device, method, materials, analyte, media and attributions.

	Combined Devices	Methods	Imprinting Materials	Analyte	Media	Attributes	Ref.
Legal/illicit drugs	LC-MS-MS	SPE	MAA and EDMA based monolith	Diazepam	Hair	LOD: 0.09	[6,7]
	EI-MS	SPE	MAA and EDMA based monolith	Benzodiazepines	Plasma	LOQ: <10 μg/L	[8]
	Electrochemical sensor	Composite NPs	PPyr sol-gel gold nanoparticles	Lorazepam	Artificial soln.	LOD: 0.09 nM	[9]
	GC-MS	SPE	MAA and EDMA based monolith	Δ9-THC-OH	Urine	LOD: 2.5-1 ng/mL	[10]
	LC-MS	SPE	EDMA and DVB based membrane	Cannabinoids	Plasma and urine	LOQ: 0.36 ng/L	[11]
	LC-MS-MS	SPE	AAm and EDMA based pill	Cannabinoids	Urine and OF	LOD: 0.75 ng/mL	[12]
	GC-FID	SPDE	MIP coated hollow stainless steel needle	MAMP, AMP, MDMA	Urine, saliva, hair	LOQ: 12 ng/mL	[17]
	LC-MS	SPE	Ground and sieved MAA based monolith	LSD	Biological fluids	LOQ: of 0.2 pg/mL	[18]
	SERS	PISA	Boronate affinity-based AuNPs	Erythropoietin	Urine	LOD: 2.9×10^{-14} M	[30]
	IMS	SPE	MAA and EDMA based column material	Metronidazole	Human serum	LOD: 10 μg/L	[72]
	Optical response without instrument	Chemosensing	Molecularly imprinted photonic hydrogels	ATR, MOR	Biological samples	LOD: 1 pg/mL (ATR) 0.1 ng/mL (MOR)	[79,81]
	Quartz crystal-TSM sensor	Chemosensing	Imprinted polymer coating	Nicotine	Human serum and urine	LOD: 2.5×10^{-8} M	[20]
Poisons	Electrochemical sensor	LSV	PPDA/SWNTs composite film	Brucine	Human serum	LOD: 2.1×10^{-7} M	[32]
	Electrochemical sensor	EIS	MAA and EDMA based NPs	Arsenic	Biological fluids	LOD: 5.0×10^{-7}	[33]
	Magnets and UV light	Magnetic response	Oleic acid modified Fe$_3$O$_4$ and Photoswitchable monomer	Caffeine	Water and beverage samples	Fast rebinding kinetics and high selectivity	[35]
DNA	Electrophoresis	GE	MIP gel as an electrophoretic matrix	Double strand DNA	Mixed-DNA sample	Simple and cost-effective detection	[36]
	No instrument	Adsorption	Hydrophobic cryogels	Plasmid DNA	E. coli lysate	Q: 45.31 mg DNA/g	[39]
	SERS	Xerogels	Spin casted xerogel films	TNT	Artificial soln.	LOD: 3 μM	[43]
Explosives and gunshot residues	SPR	Optical response	Au-NPs composite	RDX, PETN, NG, EGDN	Artificial soln.	LOD: 12 fM-20 pM	[44,45]
	RP-LC	Online SPE	Sol-Gel organosilane	Nitroaromatic explosives	Post-blast samples	Extraction recoveries higher than 90%	[46,48]
	HPLC	SPE	Polymeric microparticles	Diphenylamine (DPA)	Gunshot residues	Retention up to 90%	[52]
	IMS	SPE	AAm and EDMA based particles	Nitroaromatics	Surface water	On-site detection	[74]
Fire accelerants	Chemiresistor sensor	Chemosensing	MWCNs-PMMA composites	Ethanol vapor	Air	LOD: 0.5 ppm	[54]
	QCM	Mass sensitive	PMMA and DVB based particles	Xylene and toluene	Air	Simple and reliable	[23]
Warfare agents	Electrochemical sensor	Potensiometry	MMA, VP and EDMA based particles	MPA	Natural water	5×10^{-8} M	[56]
	Fluorescence spectrometer	Fluorescence detection	Molecularly imprinted silica particles	Ricin	Artificial soln.	2-10 μM range	[85]

AAm: Acrylamide; AMP: Amphetamine; Atropine: ATR; AuNPs: Gold nanoparticles; DVB: Divinyl benzene; EDMA: Ethylene glycol dimethacrylate; EGDN: Ethylene glycol dinitrate; EI-MS: Electrospray ionization mass spectrometry; EIS: Electrochemical impedance spectroscopy; GC-FID: Gas chromatograpy-flame ionization detector; GC-MS: Gas chromatography-mass spectrometry; GE: Gel electrophoresis; IMS: Ion mobility spectrometry; LC-MS: Liquid chromatography-mass spectrometry; LC-MS-MS: Liquid chromatography-mass spectrometry; LSV: Linear sweep voltammetry; MAA: Methacrylic acid; MAMP: Methamphetamine; MDMA: Ecstasy; Morphine: MOR; MPA: Methylphosphonic acid; MWCNs: Multi-walled carbon nanotubes; NG: Nitroglycerin; NPs: nanoparticles; OF: Oral fluid; PETN: Pentaerythritol tetranitrate; PISA: Plasmonic immunosandwich assay; PMMA: Poly(methyl methacrylate); PPDA: Poly-o-phenylenediamine; PPyr: Polypyrrole; Q: Maximum adsorption capacity; QCM: Quartz crystal microbalance; RDX: Hexahydro-1,3,5-trinitro-1,3,5-triazine; RP: Reversed phase; SERS: Surface-enhanced Raman scattering; Soln.: Solutions; SPDE: Solid phase dynamic extraction; SPE: Solid-phase extraction; SPR: Surface plasmon resonance; SWNTs: Single walled carbon nanotubes; TNT: 2,4,6-trinitrotoluene; TSM: Thickness-shear-mode; VP: 4-Vinylpyridine; Δ9-THC-OH: Δ9-tetrahydrocannabinol.

119

Conflicts of Interest: The authors declare no conflict of interest.

References

1. Uzun, L.; Turner, A.P.F. Molecularly-imprinted polymer sensors: Realising their potential. *Biosens. Bioelectron.* **2016**, *76*, 131–144. [CrossRef] [PubMed]
2. Chen, L.; Wang, X.; Lu, W.; Wu, X.; Li, J. Molecular imprinting: Perspectives and applications. *Chem. Soc. Rev.* **2016**, *45*, 2137–2211. [CrossRef] [PubMed]
3. Bayram, E.; Yilmaz, E.; Uzun, L.; Say, R.; Denizli, A. Multiclonal plastic antibodies for selective aflatoxin extraction from food samples. *Food Chem.* **2017**, *221*, 829–837. [CrossRef] [PubMed]
4. Wulff, G.; Sarhan, A. Use of polymers with enzyme-analogous structures for the resolution of racemates. *Angew. Chem. Int. Ed.* **1972**, *11*, 341–346.
5. Bui, B.T.S.; Haupt, K. Molecularly imprinted polymers: Synthetic receptors in bioanalysis. *Anal. Bioanal. Chem.* **2010**, *398*, 2481–2492.
6. Ariffin, M.M.; Miller, E.I.; Cormack, P.A.; Anderson, R.A. Molecularly imprinted solid-phase extraction of diazepam and its metabolites from hair samples. *Anal. Chem.* **2007**, *79*, 256–262. [CrossRef] [PubMed]
7. Anderson, R.A.; Ariffin, M.M.; Cormack, P.A.G.; Miller, E.I. Comparison of molecularly imprinted solid-phase extraction (MISPE) with classical solid-phase extraction (SPE) for the detection of benzodiazepines in post-mortem hair samples. *Forensic Sci. Int.* **2008**, *174*, 40–46. [CrossRef] [PubMed]
8. Figueiredo, E.C.; Sparrapan, R.; Sanvido, G.B.; Santos, M.G.; Arruda, M.A.; Eberlin, M.N. Quantitation of drugs via molecularly imprinted polymer solid phase extraction and electrospray ionization mass spectrometry: Benzodiazepines in human plasma. *Analyst* **2011**, *136*, 3753–3757. [CrossRef] [PubMed]
9. Rezaei, B.; Boroujeni, M.H.; Ensafi, A.A. A novel electrochemical nanocomposite imprinted sensor for the determination of lorazepam based on modified polypyrrole@sol-gel@gold nanoparticles/pencil graphite electrode. *Electrochim. Acta* **2014**, *123*, 332–339. [CrossRef]
10. Nestic, M.; Babic, S.; Pavlovic, D.M.; Sutlovic, D. Molecularly imprinted solid phase extraction for simultaneous determination of Δ9-tetrahydrocannabinol and its main metabolites by gas chromatography–mass spectrometry in urine samples. *Forensic Sci. Int.* **2013**, *231*, 317–324.
11. Sanchez-Gonzalez, J.; Salgueiro-Fernandez, R.; Cabarcos, P.; Bermejo, A.M.; Bermejo-Barrera, P.; Moreda-Pineiro, A. Cannabinoids assessment in plasma and urine by high performance liquid chromatography–tandem mass spectrometry after molecularly imprinted polymer microsolid-phase extraction. *Anal. Bioanal. Chem.* **2017**, *409*, 1207–1220. [CrossRef]
12. Cela-Perez, M.C.; Bates, F.; Jimenez-Morigosa, C.; Lendoiro, E.; de Castro, A.; Cruz, A.; Lopez-Rivadullab, M.; Lopez-Vilarino, J.M.; Gonzalez-Rodriguez, M.V. Water-compatible imprinted pills for sensitive determination of cannabinoids in urine and oral fluid. *J. Chromatogr. A* **2016**, *1429*, 53–64. [CrossRef] [PubMed]
13. Andersson, L.I.; Muller, R.; Vlatakis, G.; Mosbach, K. Mimics of the binding sites of opioid receptors obtained by molecular imprinting of enkephalin and morphine. *Proc. Natl. Acad. Sci. USA* **1995**, *92*, 4788–4792. [CrossRef] [PubMed]
14. Piletska, E.V.; Romero-Guerra, M.; Chianella, I.; Karim, K.; Turner, A.P.F.; Piletsky, S.A. Towards the development of multisensor for drugs of abuse based on molecular imprinted polymers. *Anal. Chim. Acta* **2005**, *542*, 111–117. [CrossRef]
15. Devanathan, S.; Salamon, Z.; Nagar, A.; Narang, S.; Schleich, D.; Darman, P.; Hruby, V.; Tollin, G. Subpicomolar sensing of δ-opioid receptor ligands by molecular-imprinted polymers using plasmon-waveguide resonance spectroscopy. *Anal. Chem.* **2005**, *77*, 2569–2574. [CrossRef] [PubMed]
16. Hu, X.; Li, G.; Li, M.; Huang, J.; Li, Y.; Gao, Y.; Zhang, Y. Ultrasensitive specific stimulant assay based on molecularly imprinted photonic hydrogels. *Adv. Funct. Mater.* **2008**, *18*, 575–583. [CrossRef]
17. Djozan, D.; Farajzadeh, M.A.; Sorouraddin, S.M.; Baheri, T. Determination of methamphetamine, amphetamine and ecstasy by inside-needle adsorption trap based on molecularly imprinted polymer followed by GC-FID determination. *Microchim. Acta* **2012**, *179*, 209–217. [CrossRef]
18. Chapuis-Hugon, F.; Cruz-Vera, M.; Savane, R.; Ali, W.H.; Valcarcel, M.; Deveaux, M.; Pichon, V. Selective sample pretreatment by molecularly imprinted polymer for the determination of LSD in biological fluids. *J. Sep. Sci.* **2009**, *32*, 3301–3309. [CrossRef] [PubMed]

19. Zhou, T.; Shen, X.; Chaudhary, S.; Ye, L. Molecularly imprinted polymer beads prepared by Pickering emulsion polymerization for steroid recognition. *J. Appl. Polym. Sci.* **2014**, *131*, 39606. [CrossRef]

20. Tan, Y.; Yin, J.; Liang, C.; Peng, H.; Nie, L.; Yao, S. A study of a new TSM bio-mimetic sensor using a molecularly imprinted polymer coating and its application for the determination of nicotine in human serum and urine. *Bioelectrochemistry* **2001**, *53*, 141–148. [CrossRef]

21. Krupadam, R.J.; Venkatesh, A.; Piletsky, S.A. Molecularly imprinted polymer receptors for nicotine recognition in biological systems. *Mol. Impr.* **2013**, *1*, 27–34. [CrossRef]

22. Zhou, T.; Jorgensen, L.; Mattebjerg, M.A.; Chronakis, I.S.; Ye, L. Molecularly imprinted polymer beads for nicotine recognition prepared by RAFT precipitation polymerization: A step forward towards multi-functionalities. *RSC Adv.* **2014**, *4*, 30292–30299. [CrossRef]

23. Matsuguchi, M.; Uno, T. Molecular imprinting strategy for solvent molecules and its application for QCM-based VOC vapor sensing. *Sens. Actuator B Chem.* **2006**, *113*, 94–99. [CrossRef]

24. Yang, J.; Hu, Y.; Cai, J.B.; Zhu, X.L.; Su, Q.D.; Hu, Y.Q.; Liang, F.X. Selective hair analysis of nicotine by molecular imprinted solid-phase extraction: An application for evaluating tobacco smoke exposure. *Food Chem. Toxicol.* **2007**, *45*, 896–903. [CrossRef] [PubMed]

25. Wu, C.T.; Chen, P.Y.; Chen, J.G.; Suryanarayanan, V.; Ho, K.C. Detection of nicotine based on molecularly imprinted TiO2-modified electrodes. *Anal. Chim. Acta* **2009**, *633*, 119–126. [CrossRef] [PubMed]

26. Jan, N.; Marclay, F.; Schmutz, N.; Smith, M.; Lacoste, A.; Castella, V.; Mangin, P. Use of forensic investigations in anti-doping. *Forensic Sci. Int.* **2011**, *213*, 109–113. [CrossRef] [PubMed]

27. Ozgur, E.; Yilmaz, E.; Sener, G.; Uzun, L.; Say, R.; Denizli, A. A new molecular imprinting-based mass-sensitive sensor for real-time detection of 17β-estradiol from aqueous solution. *Environ. Prog. Sustain. Energy* **2013**, *32*, 1164–1169. [CrossRef]

28. Zulfiqar, A.; Morgan, G.; Turner, N.W. Detection of multiple steroidal compounds in synthetic urine using comprehensive gas chromatography-mass spectrometry (GCxGC-MS) combined with a molecularly imprinted polymer clean-up protocol. *Analyst* **2014**, *139*, 4955–4963. [CrossRef] [PubMed]

29. Kellens, E.; Bove, H.; Conradi, M.; D'Olieslaeger, L.; Wagner, P.; Landfester, K.; Ethirajan, A. Improved molecular imprinting based on colloidal particles made from miniemulsion: A case study on testosterone and its structural analogues. *Macromolecules* **2016**, *49*, 2559–2567. [CrossRef]

30. Tu, X.; Muhammad, P.; Liu, J.; Ma, Y.; Wang, S.; Yin, D.; Liu, Z. Molecularly-imprinted polymer-based plasmonic immunosandwich assay for fast and ultrasensitive determination of trace glycoproteins in complex samples. *Anal. Chem.* **2016**, *88*, 12363–12370. [CrossRef] [PubMed]

31. Jackson, R.; Petrikovics, I.; Lai, E.P.; Jorn, C.C. Molecularly imprinted polymer stir bar sorption extraction and electrospray ionization tandem mass spectrometry for determination of 2-aminothiazoline-4-carboxylic acid as a marker for cyanide exposure in forensic urine analysis. *Anal. Method* **2010**, *2*, 552–557. [CrossRef]

32. Liu, P.; Zhang, X.; Xu, W.; Guo, C.; Wang, S. Electrochemical sensor for the determination of brucine in human serum based on molecularly imprinted poly-o-phenylenediamine/SWNTs composite film. *Sens. Actuator B Chem.* **2012**, *163*, 84–89. [CrossRef]

33. Alizadeh, T.; Rashedi, M.; Hanifehpour, Y.; Joo, S.W. Improvement of durability and analytical characteristics of arsenic-imprinted polymer-based PVC membrane electrode via surface modification of nano-sized imprinted polymer particles: Part 2. *Electrochim. Acta* **2015**, *178*, 877–885. [CrossRef]

34. Nakamura, Y.; Matsunaga, H.; Haginaka, J. Preparation of molecularly imprinted polymers for strychnine by precipitation polymerization and multi-step swelling and polymerization and their application for the selective extraction of strychnine from nux-vomica extract. *J. Sep. Sci.* **2016**, *39*, 1542–1550. [CrossRef] [PubMed]

35. Xu, S.; Li, J.; Song, X.; Liu, J.; Lub, H.; Chen, L. Photonic and magnetic dual responsive molecularly imprinted polymers: Preparation, recognition characteristics and properties as a novel sorbent for caffeine in complicated samples. *Anal. Method* **2013**, *5*, 124–133. [CrossRef]

36. Ogiso, M.; Minoura, N.; Shinbo, T.; Shimizu, T. Detection of a specific DNA sequence by electrophoresis through a molecularly imprinted polymer. *Biomaterials* **2006**, *27*, 4177–4182. [CrossRef] [PubMed]

37. Emir Diltemiz, S.; Hur, D.; Ersoz, A.; Denizli, A.; Say, R. Designing of MIP based QCM sensor having thymine recognition sites based on biomimicking DNA approach. *Biosens. Bioelectron.* **2009**, *25*, 599–603. [CrossRef] [PubMed]

38. Ersoz, A.; Emir Diltemiz, S.; Atilir Ozcan, A.; Denizli, A.; Say, R. 8-OHdG sensing with MIP based solid phase extraction and QCM technique. *Sens. Actuator B Chem.* **2009**, *137*, 7–11. [CrossRef]

39. Uzek, R.; Uzun, L.; Senel, S.; Denizli, A. Nanospines incorporation into the structure of the hydrophobic cryogels via novel cryogelation method: An alternative sorbent for plasmid DNA purification. *Coll. Surf. B Biointerface* **2013**, *102*, 243–250. [CrossRef] [PubMed]

40. Gao, D.; Zhang, Z.; Wu, M.; Xie, C.; Guan, G.; Wang, D. A surface functional monomer-directing strategy for highly dense imprinting of TNT at surface of silica nanoparticles. *J. Am. Chem. Soc.* **2007**, *129*, 7859–7866. [CrossRef] [PubMed]

41. Guan, G.; Zhang, Z.; Wang, Z.; Liu, B.; Gao, D.; Xie, C. Single-hole hollow polymer microspheres toward specific high-capacity uptake of target species. *Adv. Mater.* **2007**, *19*, 2370–2374. [CrossRef]

42. Li, J.; Kendig, C.E.; Nesterov, E.E. Chemosensory performance of molecularly imprinted fluorescent conjugated polymer materials. *J. Am. Chem. Soc.* **2007**, *129*, 15911–15918. [CrossRef] [PubMed]

43. Holthoff, E.L.; Stratis-Cullum, D.N.; Hankus, M.E. A nanosensor for TNT detection based on molecularly imprinted polymers and surface enhanced Raman scattering. *Sensors* **2011**, *11*, 2700–2714. [CrossRef] [PubMed]

44. Riskin, M.; Tel-Vered, R.; Willner, I. Imprinted Au-nanoparticle composites for the ultrasensitive surface plasmon resonance detection of hexahydro-1,3,5-trinitro-1,3,5-triazine (RDX). *Adv. Mater.* **2010**, *22*, 1387–1391. [CrossRef] [PubMed]

45. Riskin, M.; Ben-Amram, Y.; Tel-Vered, R.; Chegel, V.; Almog, J.; Willner, I. Molecularly imprinted Au nanoparticles composites on au surfaces for the surface plasmon resonance detection of pentaerythritol tetranitrate, nitroglycerin, and ethylene glycol dinitrate. *Anal. Chem.* **2011**, *83*, 3082–3088. [CrossRef] [PubMed]

46. Lordel, S.; Chapuis-Hugon, F.; Eudes, V.; Pichon, V. Development of imprinted materials for the selective extraction of nitroaromatic explosives. *J. Chromatogr. A* **2010**, *1217*, 6674–6680. [CrossRef] [PubMed]

47. Lordel, S.; Chapuis-Hugon, F.; Eudes, V.; Pichon, V. Selective extraction of nitroaromatic explosives by using molecularly imprinted silica sorbents. *Anal. Bioanal. Chem.* **2011**, *399*, 449–458. [CrossRef] [PubMed]

48. Lordel-Madeleine, S.; Eudes, V.; Pichon, V. Identification of the nitroaromatic explosives in post-blast samples by online solid phase extraction using molecularly imprinted silica sorbent coupled with reversed-phase chromatography. *Anal. Bioanal. Chem.* **2013**, *405*, 5237–5247. [CrossRef] [PubMed]

49. Mamo, S.K.; Gonzalez-Rodriguez, J. Development of a molecularly imprinted polymer-based sensor for the electrochemical determination of triacetone triperoxide (TATP). *Sensors* **2014**, *14*, 23269–23282. [CrossRef] [PubMed]

50. Goudsmits, E.; Sharples, G.P.; Birkett, J.W. Recent trends in organic gunshot residue analysis. *Trends Anal. Chem.* **2015**, *74*, 46–57. [CrossRef]

51. Taudte, R.V.; Beavis, A.; Blanes, L.; Cole, N.; Doble, P.; Roux, C. Detection of gunshot residues using mass spectrometry. *Biomed. Res. Int.* **2014**, *2014*, 965403. [CrossRef] [PubMed]

52. Pereira, E.; Caceres, C.; Rivera, F.; Rivas, B.; Saez, P. Preparation of molecularly imprinted polymers for diphenylamine removal from organic gunshot residues. *J. Chilean Chem. Soc.* **2014**, *59*, 2731–2736. [CrossRef]

53. Kabir, A.; Holness, H.; Furton, K.G.; Almirall, J.R. Recent advances in micro-sample preparation with forensic applications. *Trends Anal. Chem.* **2013**, *45*, 264–279. [CrossRef]

54. Alizadeh, T.; Rezaloo, F. A new chemiresistor sensor based on a blend of carbon nanotube, nano-sized molecularly imprinted polymer and poly methyl methacrylate for the selective and sensitive determination of ethanol vapor. *Sens. Actuator B Chem.* **2013**, *176*, 28–37. [CrossRef]

55. Boyd, J.W.; Cobb, G.P.; Southard, G.E.; Murray, G.M. Development of molecularly imprinted polymer sensors for chemical warfare agents. *Johns Hopkins APL Tech. Dig.* **2004**, *25*, 44–49.

56. Prathish, K.P.; Prasad, K.; Rao, T.P.; Suryanarayana, M.V.S. Molecularly imprinted polymer-based potentiometric sensor for degradation product of chemical warfare agents: Part I. Methylphosphonic acid. *Talanta* **2007**, *71*, 1976–1980. [CrossRef] [PubMed]

57. Lu, W.; Xue, M.; Xu, Z.; Dong, X.; Xue, F.; Wang, F.; Wang, Q.; Meng, Z. Molecularly imprinted polymers for the sensing of explosives and chemical warfare agents. *Curr. Org. Chem.* **2015**, *19*, 62–71. [CrossRef]

58. Mudge, S.M. Environmental forensics and the importance of source identification, Issues in environmental science and technology, No. 26. *Environ. Forensics* **2008**. [CrossRef]

59. Davis, A.; Howe, B.; Nicholson, A.; McCaffery, S.; Hoenke, K.A. Use of geochemical forensics to determine release eras of petrochemicals to groundwater, Whitehorse, Yukon. *Environ. Forensics* **2005**, *6*, 253–271. [CrossRef]

60. Alizadeh, T.; Zare, M.; Ganjali, M.R.; Norouzi, P.; Tavana, B. A new molecularly imprinted polymer (MIP)-based electrochemical sensor for monitoring 2,4,6-trinitrotoluene (TNT) in natural waters and soil samples. *Biosens. Bioelectron.* **2010**, *25*, 1166–1172. [CrossRef] [PubMed]

61. Sellergren, B.; Allender, C.J. Molecularly imprinted polymers: A bridge to advanced drug delivery. *Adv. Drug Deliv. Rev.* **2005**, *57*, 1733–1741. [CrossRef] [PubMed]

62. Li, S.; Cao, S.; Piletsky, S.A.; Turner, A.P.F. *Molecularly Imprinted Catalysts, Principles: Syntheses, and Applications*, 1st ed.; Elsevier: Waltham, MA, USA, 2016.

63. Corman, M.E.; Armutcu, C.; Uzun, L.; Denizli, A. Cryogel based molecularly imprinted composite cartridges for rapid, efficient and selective preconcentration of polycyclic aromatic hydrocarbons (PAHs) from water samples through hydrophobic interactions. *Mater. Sci. Eng. C* **2017**, *70*, 41–53.

64. Erol, K.; Kose, K.; Uzun, L.; Say, R.; Denizli, A. Polyethyleneimine assisted-two-step polymerization to develop surface imprinted cryogels for lysozyme purification. *Coll. Surf. B Biointerface* **2016**, *146*, 567–576. [CrossRef] [PubMed]

65. Ozaydin Ince, G.; Armagan, E.; Erdogan, H.; Buyukserin, F.; Uzun, L.; Demirel, G. One-dimensional surface-imprinted polymeric nanotubes for specific biorecognition by initiated chemical vapor deposition (iCVD). *ACS Appl. Mater. Interface* **2013**, *5*, 6447–6452. [CrossRef] [PubMed]

66. Osman, B.; Uzun, L.; Besirli, N.; Denizli, A. Microcontact imprinted surface plasmon resonance sensor for myoglobin detection. *Mater. Sci. Eng. C* **2013**, *33*, 3609–3614. [CrossRef] [PubMed]

67. Sener, G.; Ozgur, E.; Rad, A.Y.; Uzun, L.; Say, R.; Denizli, A. Rapid real-time detection of procalcitonin using a microcontact imprinted surface plasmon resonance biosensor. *Analyst* **2013**, *138*, 6422–6428. [CrossRef] [PubMed]

68. Zander, A.; Findlay, P.; Renner, T.; Sellergren, B.; Swietlow, A. Analysis of nicotine and its oxidation products in nicotine chewing gum by a molecularly imprinted solid-phase extraction. *Anal. Chem.* **1998**, *70*, 3304–3314. [CrossRef] [PubMed]

69. Sharma, P.S.; D'Souza, F.; Kutner, W. Molecular imprinting for selective chemical sensing of hazardous compounds and drugs of abuse. *Trends Anal. Chem.* **2012**, *34*, 59–77. [CrossRef]

70. Guan, G.; Liu, B.; Wang, Z.; Zhang, Z. Imprinting of molecular recognition sites on nanostructures and its applications in chemosensors. *Sensors* **2008**, *8*, 8291–8320. [CrossRef] [PubMed]

71. Karpas, Z. Ion Mobility Spectrometry in Forensic Science. *Encycl. Anal. Chem.* **2006**. [CrossRef]

72. Jafari, M.T.; Rezaei, B.; Zaker, B. Ion mobility spectrometry as a detector for molecular imprinted polymer separation and metronidazole determination in pharmaceutical and human serum samples. *Anal. Chem.* **2009**, *81*, 3585–3591. [CrossRef] [PubMed]

73. Rezaei, B.; Jafari, M.T.; Khademi, R. Selective separation and determination of primidone in pharmaceutical and human serum samples using molecular imprinted polymer-electrospray ionization ion mobility spectrometry (MIP-ESI-IMS). *Talanta* **2009**, *79*, 669–675. [CrossRef] [PubMed]

74. Lu, W.; Li, H.Y.; Meng, Z.H.; Liang, X.X.; Xue, M.; Wang, Q.H.; Dong, X. Detection of nitrobenzene compounds in surface water by ion mobility spectrometry coupled with molecularly imprinted polymers. *J. Hazard. Mater.* **2014**, *280*, 588–594. [CrossRef] [PubMed]

75. Homola, J. Surface plasmon resonance sensors for detection of chemical and biological species. *Chem. Rev.* **2008**, *108*, 462–493. [CrossRef] [PubMed]

76. Uzun, L.; Say, R.; Unal, S.; Denizli, A. Production of surface plasmon resonance based assay kit for hepatitis diagnosis. *Biosens. Bioelectron.* **2009**, *24*, 2878–2884. [CrossRef] [PubMed]

77. Yilmaz, E.; Majidi, D.; Ozgur, E.; Denizli, A. Whole cell imprinting based *Escherichia coli* sensors: A study for SPR and QCM. *Sens. Actuator B Chem.* **2015**, *209*, 714–721. [CrossRef]

78. Hu, X.; Li, G.; Huang, J.; Zhang, D.; Qiu, Y. Construction of self-reporting specific chemical sensors with high sensitivity. *Adv. Mater.* **2007**, *19*, 4327–4332. [CrossRef]

79. Meng, L.; Meng, P.; Tang, B.; Zhang, Q.; Wang, Y. Molecularly imprinted photonic hydrogels for fast screening of atropine in biological samples with high sensitivity. *Forensic Sci. Int.* **2013**, *231*, 6–12. [CrossRef] [PubMed]

80. Taranekar, P.; Huang, C.Y.; Advincula, R.C. Pinacolyl methyl phosphonate (PMP) detection by molecularly imprinted polymers (MIP): A labile covalent bonding approach. *Polymer* **2006**, *47*, 6485–6490. [CrossRef]

81. Meng, L.; Meng, P.; Zhang, Q.; Wang, Y. Water-compatible molecularly imprinted photonic hydrogels for fast screening of morphine in urine. *Chin. J. Anal. Chem.* **2015**, *43*, 490–496.

82. Lu, W.; Xue, F.; Huang, S.Y.; Meng, Z.H.; Xue, M. Molecularly imprinted colloidal array for detection of explosives. *Chin. J. Anal. Chem.* **2012**, *40*, 1561–1566. [CrossRef]

83. Han, C.; Shang, Z.; Zhang, H.; Song, Q. Detection of hidden drugs with a molecularly imprinted electrochemiluminescence sensor. *Anal. Method* **2013**, *5*, 6064–6070. [CrossRef]

84. Liu, M.; Lu, J.; He, Y.; Du, J. Molecular imprinting–chemiluminescence sensor for the determination of brucine. *Anal. Chim. Acta* **2005**, *541*, 97–102. [CrossRef]

85. Lulka, M.F.; Iqbal, S.S.; Chambers, J.P.; Valdes, E.R.; Thompson, R.G.; Goode, M.T.; Valdes, J.J. Molecular imprinting of Ricin and its A and B chains to organic silanes: Fluorescence detection. *Mater. Sci. Eng. C* **2000**, *11*, 101–105. [CrossRef]

86. Stringer, R.C.; Gangopadhyay, S.; Grant, S.A. Detection of nitroaromatic explosives using a fluorescent-labeled imprinted polymer. *Anal. Chem.* **2010**, *82*, 4015–4019. [CrossRef] [PubMed]

87. Stringer, R.C.; Gangopadhyay, S.; Grant, S.A. Comparison of molecular imprinted particles prepared using precipitation polymerization in water and chloroform for fluorescent detection of nitroaromatics. *Anal. Chim. Acta* **2011**, *703*, 239–244. [CrossRef] [PubMed]

88. Chianella, I.; Piletsky, S.A.; Tothill, I.E.; Chen, B.; Turner, A.P.F. MIP-based solid phase extraction cartridges combined with MIP-based sensors for the detection of microcystin-LR. *Biosens. Bioelectron.* **2003**, *18*, 119–127. [CrossRef]

89. Percival, C.J.; Stanley, S.; Braithwaite, A.; Newton, M.I.; McHale, G. Molecular imprinted polymer coated QCM for the detection of nandrolone. *Analyst* **2002**, *127*, 1024–1026. [CrossRef] [PubMed]

90. Pan, Y.; Yang, L.; Mu, N.; Shao, S.; Wang, W.; Xie, X.; He, S. A SAW-based chemical sensor for detecting sulfur-containing organophosphorus compounds using a two-step self-assembly and molecular imprinting technology. *Sensors* **2014**, *14*, 8810–8820. [CrossRef] [PubMed]

91. Tanada, N.; Kashimura, S.; Kageura, M.; Hara, K. Utility of caffeine analysis for forensic hair discrimination. *Jpn. J. Legal Med.* **1998**, *52*, 233–237. [CrossRef]

92. Liang, C.; Peng, H.; Bao, X.; Nie, L.; Yao, S. Study of a molecular imprinting polymer coated BAW bio-mimic sensor and its application to the determination of caffeine in human serum and urine. *Analyst* **1999**, *124*, 1781–1785. [CrossRef] [PubMed]

93. Uzun, L.; Tiwari, A. *Advanced Imprinting Materials*; Wiley-Scrivener: New York, NY, USA, 2016.

94. Deng, D.L.; Zhang, J.Y.; Chen, C.; Hou, X.L.; Su, Y.Y.; Wu, L. Monolithic molecular imprinted polymer fiber for recognition and solid phase microextraction of ephedrine and pseudoephedrine in biological samples prior to capillary electrophoresis analysis. *J. Chromatogr. A* **2012**, *1219*, 195–200. [CrossRef] [PubMed]

95. Peeters, M.; Csipai, P.; Geerets, B.; Weustenraed, A.; van Grinsven, B.; Thoelen, R.; Gruber, J.; De Ceuninck, W.; Cleij, T.J.; Troost, F.J.; et al. Heat-transfer-based detection of L-nicotine, histamine, and serotonin using molecularly imprinted polymers as biomimetic receptors. *Anal. Bioanal. Chem.* **2013**, *405*, 6453–6460. [CrossRef] [PubMed]

sensors

MDPI

Review

Imprinting of Microorganisms for Biosensor Applications

Neslihan Idil [1,*] and Bo Mattiasson [2,3]

[1]   Department of Biology, Faculty of Sciences, Hacettepe University, 06800 Ankara, Turkey
[2]   Department of Biotechnology, Lund University, 22362 Lund, Sweden; bo.mattiasson@biotek.lu.se
[3]   CapSenze Biosystems AB, 22363 Lund, Sweden
*   Correspondence: nsurucu@hacettepe.edu.tr; Tel.: +90-312-297-80-24

Academic Editor: Nicole Jaffrezic-Renault
Received: 31 January 2017; Accepted: 21 March 2017; Published: 29 March 2017

Abstract: There is a growing need for selective recognition of microorganisms in complex samples due to the rapidly emerging importance of detecting them in various matrices. Most of the conventional methods used to identify microorganisms are time-consuming, laborious and expensive. In recent years, many efforts have been put forth to develop alternative methods for the detection of microorganisms. These methods include use of various components such as silica nanoparticles, microfluidics, liquid crystals, carbon nanotubes which could be integrated with sensor technology in order to detect microorganisms. In many of these publications antibodies were used as recognition elements by means of specific interactions between the target cell and the binding site of the antibody for the purpose of cell recognition and detection. Even though natural antibodies have high selectivity and sensitivity, they have limited stability and tend to denature in conditions outside the physiological range. Among different approaches, biomimetic materials having superior properties have been used in creating artificial systems. Molecular imprinting is a well suited technique serving the purpose to develop highly selective sensing devices. Molecularly imprinted polymers defined as artificial recognition elements are of growing interest for applications in several sectors of life science involving the investigations on detecting molecules of specific interest. These polymers have attractive properties such as high bio-recognition capability, mechanical and chemical stability, easy preparation and low cost which make them superior over natural recognition reagents. This review summarizes the recent advances in the detection and quantification of microorganisms by emphasizing the molecular imprinting technology and its applications in the development of sensor strategies.

Keywords: microorganism imprinting; biosensor; applications

---

1. Quantifying Microbial Cells Based on Surface Properties

Antibodies have been used for the identification and quantification of microorganisms. This has been based on affinity binding to cell surface structures. As indicating mechanisms, one has been able to use microbial metabolic activity of the captured cells, or one has used labelled reagents. A problem has been that different microbial strains of a certain species may have different surface structures which leads to that not all cells will react equally to the antibodies used.

When using molecularly imprinted polymers (MIPs) for capturing microorganisms, the same options as discussed for antibodies are available, but besides those, there are other options and challenges as well. By utilizing imprinted polymers with cavities of imprinted cells of the same species/strains as shall be assayed, then size and shape are other features that are utilized in the identification of the target microorganism.

When using MIPs for quantifying microorganisms, some features that are new as compared to when protein molecules are quantified. The cell surface is huge in relation to that of a protein and many different structures might be present on the cell surface. Furthermore, a large surface fitting in a complementary imprint will be exposed to a large number of weak interactions, and maybe some stronger. However, a large number of weak interaction simultaneously will result in very strong interaction. If free cell surface antigens are available in the sample to be analysed one can expect binding of these free compounds, but if/when cells are bound protein molecules that have been captured will be displaced.

In the process of designing the MIP, a range of different monomers can be used. The strategy is normally to equilibrate the imprinting structure with monomers prior to polymerization to start. This is done in order to position monomers with specificity for certain structures so that their positions are frozen after polymerization, thereby improving the probability that the steric arrangement will be suitable for affinity capturing to take place.

The multitude of weak interactions makes it possible that other particulate matters which can establish a large number of interactions to bind even if those are not identical with the print cell. This fact highlights the need for more selectivity in the interactions. The fact that multipoint of attachment makes the binding strong also leads to problems when releasing bound material before the imprinted matter can be reused in a new assay. Many of these features discussed above will be illustrated in this review paper.

2. Use of Molecularly Imprinted Polymers in Affinity Recognition of Microbial Cells

The detection and identification of pathogenic bacterial strains present in soil, marine and estuarine waters, the intestinal tract of animals, or water contaminated with fecal matter is of great importance in many fields such as medicine, water and food safety and security [1,2]. Among various known bacterial strains, *Staphylococcus aureus* is the cause of serious infections and remains a critical threat to public health [3]. Especially, methicillin-resistant *Staphylococcus aureus* (MRSA), resistant to methicillin and other β-lactam antibiotics, have been the cause of difficult-to-treat infections in humans [4]. Another important organism in this context is *Escherichia coli* (O157:H7) which is a predominant enteropathogen and has a very low human infectious dose, since as few as ten cells are required to cause an infection [5]. It is well documented that *Listeria monocytogenes* has long been recognized as a cause of food poisoning which can lead to very serious diseases or even death. On the other hand, the identification and detection of pathogenic bacterial strains have become increasingly important in clinical diagnosis and treatment as well as prevention the spreading of responsible strains [2]. An emerging field is those bacteria with potential use in bioterrorism such as *Bacillus anthracis.*

Infectious diseases, especially those caused by life-threatening pathogenic bacterial strains, have become a great concern due to these organisms cause death and increased morbidity in hospitalized patients. Even though several attempts have been made to develop vaccines and antimicrobial agents, these infections still remain as a major challenge because of the existence of new and multi drug resistant microorganisms [2]. On the other hand, current techniques used for detecting pathogenic bacterial strains are based on plating/culturing, biochemical tests, serological and immunological assays and genotypic analysis [2,5,6]. These techniques have some disadvantages such as being time-consuming, laborious and expensive, lacking sensitivity and specificity in comparison with selective, sensitive, multi-analyte testing capacity, speed and cost-effectiveness features of new analytical methods [2,6]. Therefore, the point is that rapid, portable and reliable analytical detection is of primary importance to be applied for early and accurate diagnosis.

Biosensor technology has been introduced to be promising tools for microbial detection in complex media (blood, serum, urine, food, water, etc.) with minimum sample pre-treatment. This technology is an approach including the ability to provide specific quantifiable data related with a broad spectrum of analytes [1]. It also is suitable for high-throughput monitoring, label-free detection, short response time [7], real-time analysis, simple sample preparation, low detection limits [6]. Biosensors are comprised of a biological recognition element in close conjunction with a transducer that transforms biological response resulting from the interaction of the analyte with the bioreceptor into measurable signal [2]. Biosensor-based techniques improves their detection performance by means of considered properties, thereby holding great promise for a variety of applications in the fields of clinical diagnostics, food and water analysis, product quality and process control in the agricultural and food industry, agricultural and industrial processing, bioprocess and environmental monitoring [1,6,7].

Many efforts to create innovative methods to be used for microbial detection includes use of various materials such as silica nanoparticles [8], microfluidics [9], liquid crystals [10], carbon nanotubes [11] which could be integrated to sensor technology. Receptors, aptamers, enzymes, nucleic acids, antibodies and lectins are placed among biological recognition elements in this technology [12,13]. In many of the studies dealing with microbial detection, antibodies were used in order to recognize and separate target cells. Antibodies, serving as specific affinity ligands, contribute to generate highly selective and sensitive bio-sensing systems. However, natural receptors tend to denature in conditions outside the physiological range such as elevated/lowered temperature, strongly acidic/basic conditions, presence of organic-rich solvents etc. [3,12]. One additional limitation of biological recognition structures is the fact that they are easily degradable by e.g., proteases. This limits their use in non-sterile media. They have also additional disadvantages by the fact that usage of animals is essential for the production and there is shelf-life restrictions and special storage requirements. Moreover, handling of natural receptors is not as cost-effective as handling synthetic receptors due to these limitations [3]. The use of artificial receptors as well as natural ones coupled to biosensor technology allows the possibility of creating highly selective and sensitive sensing systems with low cost. Therefore, this alternative approach has recently attracted the interest of researchers and it has had a profound impact on the development of biosensors for various applications. With this approach, it is possible to generate well-established artificial affinity ligands and develop effective biosensing methodologies for the detection of microorganisms. In literature, it is highlighted that molecular imprinting technology is presently a wide-spread effort to fill the major gap in different fields and offers artificial recognition systems exhibiting superiority over natural ones. This technology also provides tools to form specific recognition cavities in synthetic polymer matrix which are complementary in size, shape and/or chemical functionality to the template [3]. MIPs are comprised of a tailor-made polymer matrix obtained by polymerization of functional monomers with cross-linkers in the presence of a template molecule. In addition to these, this pre-polymerization solution contains initiator in an appropriate solvent. In order to initiate polymerization, UV irradiation or temperature treatment is necessary. After the entire polymerization process is completed, the stable matrix is washed for the removal of template and unreacted monomers, by this way creating selective recognition cavities which are then able to rebind the target molecule of interest selectively and reversibly. These cavities are designed by using most often the target molecule itself or very close structural analogues of it and therefore imprinted polymer having a memory for the target molecule is produced [3]. The correct placement of the functional groups and the size-shape characteristics of binding site have considerable effect on rebinding of the target molecule. After the removal of the template, the shape stability of the polymer network could be attributed to high degree of cross-linking and the functional groups arranged in proper positions which makes it possible to rebind the template.

The corresponding sites located in imprinted polymer have been generated by: (i) covalent; (ii) non-covalent and (iii) semi-covalent interactions [14]. Binding of the template into the polymer or also removal of this molecule from the polymer matrix and rebinding of this molecule depend on same chemical interactions [15]. There are some drawbacks in covalent imprinting such as requirement of cleaving the covalent bond and a limited number interactions play a key role when the interactions between the template and polymeric structure were taken into consideration. In this imprinting method, strong covalent binding is formed which results in obtaining binding regions. However, it is not useful in most cases because of the obstacles in rebinding of the target molecules. Non-covalent imprinting is suggested to overcome these shortcomings. This approach involves interactions such as, hydrogen bonding, van der Waals forces, dipole-dipole and hydrophobic interactions. This method also contributes to produce template-free recognition sites, and thus it is the most frequently preferred due to its appropriateness on extracting the template from the rigid polymer and making it possible to rebind the target molecule with the same approach. Therefore, it has become feasible for a great number of target molecules and found wide-spread use in various applications. The concept of semi-covalent imprinting relies on the combination of covalent and non-covalent imprinting [16]. Covalent binding is formed when the template molecule is trapped during the synthesis of the polymer. On the other hand, following the removal of the template, the target molecule can bind to its complementary cavity in a reversible manner via non-covalent binding. As mentioned above, non-covalent binding approach seems rather simple and more effective in the preparation of MIP when compared to the other methods.

Molecular imprinting technology has potential applications in many fields such as catalysis, separation, purification, detection and drug delivery for a wide range of target molecules from small to macromolecules and particulate matter including microorganisms [17–19]. A number of reports focused on the synthesis of polymers via imprinting applied for aminoacids [20], peptides [21], carbohydrates [22], nucleotides [23], hormones [24], drugs [25], toxins [26], pesticides [27].

3. Imprinting of Microorganisms

In the recent literature, there are many publications describing successful imprinting of small molecules. However, from the standpoint of much larger and complex structures or molecules [28], especially imprinting of whole cells is an ongoing challenge due to some characteristics and natural structures of microorganisms [3,19]. The main complexity in imprinting of microorganisms in comparison to less complicated organic substances is the huge size resulting in some difficulties while removing the bound cells from template-shaped cavities. Furthermore, imprinted cavities have a significant effect on the recognition of microorganisms since transport of the cells into the matrix is performed slowly and resultant diffusion problem extends the response time of the system. It is important to optimize the MIP polymer composition by varying the monomers and their ratios used in polymerization and/or the method applied for designing the polymer matrix in order to overcome these limitations. For this purpose, MIPs were synthesized by increasing the degree of cross-linking and generating well-patterned surfaces which facilitate the transport of the microorganisms to and from the corresponding sites. The surfaces of microbial cells have various structures which exhibit several differences in terms of morphology and functionality. These characteristics of the surfaces present in every microorganism show extensive diversity in each species. Accordingly, a crucial prerequisite is the choice of proper functionality in the sites. Following this, the properties of functional monomers have to be taken into consideration because these monomers interact with the groups of cell surfaces having complementary functionalities. It is also necessary to manage to create functionally adapted cavities by maintaining size and shape characteristics of microorganisms in the polymer matrix. Another requirement is to arrange the right functionality in established sites [29]. Taking the imprinting of microorganisms into account, reversible and selective re-inclusion is more significant. When the imprinted cavities are occupied by microorganisms, removal of bound material becomes troublesome because of steric hindrance and therefore the property of reversible binding will be lost. It is comparatively hard to remove the template from imprinted cavities by maintaining their conformation on the polymer surface in the absence of template. A further fact that has to be kept in mind is that microbial cells have, in relation to individual protein molecules, a large number of interactions with the MIP. This leads to that also weak interactions become important due to the effects of multipoint of attachment. Therefore, new criteria to create high selectivity need to be developed.

A wide variety of functional monomers such as those forming pyrrole conducting polymers [30–32], polyurethane layers [33,34], self-assembled monolayers [35,36] and sol-gels [37,38] have been used in the design of imprints against microorganisms [12]. The imprinting of whole microorganism were successfully achieved by creating selective recognition cavities [12,39] both on polymeric beads [40–42] and in polymeric films via stamping method [34,43].

Vulfson and co-workers made an attempt dealing with bacteria-mediated lithography and published the first successful live cell imprint report. In their study, *Listeria monocytogenes* and *Staphylococcus aureus* were selected as templates. The bacterial cells attaching on the surfaces were formed due to the tendency of microorganisms to accumulate at the interface between the aqueous and organic phases. After removing the microorganism, surface imprints were patterned in size and morphology. It has to be underlined that the proposed technique addresses merely the features of size and morphology without surface chemistry. In addition, another missing side of this pioneering research is to not include rebinding studies in order to indicate imprinting selectivity [40].

In bulk imprinting, the template structure (microorganism) is added to the pre-polymerization mixture along with monomer, crosslinker, initiator and solvent. Following polymerization and template removal, it leaves behind homogenously distributed cavities in the whole polymer matrix. This approach is more appropriate for imprinting of relatively small molecules [44]. The dimensions of microorganisms limit the transport of microorganisms into the polymer matrix. Cohen et al. made an attempt in order to detect microorganisms in water. In that study, imprinting of whole cells of microorganisms with different morphologies and outer surface chemistry was performed using thin sol-gel films of organically modified silica (ORMOSILS). These imprinted films having complementary cavities did recognize the target bacterial strains, and thus high adsorption affinity was obtained along with high selectivity [38].

Another strategy is surface imprinting in which thin polymeric films were formed along with the production of template-imprinted cavities on the surface of the polymer. The template stamp is prepared, and then pressed onto the polymerizing surface. The stamping is performed on the transducer and therefore provides the formation of more robust devices easily employable for measurements. This approach is particularly successfully applied for imprinting of large-size biomolecules as well as microorganisms (i.e., yeast, bacteria). Among known imprinting techniques, surface imprinting provides a feasible route to manage successful results in the detection of microorganisms because of enabling the transport of cells to and from the cavities [3,12].

Surface imprinted polymers could be prepared by a self-assembly process. In a previous study, *E. coli* was imprinted on polymeric microspheres. The aqueous suspensions of bacterial cells were added to water containing *N*-acrylchitosan. This solution was mixed with an oil phase for the production of microspheres. Following cross-linking the chitosan matrix, the bacterial strain was extracted and analysis was performed by fluorescence microscopy and verified that the microbeads have the ability to differentiate rod-shaped *E. coli* and spherical *Micrococcus luteus* [35].

A number of publications focused on revealing alterations to the surface imprinting concept by making extra contributions to the self-assembly process. Different from previously mentioned imprinting materials, overoxidized polypyrrole (PPy) films are used in order to detect anionic species in electrochemical sensing platforms. The stable electrical conductivity property of conducting polymers makes doping of the anionic bacterial strains easy during elecropolymerization process. Tokonami et al. (2014) developed highly selective devices for the detection of both Gram-negative (*E. coli* and *Pseudomonas aeruginosa*) and Gram-positive bacteria (*Bacillus subtilis* and *S. aureus*) exploiting their surface chemistry [31].

Another strategy is molding which includes the immobilization of the template molecules on a solid substrate. For example, Qi et al. described a similar strategy in order to prepare bacteria-imprinted polymers by this technique. In this study, indium tin oxide electrodes were covered by a nanocomposite film of chitosan and reduced graphene oxide sheets by electrodeposition. Sulfate-reducing bacteria present in marine environments were selected as template microorganism and attached on this film by physical adsorption [19].

The first study to implement surface-imprinted polymers in microbial fuel cells was carried out for recognition of green algae (*Chlamydomonas reinhardtii*). The algae cells were immobilized on platinum foils coated by an algae-imprinted polymeric thin films of ethylene-covinyl alcohol (EVAL) by molding method. Sputter-coated platinum electrode was selected since the surface of the electrode could be well-patterned by sputtering on a planar plastic thin film for the production of microcontact imprinting. Fluorescence spectrometry was performed for the determination of recognition capacity of the algae-imprinted cavities [45]. Following this study, further investigations were performed to study the effects of polymer concentration on microcontact imprinting of the same microorganism. For this purpose, various concentrations of EVAL were tested due to their effect on both morphology and efficiency of imprinted films [46]. In another study, it was indicated that algal metabolism may affect the generation of electricity in biofuel cells. For this purpose, *C. reinhardtii* was microcontact imprinted onto EVAL films. The production of hydrogen catalyzed by hydrogenase of the microorganism was measured electrochemically [47].

Microcontact imprinting, an alternative approach to design well-defined surfaces, is based on the polymerization taking place between a target stamp and the surface onto which the MIP shall be formed. This method generated much attention due to offering some crucial advantages over the other molecular imprinting methods. One unique feature is that the print structure is never fully embedded in the polymer, thereby making it easier to remove the template and to rebind macromolecular and even particulate structures [48].

In a study, the effects of adsorption to bacteria imprinted polymer matrices on gene expression were studied. The purple bacteria *Rhodobacter sphaeroides* was selected as a template and it was immobilized on a glass slide for the production of the bacterial stamp and then microcontact-imprinted onto EVAL. The surfaces of the *R. sphaeroides*-imprinted (RsIPs) and non-imprinted (NIPs) EVAL thin films were analyzed by Raman spectrometry and scanning electron microscopy. The expression of the nitrogenase playing a key role in nitrogen fixation gene of *R. sphaeroides* attached on EVAL thin films was also measured by the quantitative reverse transcription polymerase chain reaction (qRT-PCR) for the confirmation [49].

4. Molecular Imprinting of Microorganisms in Biosensor Platforms

MIPs have been considered as effective alternatives to natural recognition systems in order to improve detection of microorganisms with enhanced selectivity. Molecular imprinting (MIP) technology is a powerful tool in the development of selectively working biosensor systems having polymeric MIP structure [50]. In order to create applicable sensor surfaces, the coating material has to be resistant against organic solvents and applicable for a wide range of target molecules, recognize template molecules sensitively and show high stability. MIPs open up the possibility to meet the requirements on improving adapted sensor surfaces and provide the improvement of the detection efficiency. Microorganisms with unique structures and conformations can be discriminated using molecularly imprinted materials in biosensing systems in/on which sterically and functionally complementary sites are present. In order to develop sensing tools exhibiting high operational performance, the optimization of components have to be done to generate reversible interactions between binding cavities and microorganisms.

It is not easily possible to construct well-patterned cavities during an imprinting process due to the non-rigid structure of microorganisms. Even though in principle possible, the imprinting of microorganisms are still challenging when performed by traditional MIP methodologies. In this concept, highly cross-linked polymers were formed and this makes the placement of bacterial cells complicated and restricts the access of them to the interaction regions embedded in the interior of imprinted polymers. In order to overcome intrinsic limitations, efforts to improve molecular imprinting technology were directed towards surface imprinting as a promising approach. This approach has the advantage of integration of MIPs to the sensor platforms and thus surface-molecularly imprinted sensors were designed. The polymers synthesized by this technique covered the sensor as thin MIP-films in order to fabricate selective sensing tools for different template molecules [51–53].

Biosensor technology provides the opportunity of easy miniaturisation and flexible operation for the improvement of accomplished biosensors [54]. On the other hand, several transducer devices have been proven to be suitable for on-line monitoring used in order to detect and quantify microorganisms [36,55–57]. In recent years, many attempts have been made to detect whole bacterial cells especially by using impedimetric and optical sensors [2].

Surface-plasmon resonance (SPR)-based optical biosensors are able to measure the changes in refractive index near the surface of a chip related with the interaction between the analyte and biorecognition element.

Electrochemical sensors are capable of measuring electrochemical changes which take place with the interaction between chemicals and a sensing surface of the detecting electrode. The electrical changes can be based on a change in the measured voltage between the electrodes (potentiometric transducer), a change in the measured current at a given applied voltage (amperometric transducer), or a change in the ability of the sensing material to detect changes in the electric field (Impedance including capacitance/conductance). Among electrochemical biosensors, capacitive biosensors have attractive properties such as label-free, real-time and rapid detection of macromolecules along with both high selectivity and low detection limit [58]. The capacitance measurement principle of these sensors relies on the changes in dielectric properties of the electrode surface. The interaction between analyte of interest and electrode surface leads to a decrease in capacitance [59,60].

5. Applications

Dickert and co-workers [34] were the pioneers of attempts at whole cell imprinting integrated in biosensor platforms. In their first report, a model yeast cell *Saccharomyces cerevisiae* was imprinted in polyurethane and honeycomb-like yeast imprints were formed by stamping and the measurements were performed on mass sensitive sensor (quartz microbalance—QCM—devices). The proposed sensor was prepared via sol-gel method including imprinted films composed of polyurethane and alkoxide imprinted with whole yeast cells. It was possible to detect and quantify yeast cells in cell concentrations between 10^4 and 10^6 CFU/mL at flow rates of 10 mL/min. In comparison to polyurethane, it was determined that alkoxide sol-gel layers were more robust on which low sensor responses were obtained using the template microorganism. In the second report, surface-imprinted polyurethanes were generated in order to detect *E. coli* using QCM [61].

Acrylic acid and acrylamide derivative functional monomers could be used for the imprinting of microorganisms [34]. However, when these functional monomers are used for the imprinting, exposure to UV radiation or heat treatment is necessary for initiating of the polymerization process. Other than these, the whole cells of microorganisms were imprinted on thin films of organically modified silica prepared using sol-gel technique [38]. These alternative bacteria-imprinted films produced by sol-gel method includes polycondensation of silanes and uses mild conditions in comparison to the acrylate polymerization. Crosslinked and robust MIP films obtained through sol-gel techniques are resistant to harsh conditions such as thermal and solvent stresses. These matrices show high affinity against microorganisms in terms of both morphology of the imprinted cavity and functional groups located on the surface of the microorganism. Extraction of the template results in sterically and functionally adapted binding cavities in or on a porous matrix. Resultant MIP films were capable of differentiate between bacterial species with the same geometrical form [56].

Cohen et al. used thin films of organically modified silica (ORMOSILS) formed by sol-gel technique in order to imprint whole cells of different bacterial strains. These materials have gained great interest for the creation of sensing devices with the combination of molecular imprinting [62]. They provide a platform to generate simple, versatile and long term stable biosensors exhibiting good selectivity with high sensitivity [63]. *Deinococcus radiodurans*, *E. coli*, *B. subtilis* and *Sphaerotilus natans* were selected as templates for imprinting since they have different morphologies and show size variations. The developed materials indicated high adsorption affinity towards the target microorganisms and were promising tools for rapid bacterial detection from water [38].

Apart from functional monomers such as acrylic acid and sol-gel based ones, monomers used to form conducting polymers have found wide application in order to imprint biomolecules including microorganisms [31,32,64]. The properties of easy preparation and high conductivity makes these polymers appropriate for designing biosensors. Furthermore, their desirable feature is associating anionic molecules in the structures to compensate for the cationic charges introduced by the polymer backbone. This contribution makes these biocompatible polymers being applicable for the immobilization of biomolecules (enzymes, DNA etc.) [64]. Among conducting polymers, polypyrrole (PPy) based ones are extensively implemented to the imprinting studies. These polymers also used for developing bacterial templates on the surface of an overoxidized polypyrrole film using both gram-negative and gram-positive bacteria [31]. The functional groups present on the outer surface of the bacterial cell wall such as carboxyl, phosphate and hydroxyl groups correspond to the anionic sites. By this way, docking of the bacteria into a positively charged conducting polymer easily occurs due to the presence of related functional groups of the microorganism [64]. When the entire process of microorganism imprinting is taken into account, it is obvious that capture of the target cells operates according to expectations. The binding may, however, be too firm since there are problems in releasing bound material. This is a serious limitation since the concept of biosensors is to reuse the sensor surface for repeated analyses. If it shall be used as a disposable entity, then it becomes more like a dipstick. The missing removal is one of the most urgent needs to be resolved issue. The use of conductive polymers in microorganism imprinting offers a solution to this existing problem. The template was taken out from the polymer matrix with electrochemical overoxidation protecting the shape formed on the polymer matrix [31,56].

In a previous study, *P. aeruginosa* was selected as target microorganism and was transferred on the surface of overoxidized polypyrrole (OPPy) film. Then, it was determined that proposed approach provided highly selective and rapid detection of the target bacterial strain by application of QCM even in suspensions of bacterial mixtures including *Acinetobacter calcoaceticus*, *E. coli*, and *Serratia marcescens*. The successful detection could be performed at concentrations as low as 10^3 CFU/mL within 3 min. The real sample experiments were carried out with apple juice to verify whether the prepared system works and the measurable concentration range in apple juice was pointed out between 10^7 to 10^9 CFU/mL for tested bacterial strain [30].

In another study, *B. subtilis* endospores were imprinted using conducting polymer films. The resultant imprinted films were prepared by holding endospores on the surface of glassy carbon electrodes. A polypyrrole layer presented on the top of this electrode followed by poly(3-methyl-thiophene) layer. Then, the films were sequentially deposited by electropolymerization. Measurements of the changes in the electrochemistry of the conducting polymer film depending on endospore binding were carried out in order to detect the endospores [65].

Electropolymerization offering some advantages compared to more common preparation procedures of MIP films could be effectively integrated to the biosensor platforms [53]. Qi et al. created an impedimetric biosensor with bacteria-mediated bio-imprinted films in order to develop rapid detection devices for sulfate reducing bacteria. Chitosan (CS) was preferred due to its unique features such as biodegradability, low toxicity and good biocompatibility which makes them appealing alternatives in designing biosensors. A nanocomposite film with reduced graphene sheets (RGSs)/CS was electrodeposited on the surface of an indium tin oxide (ITO) electrode, and the RGSs/CS hybrid film behaved as a recognition element for the binding of target microorganism. The role of RGSs was increasing the conductivity of the film. Following bacterial attachment on the RGSs-CS nanocomposite films, a layer of CS was electrodeposited around the immobilized microorganism. After template removal, binding studies were performed by monitoring the impedance change of the biosensor with a detection limit of 0.7×10^4 CFU/mL. The selectivity studies carried out using bacteria-mediated bioimprinted film with *M. luteus*, *S. aureus*, *Vibrio anguillarum* and *Vibrio alginolyticus* confirmed the selectivity of the developed sensor [19].

Roy et al. developed an electrochemical sensor with Ag-ZnO bimetallic nanoparticle and graphene oxide nanocomposite using stamping method. The purpose of this study was to detect, remove and kill *E. coli* cells. The nanocomposite was selected as a platform in order to imprint target microorganism. The MIP-modified sensor was capable of detecting *E. coli* with limit of detection 5.9 CFU/mL, removing

98% of cells. In addition, 16 cm^2 modified area of the plate is sufficient to kill bacteria in a concentration of 10^5 CFU/mL. The real-sample experiments were performed with waste water samples. It was demonstrated that *E. coli* cells could be detected in the complicated matrix of waste water with the recovery of 92%–98%. The generated sensor offered simple, low-cost, short time analysis [5].

In our previous study, a label-free, selective and sensitive microcontact imprinted capacitive biosensor was generated in order to detect *E. coli*. The recognition of the target cells was successfully achieved by this sensor prepared with the combination of microcontact imprinting method and capacitive biosensor technology. After preparation of bacterial stamps, microcontact-*E. coli* imprinted gold electrodes were produced using an amino acid based recognition element, N-methacryloyl-L-histidine methylester (MAH) and 2-hydroxyethyl methacrylate (HEMA) as monomers and ethyleneglycol dimethacrylate (EGDMA) as crosslinker under UV-polymerization conditions. *E. coli*-imprinted capacitive electrodes were prepared using microcontact imprinting by bringing the modified surface with the vinyl groups of the electrode and the monomer solution in contact with the *E. coli*-immobilized on the glass slide like a sandwich (Figure 1).

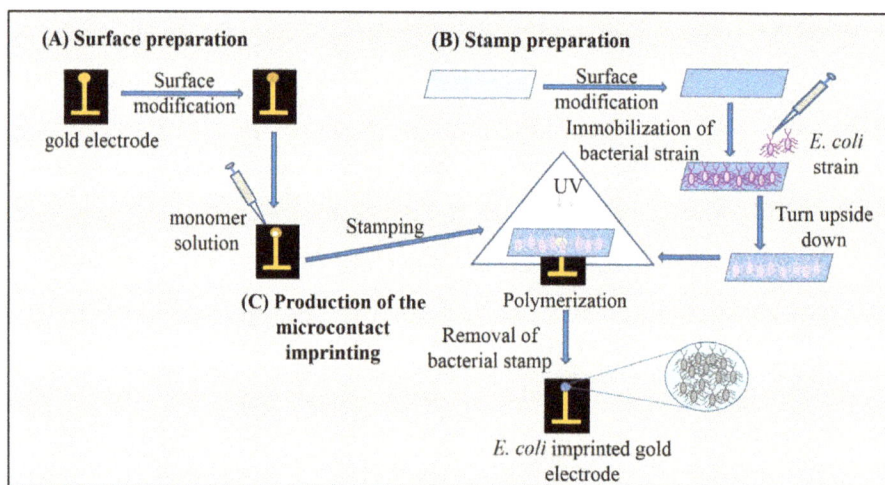

Figure 1. Schematic representation of microcontact imprinting of *E. coli* onto the polymer modified surfaces. (A) preparation of electrode surface; (B) preparation of bacteria stamps; (C) production of the microcontact imprinting (reproduced from [60] with permission).

SEM images in Figure 2 also show the changes in the morphology of the polymeric surface on the *E. coli*-MIP electrode in different magnifications, Figure 2B–D also show the captured cells. The unique combination of these two techniques provides selective detection with a detection limit of 70 CFU/mL. Capacitive sensorgram of the *E. coli* imprinted electrode and calibration curve of *E. coli* obtained at a concentration range of (1.0×10^2–1.0×10^7 CFU/mL) were given in Figure 3A,B, respectively. The recognition of *E. coli* took place by the functional group of MAH in the imprinted film and shape-complementary cavities to *E. coli*. The generated capacitive sensor has high selectivity and was able to distinguish *E. coli* when present together with competing bacterial strains which are known to have similar shape. In addition, the prepared sensor has the ability to detect *E. coli* with a recovery of 81%–97% in e.g., river water [60].

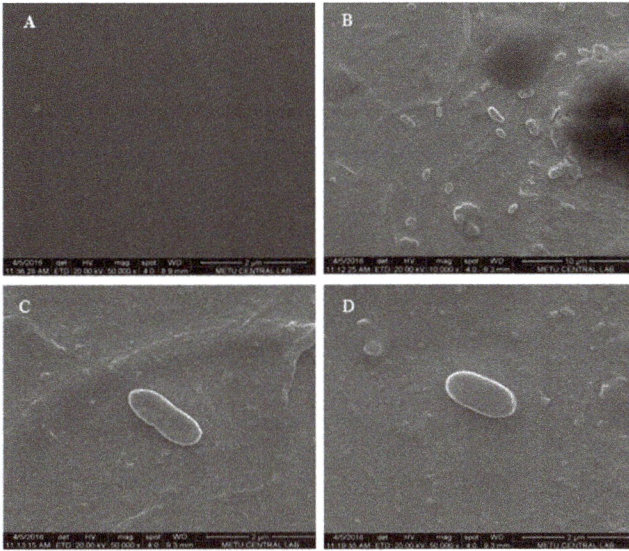

Figure 2. SEM images of bare gold electrode (**A**) −50,000×, and *E. coli* imprinted electrode (**B**) −10,000×, (**C,D**) −50,000× magnifications (reproduced from [60] with permission).

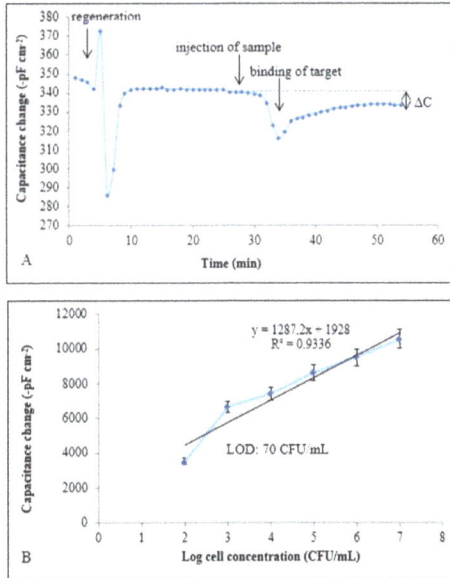

Figure 3. (**A**) Capacitive sensorgram of the *E. coli* imprinted electrode (concentration of *E. coli*: 1.0×10^7 CFU/mL), ΔC: change in capacitance; (**B**) calibration curve of *E. coli* obtained at a concentration range of (1.0×10^2–1.0×10^7 CFU/mL) under optimum conditions, LOD: limit of detection, flow rate: 100 µL/min; sample volume: 500 µL; running buffer: 10 mM sodium phosphate, pH: 7.4; regeneration buffer: pure ethyl alcohol, 10 mg/mL lysozyme solution (in 10 mM Tris-HCl buffer, pH 8.0, with 1 mM EDTA) and 50 mM glycine-HCl, pH 2.5; T 25 °C. (Reproduced from [60] with permission).

Yilmaz et al. developed microcontact imprinted SPR and QCM sensor surfaces for the detection of *E. coli*. They used an amino acid-based recognition element, MAH, as functional monomer. Recognition occurred via of functional group (MAH) in imprinted film and shape-complementary cavities belonging to *E. coli* on the imprinted film surface. The schematic representation of microcontact imprinted SPR and QCM sensor surfaces developed in this study. Characterization studies were performed with SEM and AFM. In that study, the limit of detection (LOD) and the limit of quantification (LOQ) of QCM system were found as 3.72×10^5 CFU/mL and 1.24×10^6 CFU/mL, respectively. LOD and LOQ of SPR system were also calculated and were found as 1.54×10^6 CFU/mL and 5.13×10^6 CFU/mL, respectively. Real-time responses and linear regions when plotting the relation of log cell conc. vs. the signal registered of SPR and QCM. The researchers preferred the apple juice spiked with *E. coli* at different concentrations in the range of 0.5–4.0 McFarland (approx. 1.5×10^8 to 12×10^8 CFU/mL) as real sample in order to verify the proposed SPR and QCM sensors for detecting *E. coli* [36].

There are several publications in which bio-sensing systems are based on application of molecular imprinting. In recent years, imprinted polymers have gained increasing attention of researchers in different fields. In these platforms, imprinted materials are used as artificial recognition elements integrated with a transducer component. The real time detection of an interaction occuring between the unique binding sites and the target molecule was a stimulating feature in order to improve such systems as options to conventional methods. Due to some difficulties with proper selective detection of microorganism with cell surface imprinted materials, the development of effective sensing systems still remains to be addressed. The representative examples of microorganism detection using MIP based materials in sensor applications in literature were shown in Table 1.

Table 1. Representative examples of microorganism detection using MIP based materials in sensor applications.

Template	Functional Monomer	Imprinting Technique	Determination Technique	LOD/LOQ	Ref.
Saccharomyces cerevisiae	Polyurethane	Surface imprinting	QCM	10^4–10^6 CFU/mL	[34]
Saccharomyces cerevisiae	Titanium ethylate	Surface imprinting	QCM	10^4–10^6 CFU/mL	[34]
Escherichia coli	Polyureathane	Surface Imprinting	QCM	0.1 mg/mL	[61]
Deinococcus radiodurans	Tetraehoxysilane	Bulk imprinting	Fluorescence	10^7–10^9 CFU/mL	[38]
Escherichia coli	Tetraehoxysilane	Bulk imprinting	Fluorescence	10^7–10^9 CFU/mL	[38]
Bacillus subtilis	Tetraehoxysilane	Bulk imprinting	Fluorescence	10^7–10^9 CFU/mL	[38]
Sphaerotilus natans	Tetraehoxysilane	Bulk imprinting	Fluorescence	10^7–10^9 CFU/mL	[38]
Pseudomonasaeruginosa	Pyrrole	Electro polymerization	Fluorescence	-	[32]
Escherichia coli	Pyrrole	Electro polymerization	QCM	10^3–10^9 CFU/mL	[32]
Pseudomonasaeruginosa	Pyrrole	Electro polymerization	QCM	10^3–10^9 CFU/mL	[32]
A. calcoaceticus	Pyrrole	Electro polymerization	QCM	10^3–10^9 CFU/mL	[32]
S. marcescens	Pyrrole	Electro polymerization	QCM	10^3–10^9 CFU/mL	[32]
Escherichia coli	Tetraehoxysilane	Bulk imprinting	QCM	10^2 CFU/mL	[37]
Sulphate reducing bacteria	Chitosan reduced graphene	Electro polymerization	EIS	10^4–10^8 CFU/mL	[19]
Bacillus subtilis endospore	Pyrrole	Electro polymerization	EIS	10^4–10^7 CFU/mL	[65]
Escherichia coli	Silane	Stamping	Electrochemical sensor	5.9 CFU/mL	[5]
Escherichia coli	MAH	Microcontact Imprinting	Capacitive biosensor	70 CFU/mL	[60]
Escherichia coli	MAH	Microcontact Imprinting	QCM	1.54×10^6 CFU/mL	[36]
Escherichia coli	MAH	Microcontact Imprinting	SPR	3.72×10^5 CFU/mL	[36]

CFU: colony forming unit, EIS: electrochemical impedance spectroscopy, MAH: *N*-methacryloyl-L-histidine methylester, QCM: quartz crystal microbalance, SPR: surface plasmon resonance.

MIPs show selectively high affinity towards target molecule/cell selected to be used in the imprinting process [66]. These polymers have higher physical robustness, strength, resistance to elevated temperature and inertness against organic solvents in comparison with natural systems [66,67]. Furthermore, their production is less expensive and storage life is higher with preserving their capacity [66].

When the limitations were taken into consideration, formation of new MIPs has complicated challenges which result mainly from the fact that all target molecules/cells are different from each other with the requirement of various functional monomer and cross-linker combinations [68]. The accurate choice and adjustment of the polymer concentration composition provides enough binding regions complementary to the functional groups of the target molecule/cell. The production of favorably developed MIPs is restricted because of inefficient removal of stamp from the polymer surface. In some cases, strong removal methods can be necessary and this sometimes leads to damage binding region of target molecule/cell while fully removal of the template.

6. Concluding Remarks

The present review summarizes the status with regard to biosensor-based assays of microorganisms by registration of their binding to artificial affinity sites created via use of molecularly imprinted polymers. Several of the assays are direct binding assays, i.e., one can register the captured cells directly and thereby get an opinion about the number of cells (often as CFU) per mL. The assays are quick and that is certainly a strong feature. A limitation is still the relatively poor knowledge base on how to design the best affinity cavity to selectively capture and quantify particular microbial cells. Thus, the data so far generated give clear indications of possibilities for an interesting future development of this relatively new approach to quantify microorganisms (most often pathogenic) in our environment.

When comparing microorganism analysis to that of small molecules or even proteins, studies on microorganisms lag far behind with regard to development of the particular analytical field. At the same time it is encouraging to follow how the protein assays have developed from a very low level with an accelerated pace into advanced analytical systems. It is therefore a not too brave guess that analytical devices based on MIPs will undergo a similar development as that of protein analysis.

Acknowledgments: I would like to thank Hacettepe University, Ankara, Turkey.

Conflicts of Interest: The authors declare no conflict of interest.

References

1. Ivnitski, D.; Abdel-Hamid, I.; Atanasov, P.; Wilkins, E. Biosensors for detection of pathogenic bacteria. *Biosens. Bioelectron.* **1999**, *14*, 599–624. [CrossRef]
2. Ahmed, A.; Rushworth, J.V.; Hirst, N.A.; Millner, P.A. Biosensors for whole-cell bacterial detection. *Clin. Microbiol. Rev.* **2014**, *27*, 631–646. [CrossRef] [PubMed]
3. Bole, A.L.; Manesiotis, P. Advanced Materials for the Recognition and Capture of Whole Cells and Microorganisms. *Adv. Mater.* **2016**, 5349–5366. [CrossRef] [PubMed]
4. David, M.Z.; Daum, R.S. Community-associated methicillin-resistant *Staphylococcus aureus:* Epidemiology and clinical consequences of an emerging epidemic. *Clin. Microbiol. Rev.* **2010**, *23*, 616–687. [CrossRef] [PubMed]
5. Roy, E.; Patra, S.; Tiwari, A.; Madhuri, R.; Sharma, P.K. Single cell imprinting on the surface of Ag-ZnO bimetallic nanoparticle modified graphene oxide sheets for targeted detection, removal and photothermal killing of *E. Coli. Biosens. Bioelectron.* **2017**, *89*, 620–626. [CrossRef] [PubMed]
6. Lazcka, O.; Del Campo, F.J.; Muñoz, F.X. Pathogen detection: A perspective of traditional methods and biosensors. *Biosens. Bioelectron.* **2007**, *22*, 1205–1217. [CrossRef] [PubMed]
7. Leonard, P.; Hearty, S.; Brennan, J.; Dunne, L.; Quinn, J.; Chakraborty, T.; Kennedy, R. O'Advances in Biosensors for Detection of Pathogens in Food and Water. *Enzyme Microb. Technol.* **2003**, *32*, 3–13. [CrossRef]

8. Science, S.; Tsing, N. Nanoparticles-Based Electrochemical Biosensor for Single Bacterium Detection by Redox Signal Amplification. In Proceedings of the 17th International Conference on Miniaturized Systems for Chemistry and Life Sciences, Freiburg, Germany, 27–31 October 2013; pp. 239–241.
9. Boehm, D.A.; Gottlieb, P.A.; Hua, S.Z. On-chip microfluidic biosensor for bacterial detection and identification. *Sens. Actuators B Chem.* **2007**, *126*, 508–514. [CrossRef]
10. Sivakumar, S.; Wark, K.L.; Gupta, J.K.; Abbott, N.L.; Caruso, F. Liquid crystal emulsions as the basis of biological sensors for the optical detection of bacteria and viruses. *Adv. Funct. Mater.* **2009**, *19*, 2260–2265. [CrossRef]
11. Andrade, C.A.S.; Nascimento, J.M.; Oliveira, I.S.; De Oliveira, C.V.J.; De Melo, C.P.; Franco, O.L.; Oliveira, M.D.L. Colloids and Surfaces B: Biointerfaces Nanostructured sensor based on carbon nanotubes and clavanin A for bacterial detection. *Colloids Surf. B Biointerfaces* **2015**, *135*, 833–839. [CrossRef] [PubMed]
12. Iskierko, Z.; Sharma, P.S.; Bartold, K.; Pietrzyk-Le, A.; Noworyta, K.; Kutner, W. Molecularly imprinted polymers for separating and sensing of macromolecular compounds and microorganisms. *Biotechnol. Adv.* **2016**, *34*, 30–46. [CrossRef] [PubMed]
13. Chambers, J.P.; Arulanandam, B.P.; Matta, L.L.; Weis, A.; Valdes, J.J. Biosensor recognition elements. *Curr. Issues Mol. Biol.* **2008**, *10*, 1–12. [PubMed]
14. Whitcombe, M.J.; Kirsch, N.; Nicholls, I.A. Molecular imprinting science and technology: A survey of the literature for the years 2004–2011. *J. Mol. Recognit.* **2014**, *27*, 297–401. [PubMed]
15. Wulff, G. Molecular Imprinting in Cross-Linked Materials with the Aid of Molecular Templates—A Way towards Artificial Antibodies. *Angew. Chem. Int. Ed. Engl.* **1995**, *34*, 1812–1832.
16. Qi, P.; Wang, J.; Wang, L.; Li, Y.; Jin, J.; Su, F.; Tian, Y.; Chen, J. Molecularly imprinted polymers synthesized via semi-covalent imprinting with sacrificial spacer for imprinting phenols. *Polymer* **2010**, *51*, 5417–5423. [CrossRef]
17. Chen, L.X.; Xu, S.F.; Li, J.H. Recent advances in molecular imprinting technology: Current status, challenges and highlighted applications. *Chem. Soc. Rev.* **2011**, *40*, 2922–2942. [CrossRef] [PubMed]
18. Haupt, K.; Mosbach, K. Molecularly imprinted polymers and their use in biomimetic sensors. *Chem. Rev.* **2000**, *100*, 2495–2504. [CrossRef] [PubMed]
19. Qi, P.; Wan, Y.; Zhang, D. Impedimetric biosensor based on cell-mediated bioimprinted films for bacterial detection. *Biosens. Bioelectron.* **2013**, *39*, 282–288. [CrossRef] [PubMed]
20. Yu, C.; Mosbach, K. Molecular Imprinting Utilizing an Amide Functional Group for Hydrogen Bonding Leading to Highly Efficient Polymers. *J. Org. Chem.* **1997**, *62*, 4057–4064. [CrossRef]
21. Ramström, O.; Nicholls, I.A.; Mosbach, K. Synthetic peptide receptor mimics: Highly stereoselective recognition in non-covalent molecularly imprinted polymers. *Tetrahedron Asymmetry* **1994**, *5*, 649–656. [CrossRef]
22. Striegler, S. Investigation of disaccharide recognition by molecularly imprinted polymers. *Bioseparation* **2001**, *10*, 307–314. [CrossRef] [PubMed]
23. Tsunemori, H.; Araki, K.; Uezu, K.; Goto, M.; Furusaki, S. Surface imprinting polymers for the recognition of nucleotides. *Bioseparation* **2001**, *10*, 315–321. [CrossRef] [PubMed]
24. Özgür, E.; Yilmaz, E.; Şener, G.; Uzun, L.; Say, R.D.L.; Denizli, A. A new molecular imprinting-based mass-sensitive sensor for real-time detection of 17β-estradiol from aqueous solution. *Environ. Prog. Sustain. Energy* **2013**, *32*, 1164–1169. [CrossRef]
25. Panahi, H.A.; Feizbakhsh, A.; Khaledi, S.; Moniri, E. Fabrication of new drug imprinting polymer beads for selective extraction of naproxen in human urine and pharmaceutical samples. *Int. J. Pharm.* **2013**, *441*, 776–780. [CrossRef] [PubMed]
26. Yu, J.C.C.; Lai, E.P.C. Molecularly imprinted polymers for ochratoxin A extraction and analysis. *Toxins* **2010**, *2*, 1536–1553. [CrossRef] [PubMed]
27. Garcia, R. Application of Molecularly Imprinted Polymers for the Analysis of Pesticide Residues in Food—A Highly Selective and Innovative Approach. *Am. J. Anal. Chem.* **2011**, *2*, 16–25. [CrossRef]
28. Blanco-López, M.C.; Gutiérrez-Fernández, S.; Lobo-Castañón, M.J.; Miranda-Ordieres, A.J.; Tuñón-Blanco, P. Electrochemical sensing with electrodes modified with molecularly imprinted polymer films. *Anal. Bioanal. Chem.* **2004**, *378*, 1922–1928. [CrossRef] [PubMed]
29. Sharma, P.S.; Iskierko, Z.; Pietrzyk-Le, A.; D'Souza, F.; Kutner, W. Bioinspired intelligent molecularly imprinted polymers for chemosensing: A mini review. *Electrochem. Commun.* **2015**, *50*, 81–87. [CrossRef]

30. Tokonami, S.; Nakadoi, Y.; Takahashi, M.; Ikemizu, M.; Kadoma, T.; Saimatsu, K.; Dung, L.Q.; Shiigi, H.; Nagaoka, T. Label-free and selective bacteria detection using a film with transferred bacterial configuration. *Anal. Chem.* **2013**, *85*, 4925–4929. [CrossRef] [PubMed]

31. Tokonami, S.; Nakadoi, Y.; Nakata, H.; Takami, S.; Kadoma, T.; Shiigi, H.; Nagaoka, T. Recognition of gram-negative and gram-positive bacteria with a functionalized conducting polymer film. *Res. Chem. Intermed.* **2014**, *40*, 2327–2335. [CrossRef]

32. Tokonami, S.; Saimatsu, K.; Nakadoi, Y.; Furuta, M.; Shiigi, H.; Nagaoka, T. Vertical immobilization of viable bacilliform bacteria into polypyrrole films. *Anal. Sci.* **2012**, *28*, 319–321. [CrossRef] [PubMed]

33. Dickert, F.L.; Hayden, O.; Halikias, K.P. Synthetic receptors as sensor coatings for molecules and living cells. *Analyst* **2001**, *126*, 766–771. [CrossRef] [PubMed]

34. Dickert, F.L.; Hayden, O. Bioimprinting of polymers and sol-gel phases. Selective detection of yeasts with imprinted polymers. *Anal. Chem.* **2002**, *74*, 1302–1306. [CrossRef] [PubMed]

35. Shen, X.; Svensson Bonde, J.; Kamra, T.; Bülow, L.; Leo, J.C.; Linke, D.; Ye, L. Bacterial imprinting at pickering emulsion interfaces. *Angew. Chem. Int. Ed.* **2014**, *53*, 10687–10690. [CrossRef] [PubMed]

36. Yilmaz, E.; Majidi, D.; Ozgur, E.; Denizli, A. Whole cell imprinting based *Escherichia coli sensors*: A study for SPR and QCM. *Sens. Actuators B Chem.* **2015**, *209*, 714–721. [CrossRef]

37. Starosvetsky, J.; Cohen, T.; Cheruti, U.; Bilanovi, D.; Armon, R. Effects of Physical Parameters on Bacterial Cell Adsorption onto Pre-Imprinted Sol-Gel Films. *J. Biomater. Nanobiotechnol.* **2012**, *2012*, 499–507. [CrossRef]

38. Cohen, T.; Starosvetsky, J.; Cheruti, U.; Armon, R. Whole cell imprinting in sol-gel thin films for bacterial recognition in liquids: Macromolecular fingerprinting. *Int. J. Mol. Sci.* **2010**, *11*, 1236–1252. [CrossRef] [PubMed]

39. Borovicka, J.; Stoyanov, S.D.; Paunov, V.N. Shape recognition of microbial cells by colloidal cell imprints. *Nanoscale* **2013**, *57*, 8560–8568. [CrossRef] [PubMed]

40. Aherne, A.; Alexander, C.; Payne, M.J.; Perez, N.; Vulfson, E.N. Bacteria-mediated lithography of polymer surfaces. *J. Am. Chem. Soc.* **1996**, *118*, 8771–8772. [CrossRef]

41. Alexander, C.; Vulfson, E.N. Spatially functionalized polymer surfaces produced via cell-mediated lithography. *Adv. Mater.* **1997**, *9*, 751–755. [CrossRef]

42. Harvey, S.D.; Mong, G.M.; Ozanich, R.M.; McLean, J.S.; Goodwin, S.M.; Valentine, N.B.; Fredrickson, J.K. Preparation and evaluation of spore-specific affinity-augmented bio-imprinted beads. *Anal. Bioanal. Chem.* **2006**, *386*, 211–219. [CrossRef] [PubMed]

43. Hayden, O.; Dickert, F.L. Selective microorganism detection with cell surface imprinted polymers. *Adv. Mater.* **2001**, *13*, 1480–1483. [CrossRef]

44. Kempe, M.; Mosbach, K. Separation of amino acids, peptides and proteins on molecularly imprinted stationary phases. *J. Chromatogr. A* **1995**, *691*, 317–323. [CrossRef]

45. Chen, W.J.; Lee, M.H.; Thomas, J.L.; Lu, P.H.; Li, M.H.; Lin, H.Y. Microcontact imprinting of algae on poly(ethylene-Co-Vinyl alcohol) for biofuel cells. *ACS Appl. Mater. Interfaces* **2013**, *5*, 11123–11128. [CrossRef] [PubMed]

46. Lee, M.; Thomas, J.L.; Lai, M.; Shih, C.; Lin, H. Microcontact Imprinting of Algae for Biofuel Systems: The effects of the polymer concentration. *Langmuir* **2014**, *30*, 14014–14020. [CrossRef] [PubMed]

47. Lee, M.; Thomas, J.L.; Lai, M.; Lin, H. Advances Recognition of algae by microcontact-imprinted polymers modulates hydrogenase expression. *RSC Adv.* **2014**, *4*, 61557–61563. [CrossRef]

48. Sener, G.; Ozgur, E.; Rad, A.Y.; Uzun, L.; Say, R.; Denizli, A. Rapid real-time detection of procalcitonin using a microcontact imprinted surface plasmon resonance biosensor. *Analyst* **2013**, *138*, 6422–6428. [CrossRef] [PubMed]

49. Lee, M.H.; Thomas, J.L.; Li, M.H.; Shih, C.P.; Jan, J.S.; Lin, H.Y. Recognition of *Rhodobacter sphaeroides* by microcontact-imprinted poly(ethylene-co-vinyl alcohol). *Colloids Surf. B Biointerfaces* **2015**, *135*, 394–399. [PubMed]

50. Lieberzeit, P.A.; Schirk, C.; Glanznig, G.; Gazda-Miarecka, S.; Bindeus, R.; Nannen, H.; Kauling, J.; Dickert, F.L. From nanopatterning to functionality—Surface and bulk imprinting for analytical purposes. *Superlattices Microstruct.* **2004**, *36*, 133–142. [CrossRef]

51. Eersels, K.; Lieberzeit, P.; Wagner, P. A Review on Synthetic Receptors for Bioparticle Detection Created by Surface-Imprinting Techniques—From Principles to Applications. *ACS Sens.* **2016**, *1*, 1171–1187. [CrossRef]

52. Malitesta, C.; Mazzotta, E.; Picca, R.A.; Poma, A.; Chianella, I.; Piletsky, S.A. MIP sensors—The electrochemical approach. *Anal. Bioanal. Chem.* **2012**, *402*, 1827–1846. [PubMed]
53. Sharma, P.S.; Pietrzyk-Le, A.; D'Souza, F.; Kutner, W. Electrochemically synthesized polymers in molecular imprinting for chemical sensing. *Anal. Bioanal. Chem.* **2012**, *402*, 3177–3204. [CrossRef] [PubMed]
54. Sokolov, A.N.; Roberts, M.E.; Bao, Z. Fabrication of low-cost electronic biosensors. *Mater. Today* **2009**, *12*, 12–20. [CrossRef]
55. Farka, Z.; Kovář, D.; Skládal, P. Rapid detection of microorganisms based on active and passive modes of QCM. *Sensors* **2015**, *15*, 79–92. [CrossRef] [PubMed]
56. Sciences, A. *Biosensors for Security and Bioterrorism Applications*; Springer: New York, NY, USA, 2016.
57. Dudak, F.C.; Boyaci, I.H. Rapid and label-free bacteria detection by surface plasmon resonance (SPR) biosensors. *Biotechnol. J.* **2009**, *4*, 1003–1011. [CrossRef] [PubMed]
58. Lebogang, L.; Mattiasson, B.; Hedström, M. Capacitive sensing of microcystin variants of *Microcystis aeruginosa* using a gold immunoelectrode modified with antibodies, gold nanoparticles and polytyramine. *Microchim. Acta* **2014**, *181*, 1009–1017. [CrossRef]
59. Berggren, C.; Bjarnason, B.; Johansson, G. Capacitive biosensors. *Electroanalysis* **2001**, *13*, 173–180. [CrossRef]
60. Idil, N.; Hedström, M.; Denizli, A.; Mattiasson, B. Whole cell based microcontact imprinted capacitive biosensor for the detection of *Escherichia coli*. *Biosens. Bioelectron.* **2016**, *87*, 807–815. [CrossRef] [PubMed]
61. Findeisen, A.; Wackerlig, J.; Samardzic, R.; Pitkänen, J.; Anttalainen, O.; Dickert, F.L.; Lieberzeit, P.A. Artificial receptor layers for detecting chemical and biological agent mimics. *Sens. Actuators B Chem.* **2012**, *170*, 196–200. [CrossRef]
62. Díaz-García, M.E.; Laíñño, R.B. Molecular Imprinting in Sol-Gel Materials: Recent Developments and Applications. *Microchim. Acta* **2005**, *149*, 19–36. [CrossRef]
63. Fireman-Shoresh, S.; Turyan, I.; Mandler, D.; Avnir, D.; Marx, S. Chiral electrochemical recognition by very thin molecularly imprinted sol-gel films. *Langmuir* **2005**, *21*, 7842–7847. [CrossRef] [PubMed]
64. Nagaoka, T.; Shiigi, H.; Tokonami, S.; Saimatsu, K. Entrapment of Whole Cell Bacteria into Conducting Polymers. *J. Flow Inject. Anal.* **2012**, *29*, 7–10.
65. Namvar, A.; Warriner, K. Microbial imprinted polypyrrole/poly(3-methylthiophene) composite films for the detection of *Bacillus* endospores. *Biosens. Bioelectron.* **2007**, *22*, 2018–2024. [CrossRef] [PubMed]
66. Ertürk, G.; Mattiasson, B. Molecular Imprinting Techniques Used for the Preparation of Biosensors. *Sensors* **2017**, *17*, 288.
67. Vasapollo, G.; Del Sole, R.; Mergola, L.; Lazzoi, M.R.; Scardino, A.; Scorrano, S.; Mele, G. Molecularly imprinted polymers: Present and future prospective. *Int. J. Mol. Sci.* **2011**, *12*, 5908–5945. [CrossRef] [PubMed]
68. Chen, L.; Wang, X.; Lu, W.; Wu, X.; Li, J. Molecular imprinting: Perspectives and applications. *Chem. Soc. Rev.* **2016**, *45*, 2137–2211. [CrossRef] [PubMed]

sensors

MDPI

Review

Molecular Imprinting of Macromolecules for Sensor Applications

Yeşeren Saylan [1], Fatma Yilmaz [2], Erdoğan Özgür [1], Ali Derazshamshir [1], Handan Yavuz [1] and Adil Denizli [1,*]

[1] Department of Chemistry, Division of Biochemistry, Hacettepe University, 06800 Ankara, Turkey; yeseren@hacettepe.edu.tr (Y.S.); erdoganozgur@hacettepe.edu.tr (E.Ö.); tural@hacettepe.edu.tr (A.D.); handany@hacettepe.edu.tr (H.Y.)

[2] Department of Chemistry Technology, Abant Izzet Baysal University, 14900 Bolu, Turkey; fyilmaz71@gmail.com

* Correspondence: denizli@hacettepe.edu.tr; Tel.: +90-312-297-7963

Academic Editors: Gizem Ertürk and Bo Mattiasson
Received: 31 January 2017; Accepted: 7 April 2017; Published: 19 April 2017

Abstract: Molecular recognition has an important role in numerous living systems. One of the most important molecular recognition methods is molecular imprinting, which allows host compounds to recognize and detect several molecules rapidly, sensitively and selectively. Compared to natural systems, molecular imprinting methods have some important features such as low cost, robustness, high recognition ability and long term durability which allows molecularly imprinted polymers to be used in various biotechnological applications, such as chromatography, drug delivery, nanotechnology, and sensor technology. Sensors are important tools because of their ability to figure out a potentially large number of analytical difficulties in various areas with different macromolecular targets. Proteins, enzymes, nucleic acids, antibodies, viruses and cells are defined as macromolecules that have wide range of functions are very important. Thus, macromolecules detection has gained great attention in concerning the improvement in most of the studies. The applications of macromolecule imprinted sensors will have a spacious exploration according to the low cost, high specificity and stability. In this review, macromolecules for molecularly imprinted sensor applications are structured according to the definition of molecular imprinting methods, developments in macromolecular imprinting methods, macromolecular imprinted sensors, and conclusions and future perspectives. This chapter follows the latter strategies and focuses on the applications of macromolecular imprinted sensors. This allows discussion on how sensor strategy is brought to solve the macromolecules imprinting.

Keywords: macromolecule; molecular imprinting; sensor

1. Molecular Imprinting Methods

One of the first reports about molecular imprinting was published by Wulff and Sarhan in 1972 [1]. Molecular imprinting is one of the most popular methods to present molecular recognition sites and has attracted growing attempts for the preparation of complementary parts of the target molecules [2]. This method mainly relies on the molecular identification reaction that occurs at the surrounding of the target molecule called as a template. As seen in Figure 1, the pre-complex was first formed by the template and functional monomers. The polymerization was completed after the cross-linker and pre-complex interactions that keep the position of the functional groups to bind the template able to produce the molecular recognition sites. At the end of the polymerization stage, the polymeric matrix has specific recognition sites after removing template with suitable desorption agents, so molecularly imprinted polymers (MIPs) are able to bind the template with high selectivity as compared to other competing molecules [3]. MIPs can be produced by several combinations of cross-linkers, functional monomers and solvents [4–6]. The quality of the MIPs and their binding features are changed not only via the combination of the mixture, but also the experimental circumstances, such as the initiator type and amount, polymerization temperature, interaction mechanisms, and so on [7–13].

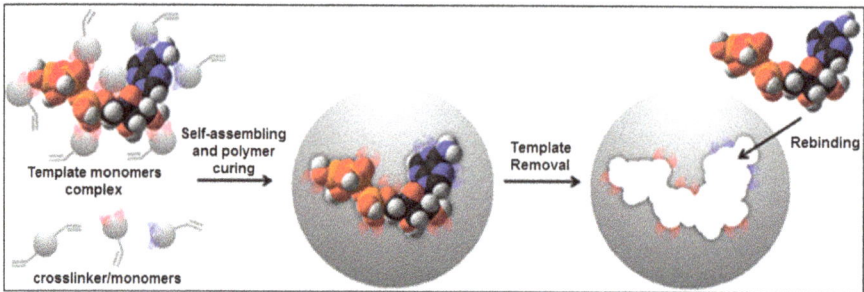

Figure 1. A schematic representation of the molecular imprinting method [14].

It is generally supposed that the template acts as a critical molecule and the other compounds (cross-linkers, functional monomers, and solvents) should be selected based on the chemical and physical features of the template [15]. Furthermore, the stability as binding strength of the target molecule of the imprinted polymers is optimized by varying the monomer and cross-linker composition [16].

On the basis of these facts, Baggiani et al. offered an alternative perspective to the molecular imprinting method [17]. As illustrated in Figure 2, the existence of the template in the pre-complex mixture helps to improve interactions that pre-exist in a non-imprinted polymer (NIP). As a result, if the NIP does not bind the target molecule, the MIP will display a weak imprinting effect. On the other hand, if the NIP binds the target, the MIP will display a strong imprinting effect.

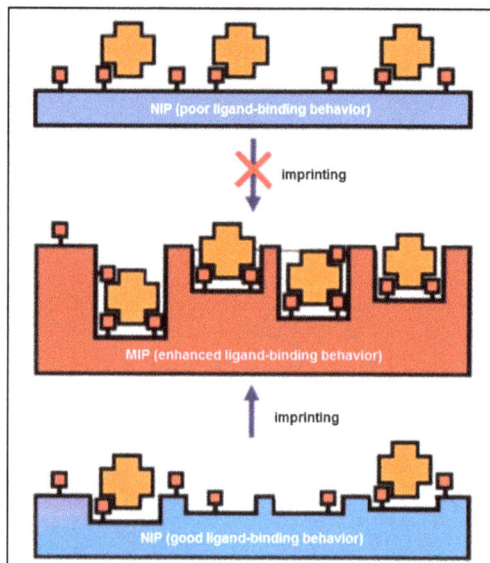

Figure 2. A schematic illustration of the alternative imprinting hypothesis [17].

They provided a library in the absence of any template to verify their hypothesis. This library was screened for various possible ligands, and the composition of the best-binding NIP yielded the MIP with outstanding binding features. The scanning of the numerous polymers authorized a definite correlation between the binding features of the NIP and MIP libraries [17]. MIPs have been synthesized successfully using different types of molecular imprinting methods including surface imprinting, micro-contact imprinting and epitope imprinting and these polymers have been used in many applications such as purification [18], isolation [19], chiral separation [20], catalysis [21] and in sensors [22].

2. Developments in Macromolecular Imprinting Methods

Biological molecules such as amino acids, nucleic acids etc. are macromolecules which display significant biological activities and they can be classified as organic molecules [23,24]. The structures of macromolecules are often uninformative about function. Owing to this wide range of functions, detection of proteins, enzymes, nucleic acids, and cells has attracted a huge interest in the high-speed development of biomedicine and proteomics, and other studies as well [25].

Even though the use of molecular imprinting method for the detection of macromolecules has a number of advantages, however it also has some drawbacks as well. Firstly, desorption and recognition of the template is the earliest point in the macromolecular imprinting method because rebinding capacity of the extremely cross-linked forms of the MIPs with asymmetric configuration is reduced. Then, the template cannot be removed smoothly from the imprinted cavities. Thus, a decrease of the adsorption and desorption processes occurs. Desorption efficiency of the template in the case of imprinting of macromolecules is low due to the large size of the macromolecule. Nanomaterials can be used to overcome these problems, because nanomaterials have high surface area and volume ratio and also most of the interacting sites are exposed on the surface. Thus, after easy removal of the template, higher rates of adsorption can be achieved [26–28]. Different types of imprinting methods also help to figure out any kind of problems in the imprinting of macromolecules, which are summarized below.

2.1. Surface Imprinting Methods

Surface imprinting methods help reduce the mass transfer resistance in macromolecule imprinting processes [29]. In recent years, surface imprinting, as a significant progress in macromolecular imprinting methods, has gained great attention, especially in separation, sensor and diagnostic applications [30].

These methods can be divided into two sub-classes—top-down and bottom-up—based on the location where the polymerization occurs. In a top-down process, the template is bound to a support, which is removed after formation and the interaction areas on the polymer surface retreat [31,32]. There are some studies that show the production of surface imprinted nanomaterials [33–35]. The support utilize to immobilize the template enhances the substrate on which the polymer is inserted and improves the substrate in the bottom-up process. In this method, the removal of the template has to be accomplished impressively to obtain imprinted materials. The bottom-up process has been employed for the preparation of macromolecule imprinted microparticles [36–38], quantum dots [39], magnetic nanoparticles [40,41], carbon nanotubes [42], and gold electrodes [43,44]. A comparable method has been employed to produce sites imprinted with macromolecules [45,46].

Recently, Wang et al. [47] offered a surface imprinted model to detect glycoproteins, as illustrated in Figure 3. Dopamine and *m*-aminophenyl boronic acid were deposited on the sensor surface. After template removal, the imprinted cavities demonstrated recognition ability with high affinity in basic medium. The dissociation constant was calculated as 6.6×10^{-9} at pH 9.0 and 2.7×10^{-7} at pH 3.0 for horseradish peroxidase (HRP).

Further examples are the ionic liquid-modified graphene-based sensor [48], quartz crystal microbalance (QCM)-based sensor [49] and molecularly imprinted micro-particles for macromolecule detection [50].

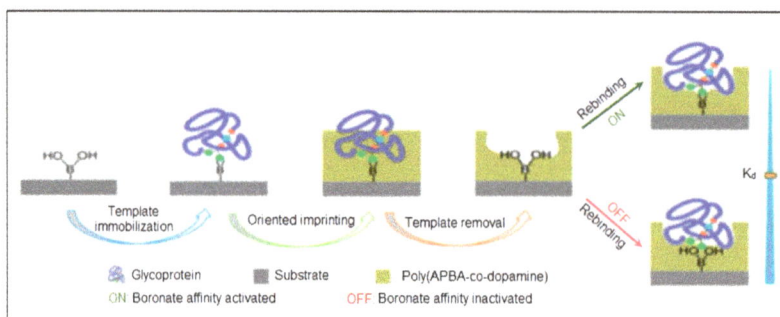

Figure 3. A schematic illustration of the surface imprinting of glycoproteins [47].

2.2. Micro-Contact Imprinting Method

Another solution for the problem of imprinting of large, labile and non-rigid macromolecules has been figured out by researchers using a micro-contact imprinting method. The micro-contact imprinting method needs a small amount of template mass, which is efficiently used as a monolayer. Some scientists have applied a stamp to place the template moiety during the imprinting process [51–53]. They fabricated a single layer of the template with the functional monomer on a glass slide and then attached this layer onto a glass slide that has the cross-linker. At the end of the UV polymerization, the glass slide was smoothly removed from the surface of the polymer, which demonstrated the desired recognition features.

The operative site of the template that helps to produce interactions with molecules in the polymer was decreased in the imprinted polymeric films [54,55]. Silicon wafers, glass slides, gold surfaces were employed as a stamp to fabricate imprinted polymeric matrices. Micro-contact imprints of some templates such as creatine kinase, lysozyme, RNase A, myoglobin and C-reactive protein were formed by micro-contact imprinting methods [56–60].

2.3. Epitope Imprinting Methods

Epitope imprinting methods have become a new method for the identification and separation of target molecules. According to this approach, a region of a macromolecule is employed as a template instead of the whole macromolecule during the imprinting process. Thus, small peptide sequences could recognize a whole protein [61–64] or F_{ab} fragments were used to detect human immunoglobulin G as a model of the protein fragment [65]. Epitope imprinting methods have several advantages such as organic solvents can be used, and the costs of the peptides are lower than proteins and they can be produced in very pure form, so increased selectivity can be achieved [66].

Thermodynamic considerations imply that the non-rigid template usage leads to less clear recognition sites in MIPs [67], so the composition includes the macromolecule in the polymerization by framing and also they have some disadvantages such as low specificity and insufficient reproducibility [68–71]. Mosbach et al. also proved the efficient recognition of short oligopeptides by MIPs [72–74]. In order to resolve the recognition problem, Rachkov and Minoura suggested imprinting only fixed limited areas of recognition sites [75]. Epitope imprinting and a temperature- dependent capture and release process were performed by Li and coworkers [76]. In their study, SiO_2 nanoparticles were immobilized by glycidoxypropyltrimethoxysilane-iminodiacetic acid and 3-(trimethoxysilyl)propyl methacrylate to promote the template modification, as reflected in Figure 4. After immobilization of a His-tag-anchored epitope of human serum albumin, polymerization was conducted using N-isopropylacrylamide as a monomer to obtain a thermosensitive imprinted shell. Finally, the template was removed by ethylenediaminetetraacetic acid, and the formed epitope imprinted sites could capture template at 45 °C and release it at 4 °C.

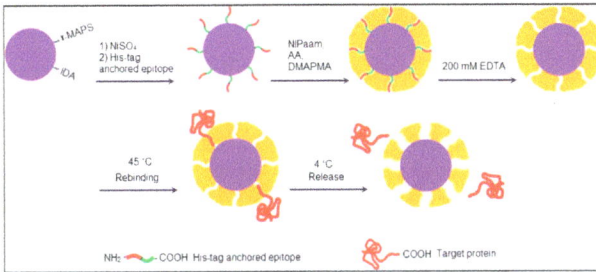

Figure 4. A schematic representation of the epitope imprinting method [76].

3. Macromolecular Imprinted Sensors

Sensors are tools for the analysis of molecules to obtain their composition, structure and function by converting biological responses into electrical signals [77]. They should fundamentally comprise a transducer (electrochemical [78], piezoelectric [79], or optical [80]) and a recognition molecule, which interacts with an analyte. A number of optical sensing studies have been performed on sensors involving chemiluminescence [81], fluorescence [82], light absorption [83], and reflectance [84] that can be classified as label-based and label-free. While label-based sensing is highly sensitive, and the limit of detection is also very low, it is limited to hard labelling processes that may also hinder the molecule function [85]. On the contrary, label-free sensing reacts in the natural forms and depends on the measurement of refractive index changes. As a result, label-free sensing is comparatively simple and cheap, and also permits one to perform quantitative and kinetic analyses for molecular recognition [86]. The recognition molecules are significant components of sensors because they are responsible for the capture of the target molecules, so the recognition molecule selection is completely based on the target molecule. The recognition molecules are selected according to their high affinity and also stability to the target molecule. Chemical sensors can be sorted according to the recognition molecules used. The advantages and disadvantages of some recognition molecules in sensors are as listed in Table 1 [87].

Table 1. Advantages and disadvantages of some recognition molecules in sensors.

Recognition Molecule	Sensor Identification	Advantages	Disadvantages
Enzyme	Enzymatic sensor	- Specificity - Basic equipment and procedures	- High cost and time-wasting purification - Low stability - Efficiency only at an optimum pH and temperature
Antibody	Immunosensor	- Extreme affinity - Specificity	- Minimal target - Hard production - Use of animals requirement - Stability deficiency
Nucleic acid	Genosensor	- Stability	- Minimal target
Cell	Cell sensor	- Low cost preparation - Low purification needs	
Aptamer	Aptasensor	- Easy to modify - Possibility of structure design - Possibility of denaturalization and rehybridization - Possibility of distinguishing targets - Thermal stability - In-vitro synthesis	
MIPs	MIP sensor	- Extreme thermal, chemical and mechanical resistance - Reusability	- Complicated fabrication methods - Time-wasting procedure - Incompatibility with aqueous solutions - Template release - Lower specificity

3.1. Enzyme Imprinted Sensors

Molecular imprinting processes lock the enzyme into a certain conformation that is favorable for catalysis. In this rigid form, enzymes remain more active and selective in the presence of organic solvents than they are in aqueous media. This makes a large number of applications in chemical, pharmaceutical and polymer industries possible [88].

A surface plasmon resonance (SPR) sensor with lysozyme-imprinted nanoparticles was designed by Sener et al. as a recognition element to detect lysozyme [89]. They immobilized lysozyme imprinted nanoparticles onto the SPR sensor surface. As shown in Figure 5, this SPR sensor could perform in both aqueous and natural solutions. The concentration of lysozyme was as low as 32.2 nM. They also calculated that the limit of detection (LOD), association (K_a) and dissociation (K_d) values as 84 pM, 108.71 nM^{-1} and 9.20 pM, respectively.

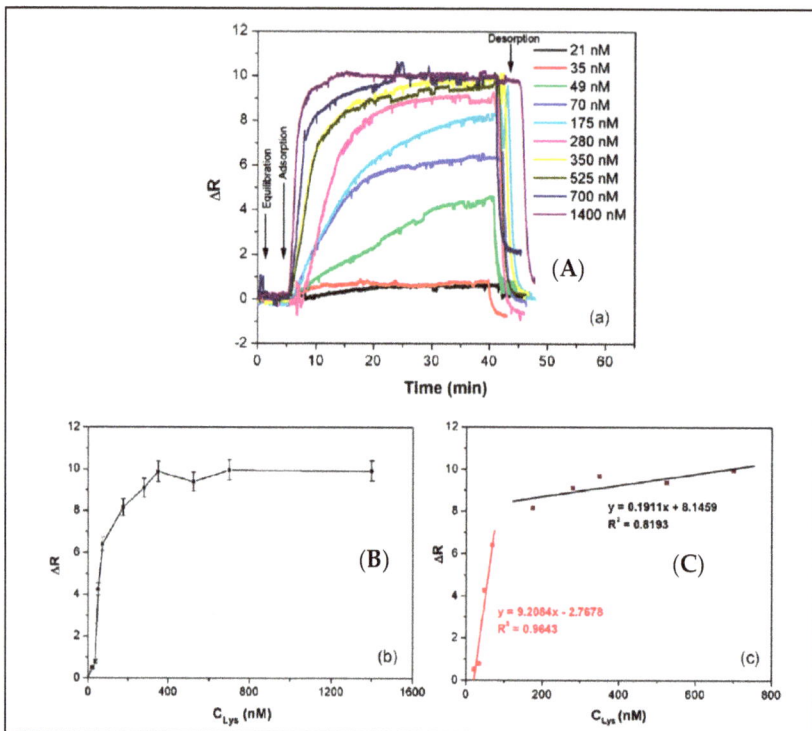

Figure 5. The detection of lysozyme with lysozyme imprinted SPR sensor: (**A**) concentration dependence of lysozyme imprinted SPR sensor; (**B**) concentration versus SPR sensor response; (**C**) linear regions [89].

Saylan and colleagues developed an SPR-based sensor to detect lysozyme with hydrophobic poly(N-methacryloyl-L-phenylalanine) nanoparticles [90]. Various concentrations of lysozyme solutions were used to calculate kinetic and affinity coefficients (Figure 6A). The equilibrium and adsorption isotherm models of interactions between the lysozyme solutions and the SPR sensor were determined and the maximum reflection, association and dissociation constants were calculated by a Langmuir model as 4.87 nM, 0.019 nM and 54 nM, respectively. Selectivity studies of the SPR sensor were performed with competitive agents like hemoglobin and myoglobin (Figure 6B). The results showed that the SPR sensor could detect lysozyme in lysozyme solutions with high accuracy, good sensitivity, in real-time, label-free, and with a low LOD value of 0.66 nM.

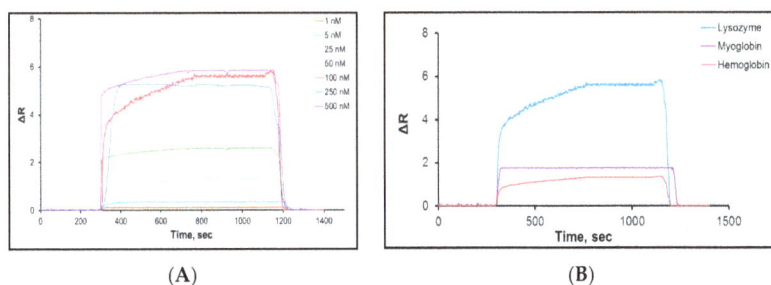

Figure 6. The (**A**) concentration dependency and (**B**) selectivity experiments of SPR sensor [90].

Sunayama et al. reported a new functional monomer which could convert the macromolecule identification signal event into a fluorescent signal [91]. They prepared lysozyme-imprinted polymers which were organized on glass substrates by copolymerization of a functional monomer, and cross-linker, in the presence of lysozyme (Figure 7A,B).

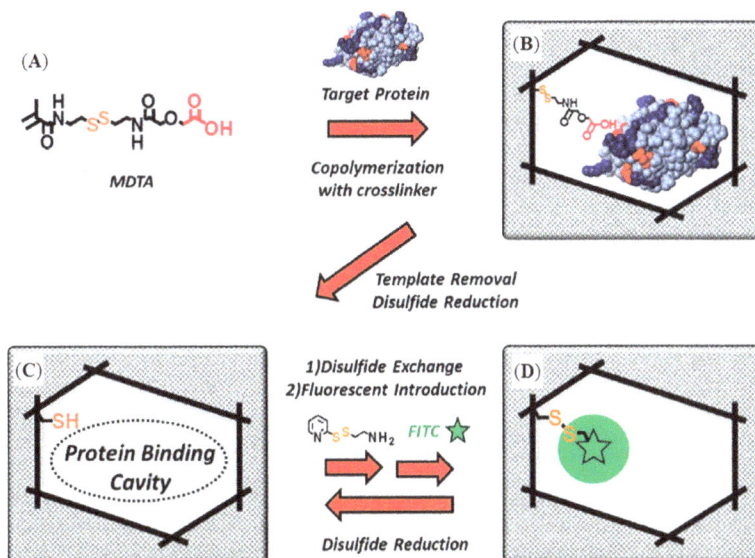

Figure 7. Schematic illustration of the preparation of lysozyme-imprinted polymer: (**A**) ([[2-(2-methacrylamido)ethyldithio]ethylcarbamoyl]methoxy)acetic acid structure; (**B**) protein-imprinted polymer preparation; (**C**) binding cavity created by the disulfide linkage reduction and (**D**) fluorophore introduction by the disulfide linkage reformation [91].

In the first post-imprinting modifications after the removal of lysoyzme resulted in the creation of the lysozyme-binding cavities, the residual (ethylcarbamoylmethoxy)acetic acid moiety within the cavities was removed by reduction (Figure 7C). In the second post-imprinting modification, the disulfide linkage was reformed using aminoethylpyridyl disulfide to introduce aminoethyl groups (Figure 7D), followed by treatment with fluorescein isothiocyanate to label the amino groups within the cavities, in the third post-imprinting modification. The reusability and tunability of the prepared MIPs were evaluated by fluorescence measurements to demonstrate the effectiveness of the proposed method. All studies about enzyme detection were summarized in Table 2 according to the different parameters.

Table 2. Comparison of the sensor studies for enzyme detection.

Parameters	Reference		
	[89]	[90]	[91]
Target	Lysozyme	Lysozyme	Lysozyme
Linear dynamic range	21–1400 nM	1–500 nM	0–3 μM
Buffers (Ads, Des)	pH 7.4, 1 M NaCl (pH 8.0)	pH 7.0, ethylene gycol	10 mM Tris/HCl (pH 7.4), NA
Time	45 min	23 min	NA
Limit of detection	32.2 nM	0.66 nM	NA

NA: not available.

3.2. Antibody/Antigen Imprinted Sensors

Molecular imprinting methods have become a straightforward and versatile way of making synthetic receptors that can recognize target molecule with affinity and selectivity. This has led to them being called antibody mimics, and they can be used in sensor systems to detect antigens and also antibody molecules with higher physical and chemical stability than their biomolecule counterparts, which are restricted in stability under external conditions besides being expensive. Some studies in which clinically significant antigen/antibody molecules were selected as target molecules are summarized below.

A cyclic citrullinated peptide antibody-imprinted SPR sensor was prepared by Dibekkaya and collaborators to detect cyclic citrullinated peptide antibodies [92]. They used different concentrations of cyclic citrullinated peptide antibodies for real-time detection. They also calculated LOD, K_a and K_d constants of 0.177, 0.589 RU/mL and 1.697 mL/RU, respectively. According to their results, the cyclic citrullinated peptide antibody-imprinted SPR sensor can be employed several times for cyclic citrullinated peptide antibodies detection with remarkable selectivity and sensitivity.

Uzun et al. developed a hepatitis B surface antibody imprinted film on a SPR sensor and measured hepatitis B surface antibody concentrations in human serum [93]. They first characterized the SPR sensor by atomic force microscopy (Figure 8) and then performed kinetic studies with different concentrations of hepatitis B surface antibody positive human serum samples (Figure 9). The LOD value was found to be 208.2 mIU/mL.

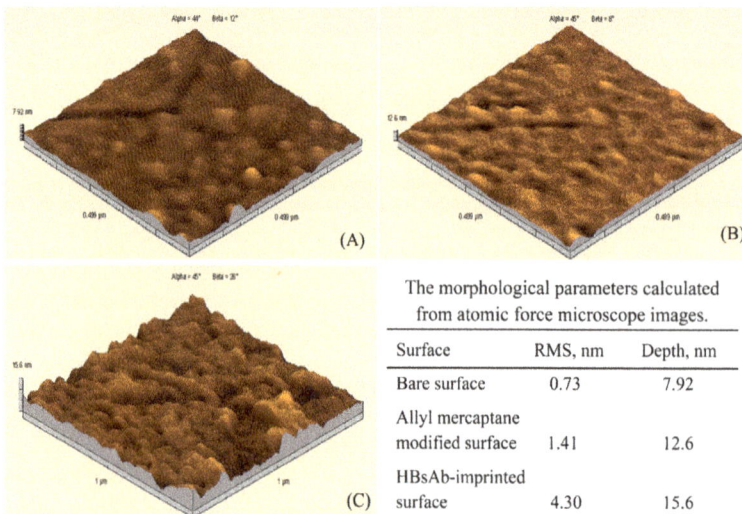

The morphological parameters calculated from atomic force microscope images.

Surface	RMS, nm	Depth, nm
Bare surface	0.73	7.92
Allyl mercaptane modified surface	1.41	12.6
HBsAb-imprinted surface	4.30	15.6

Figure 8. Atomic force microscopy pictures of (**A**) non-modified; (**B**) allyl mercaptan-modified; (**C**) hepatitis B surface antibody-imprinted SPR sensor [93].

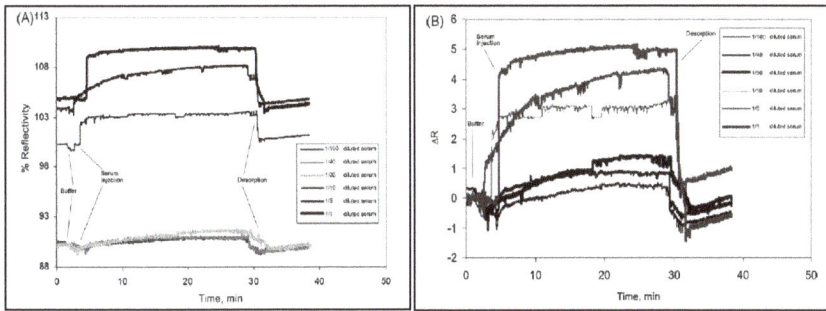

Figure 9. Sensorgrams for the interaction between hepatitis B surface antibody positive human serum and hepatitis B surface antibody imprinted SPR sensor (**A**) reflectivity and (**B**) ΔR vs. time [93].

A micro-contact imprinting-based SPR sensor to detect prostate specific antigen (PSA) was developed by Ertürk et al. [94]. As shown in Figure 10A–F, they prepared the SPR sensor by UV polymerization based on the micro-contact imprinting method. They detected PSA in a 0.1–50 ng/mL concentration range with a LOD value of 91 pg/mL. They also analysed 10 clinical samples using their PSA-imprinted SPR sensor and indicated around 98% accuracy between their results and those obtained by a commercial ELISA method.

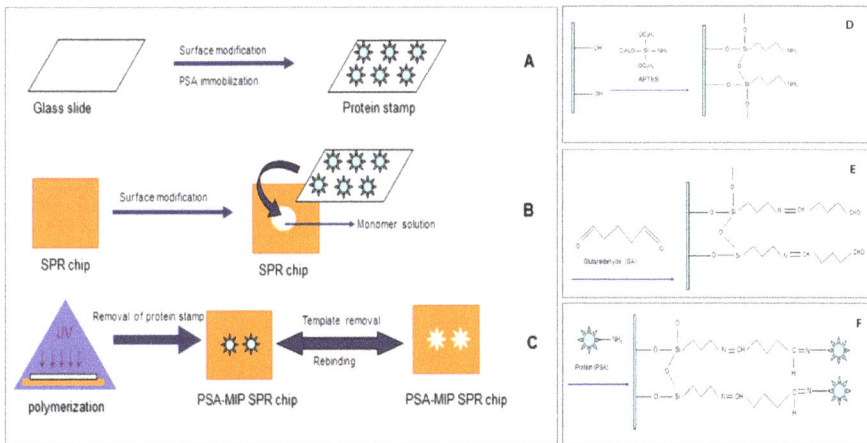

Figure 10. (**A**) Glass slides preparation, (**B**) surface modification of SPR sensor, (**C**) micro-contact imprinting of prostate specific antigen, (**D**) surface modification of glass slides, (**E**) amino groups activation on glass slides and (**F**) prostate specific antigen immobilization onto the glass slides [94].

Uludag et al. described a system to detect total prostate-specific antigen in human serum samples with SPR and QCM sensors [95]. They performed a sandwich assay using antibody- based nanoparticles and detected 2.3 ng/mL and 0.29 ng/mL total prostate-specific antigen concentrations in human serum (Figure 11). They also showed that their results correlated well with those of a QCM sensor.

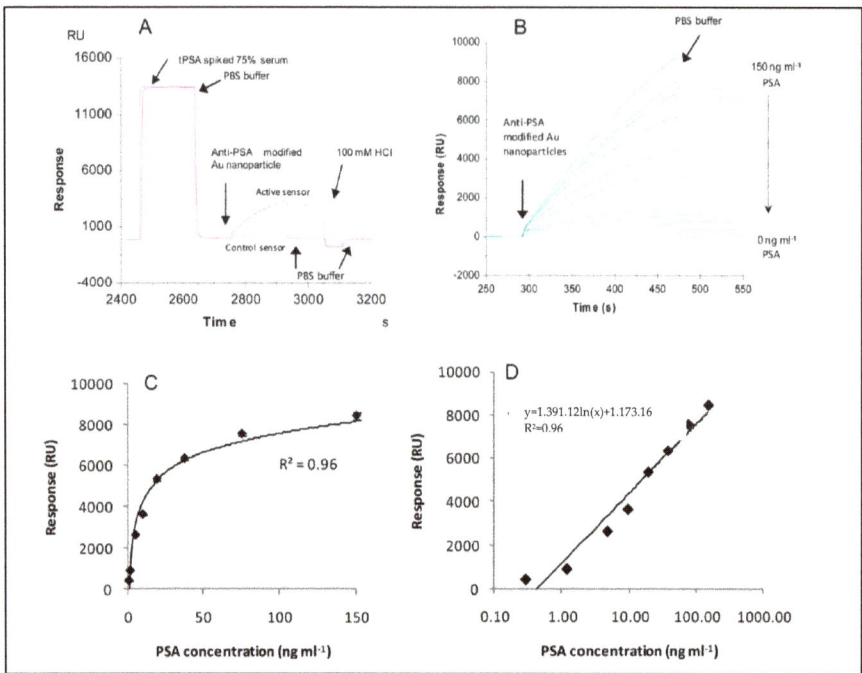

Figure 11. The (**A**) detection of total PSA antibody-modified gold nanoparticles after the PSA injection; (**B**) 4.69, 1.17, 0.29, 0 ng/mL on total PSA antibody immobilized surface and 150 ng/mL on IgG-immobilized surface; (**C**, linear, **D**, log scales) the calibration curves that came by from the assay [95].

Ertürk and coworkers prepared a F_{ab} fragments-imprinted SPR sensor to detect human immunoglobulin G (IgG) [65]. They digested IgG molecules with papain and concentrated them by fast protein liquid chromatography. They formed a complex between F_{ab} fragments and the specific monomer, and then prepared nanofilms on the SPR sensor surface using a cross-linker and functional monomer. The group carried out IgG detection studies using different concentrations of aqueous IgG solutions. They also performed experiments to verify the selectivity of the F_{ab}-imprinted SPR sensor by using bovine serum albumin, IgG, F_{ab} and F_c fragments. The SPR sensor has no response to bovine serum albumin (Figure 12D) and F_c (Figure 12A) solutions, while has specific responses to F_{ab} fragments (Figure 12B) and IgG (Figure 12E) with higher affinity. They also performed with pre-mixed protein solutions that contained $F_{ab}/F_c/BSA$ (Figure 12C) and $IgG/F_c/BSA$ (Figure 12F) for a second confirmation of SPR sensor selectivity.

Chianella et al. modified the common ELISA test by replacing the antibodies with molecularly imprinted nanoparticles as depicted in Figure 13 [96]. They achieved detection of vancomycin in studies on competition with a horseradish peroxidase–vancomycin conjugate. Their study was able to detect vancomycin in aqueous and blood plasma solutions in the range of 0.001–70 nM with LOD value of 0.0025 nM. They showed that the sensitivity of the study was three times higher than the ELISA predicated on antibodies.

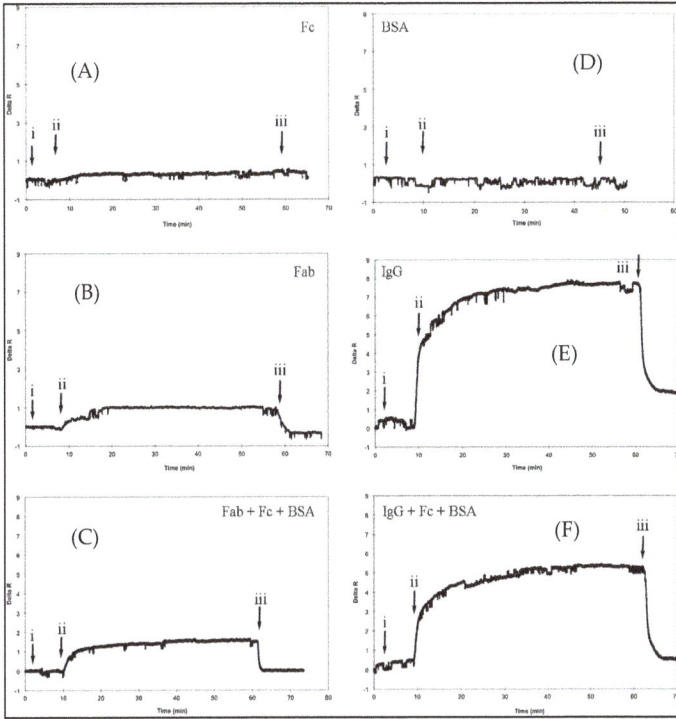

Figure 12. The selectivity SPR sensor (**A–F**): (i) Equilibrium by phosphate buffer, (ii) the analyte solutions application, and (iii) desorption with phosphate buffer that has 1 M NaCl [65].

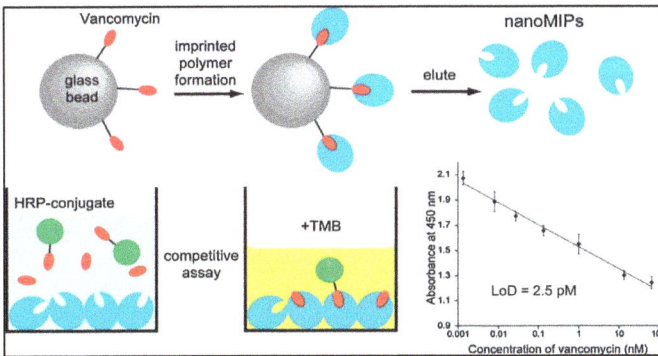

Figure 13. The schematic model of solid phase imprinting to produce artificial antibodies [96].

As indicated by Türkoğlu et al. polymeric nanoparticles were fixed to a SPR sensor to detect human IgG in human serum [97]. They performed detection studies by utilizing aqueous IgG solutions at different concentrations. They found a K_a value as 1.810 µg/mL with a high correlation coefficient ($R^2 = 0.9274$). They also showed that the best adsorption model showing the interaction between the SPR sensor and IgG was the Langmuir adsorption isotherm. All studies about antibody and antigen detection were summarized in Table 3 according to the different parameters.

Table 3. Comparison of the sensor studies for antibody and antigen detection.

Parameters				Reference			
	[65]	[92]	[93]	[94]	[95]	[96]	[97]
Target	Fab fragment	Cyclic citrullinated peptide antibody	Hepatitis B surface antibody	Prostate specific antigen	Prostate specific antigen	Vancomycin	Immunoglobulin G
Linear dynamic range	2–15 mg/mL	1–200 RU/mL	0–120 mIU/mL	0.1–50 ng/mL	0.29–5000 ng/mL	0.001–70 nM	0.05–2.0 mg/mL
Buffers (Ads, Des)	pH 7.4, 1 M NaCl	pH 7.0, Acetic acid with Tween 20	NA, 1 M Ethylene glycol	pH 7.4, Glycine HCl (pH 2.5)	PBS, 100 mM HCl	PBS (pH 7.4), NA	pH 7.4, 1 M NaCl
Time	40 min	40 min	40 min	50 min	53 min	NA	52 min
Limit of detection	56 ng/mL	0.177 RU/mL	208.2 mIU/mL	91 pg/mL	0.29 ng/mL	2.5 pM	NA

NA: not available.

3.3. Protein Imprinted Sensors

Molecular imprinting techniques are proven to work well with small molecules but they are challenging for larger molecules like proteins because of their complexity, conformational flexibility and solubility. However it has been shown that it is possible to prepare protein-imprinted films or nanoparticles for various applications including sensors, by i.e. surface imprinting, epitope imprinting, metal-ion coordination, or using natural nontoxic and biocompatible polymers as facile and green approaches [98].

Osman et al. developed a SPR sensor combined with molecularly imprinted synthetic receptors. A myoglobin-imprinted polymeric film was integrated onto the SPR sensor and characterized by atomic force microscopy and contact angle measurements (Figure 14). Then they evaluated the detection behaviors of the developed sensor for myoglobin with myoglobin solutions in different concentrations scale in phosphate buffer and serum. The LOD value of the SPR sensor was calculated as 26.3 ng/mL [60].

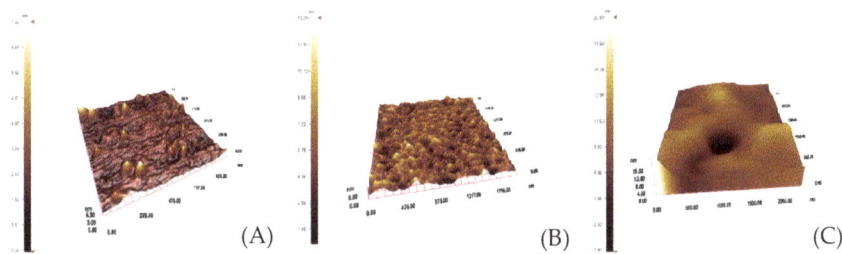

Figure 14. The characterization of the glass and SPR sensor: The atomic force microscopy images of (**A**) myoglobin immobilized glass, (**B**) bare SPR sensor, (**C**) myoglobin imprinted SPR sensor [60].

Moreira et al. described a novel use of the polymeric film poly(*o*-aminophenol) that was made responsive to a specific protein [99]. This was accomplished through electropolymerization of aminophenol with protein. Proteins embedded in the outer surface of the polymeric film were digested by proteinase K and then washed out thereby creating empty sites. The films acted as biomimetic artificial antibodies and were produced on a gold screen printed electrode, as a step towards disposable sensors. The sensors displayed linear responses to myoglobin down to 4.0 and 3.5 g/mL with LOD values of 1.5 and 0.8 g/mL.

Chunta and co-workers synthesized MIPs for the detection of low-density lipoprotein [100]. They examined that the ratios of monomers acrylic acid, methacrylic acid, and N-vinylpyrrolidone and analyzed by samples by QCM (Figure 15). The QCM sensor had an accuracy of 95−96% at the 95% confidence interval with 6−15% precision. Their results showed that QCM sensor responses were in agreement with those of standard methods.

Reddy et al. studied hydrophilic molecularly imprinted hydrogels to detect bovine hemoglobin and insulin [101]. They investigated the specific binding capacity of four different functional monomers. After finding the optimal conditions, they synthesized the hydrogel from acrylamide functional monomer that was found to have the best specific adsorption capacity.

Bakhshpour and coworkers detected protein C in human serum by QCM [102]. The protein C micro-contact imprinted polymeric film was prepared on a glass surface. Then, the QCM sensor was prepared using suitable functional monomers and cross-linker with copper (II) ions. The polymerization was performed under UV light for 20–25 min (Figure 16). Detection of protein C was studied in a concentration range of 0.1–30 μg/mL. The LOD value for protein C analysis was determined as 0.01 μg/mL.

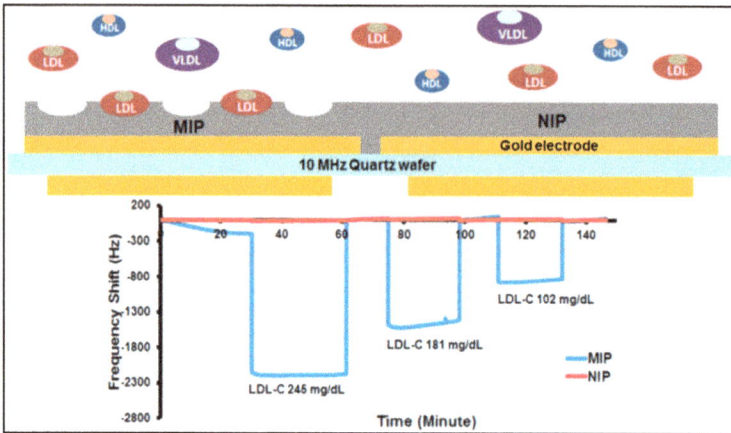

Figure 15. The scheme and recovery rates of the QCM sensor [101].

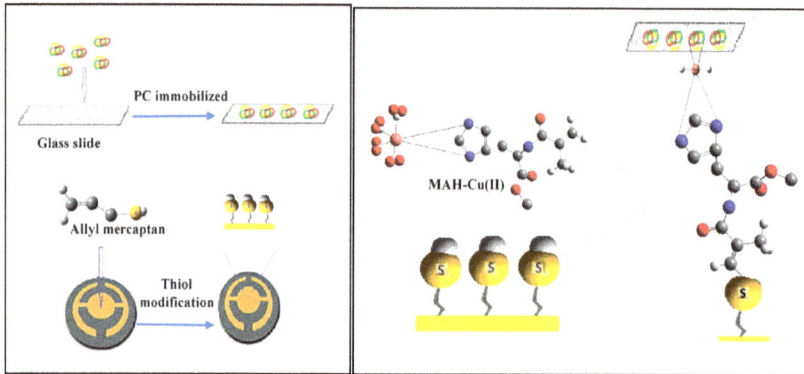

Figure 16. Schematic representation of the protein C-imprinted QCM sensor [102].

Another study was published by Wang et al. was based on the use of a carbon electrode to fabricate an imprinted electrochemical sensor for bovine hemoglobin detection [103]. They examined the fabrication parameters such as the concentration of pyrrole, scan cycles and rates of the imprinted electrochemical sensor. After optimization, they showed that the imprinted electrochemical sensor had fast rebinding features that helped detect bovine hemoglobin with a high correlation coefficient (R = 0.998) and low LOD value (3.09×10^{-11} g/L). They also revealed that the imprinted electrochemical sensor had a very good selectivity and stability, and can be used for immunoassays and clinical applications.

Wang and collaborators also constructed a potentiometric myoglobin or hemoglobin sensor using self-assembled monolayers [104]. They produced a sensing surface on the sensor by using self-assembled monolayers and template molecules. Their results showed that the sensor could detect myoglobin or hemoglobin proteins in the presence/absence of other proteins in aqueous solution.

Moreira and co-workers built a self-assembled monolayer sensor for myoglobin detection [105]. They found the optimum conditions in HEPES buffer. The LOD value was found to be 1.3×10^{-6} mol/L. They also showed the imprinting effect by non-imprinted particles with no ability to detect myoglobin. The sensor had a good selectivity towards other molecules such as creatinine, sacarose, fructose, galactose, sodium glutamate, and alanine.

Çiçek et al. prepared bilirubin-imprinted polymeric films on a QCM sensor [106]. They prepared sample solutions in the concentration range of 1–50 g/mL to determine the relationship between the bilirubin sample solutions and the QCM sensor response. They also calculated the LOD and LOQ values as 0.45 and 0.9 g/mL, respectively.

Dechtrirat and collaborators prepared a MIP film on a gold-based transducer surface for cytochrome C detection in aqueous solution [107]. They used a combination of the epitope and surface imprinting approaches. They characterized the recognition capabilities of the films and confirmed that the MIP film was able to detect cytochrome C selectively. They also showed that the MIP film could discriminate even single amino acid mismatched sequences on the target peptide. All studies about protein detection were summarized in Table 4 according to the different parameters.

3.4. Nucleic Acid Imprinted Sensors

Nucleic acids are suitable molecular recognition elements for the detection of DNA and RNA molecules owing to their potential to form Watson-Crick pairs. Beyond this, the use of nucleic acids as biorecognition elements was extended to aptamers, which bind non-nucleotide molecules as well with a broad scope. The nucleic acid or aptamer-imprinted systems offer several advantages in terms of improved selectivity with a protection against enzymatic and chemical degradation [108].

Diltemiz et al. developed a SPR sensor by using methacrylamidohistidine-platinum (II) for recognition of DNA [109]. They reported that not only the detection of guanosine guanine and but also assay DNA sequencing is possible with the MIP SPR sensor. The formation of guanosine-imprints on SPR sensor surface is shown in Figure 17. Briefly, they cleaned the SPR sensor surface and dipped it into monomer solution for 24 h. After that, immobilized surface was dipped in the mixture that included the metal-chelate monomer, crosslinker and initiator for the polymerization. They carried out the polymerization at room temperature by using UV light for 4 h.

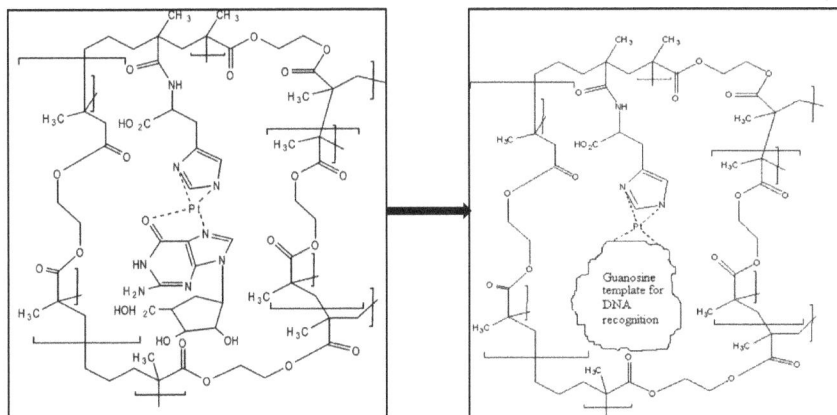

Figure 17. Scheme of the formation of guanosine-imprints on SPR sensor surface [109].

Table 4. Comparison of the sensor studies for protein detection.

Parameters	[60]	[99]	[100]	[101]	[102]	[103]	[104]	[105]	[106]	[107]
						Reference				
Target	Myoglobin	Myoglobin	Lipoprotein	Bovine hemoglobin and trypsin	Protein C	Bovine hemoglobin	Hemoglobin and myoglobin	Myoglobin	Bilirubin	Cytochrome C
Linear dynamic range	0.1–10 µg/mL	0.5–53.3 µg/mL	4–400 mg/dL	3 mg/mL	0.1–30 µg/mL	1×10^{-3}– 1×10^{-10} g/L	0–250 µg/mL	1.24×10^{-6}– 3.71×10^{-7} mol/L	1.71–85.51 µM	1–500 µg/mL
Buffers (Ads, Des)	pH 7.4, 1 M Ethylene glycol	pH 5.0, Proteinase K (pH 7.4)	pH 7.4, Acetic acid (10%) nd SDS (0.1%)	NA	pH 5.0, 0.5 M NaCl (pH 5.0)	pH 7.0, NA	Dulbecco's PBS (pH 7.15), NA	HEPES (pH 4.0), oxalic acid	pH 11.0, 2 M NaOH, 1 M Na$_2$CO$_3$, 25 mM EDTA	pH 7.4, Tween-20 (0.1%)
Time	27 min	NA	60 min	50 min	30 min	120 min	2–10 min	<15 s	25 min	7.5 min
Limit of detection	87.6 ng/mL	0.827 µg/mL	4 mg/mL	NA	0.01 µg/mL	3.09×10^{-11} g/L	NA	1.3×10^{-6} mol/L	0.45 µg/mL	NA

NA: not available.

Ersöz and colleagues prepared an imprinted QCM sensor using methacryloylaminoantipyrine-Fe(III) as a metal-chelating monomer to detect 8-hydroxy-2′-deoxyguanosine [110]. They used a photograft surface polymerization technique to synthesize the imprinted film by UV light. They determined the K_a value of the imprinted QCM sensor as 78,760 M^{-1}. They also analysed 8-hydroxy-2′-deoxyguanosine levels in the blood serum of a breast cancer patient with the imprinted QCM sensor and found a 8-hydroxy-2′-deoxyguanosine level of 0.297 µM could be detected in 20 min.

Diltemiz and her research group also developed QCM sensors to determine thymine by using methacryloylamidoadenine monomer [111]. They investigated the binding affinity of the QCM sensor by using the Langmuir isotherm and calculated K_a value as 1.0×10^5 M^{-1}. Their results showed that the QCM sensor had homogeneous binding sites for thymine. They also reported the selectivity of the QCM sensor according to single-stranded DNA, uracil and single-stranded RNA binding experiments. They obtained the initial slope for single-stranded DNA, uracil and single-stranded RNA as −9400 Hz mmol/L, −4700 Hz mmol/L and −2200 Hz mmol/L, while the initial slope of the curve was −11,800 Hz mmol/L.

Taghdisi et al. developed an electrochemical sensor that depended on the form of a dual-aptamer-complementary strand of aptamer conjugate, gold electrode and exonuclease I to detect myoglobin [112]. They observed a weak electrochemical signal in the absence of myoglobin due to the intact form. After the addition of myoglobin, they also observed a strong electrochemical signal because the dual-aptamer left the complementary strand of aptamer and bound to myoglobin. The electrochemical sensor displayed the high selectivity for myoglobin with low LOD value (27 pM).

Li and colleagues also fabricated an electrochemical sensor to detect myoglobin [113]. They built the electrochemical sensor by grafting myoglobin-binding-aptamer onto the surface of a gold nanoparticles/arginine-glycine-aspartic/carboxylatedgraphene/glassy carbon electrode. Their results showed the electrochemical sensor has good reproducibility and stability. Their electrochemical sensor was also used in biochemical assays with satisfactory results. All studies about nucleic acid detection were summarized in Table 5 according to the different parameters.

Table 5. Comparison of the sensor studies for nucleic acid detection.

Parameters	Reference				
	[109]	[110]	[111]	[112]	[113]
Target	Guanosine and guanine	8-Hydroxy-2-deoxyguanosine	Thymine	Myoglobin	Myoglobin
Linear dynamic range	20–400 µmol/L	0.1–1 mM	10–100 µM	0–80 nM	0.0001–0.2 g/L
Buffers (Ads, Des)	PBS, 0.1 M glycine–HCl (pH 2.2)	pH 10.0, 0.1 M Glycine–HCl	pH 7.4 HBS, 0.5 M NaCl	pH 7.4, NA	PBS, NA
Time	NA	30 min	60 min	45 min	120 min
Limit of detection	NA	0.0125 µM		27 pM	26.3 ng/mL

NA: not available.

3.5. Cell Imprinted Sensors

Recognition and isolation of cells have particular importance in clinical, diagnostic, environmental and security applications. Their size and density limit the use of traditional methods in complex sample mixtures. Molecularly imprinted sensors prepared with whole cells or epitopes provide rapid cell detection platforms [114].

Ren and co-workers covered glass with bacteria that was pushed into another glass coated with polydimethylsiloxane. As illustrated in Figure 18, the cell-imprinted polydimethylsiloxane created the resulting surface to favourably capture the imprinted bacteria. They also interpreted this result as powerful proof that chemical interaction acts a superior function in cell sorting with cell-imprinted polydimethylsiloxane films [115].

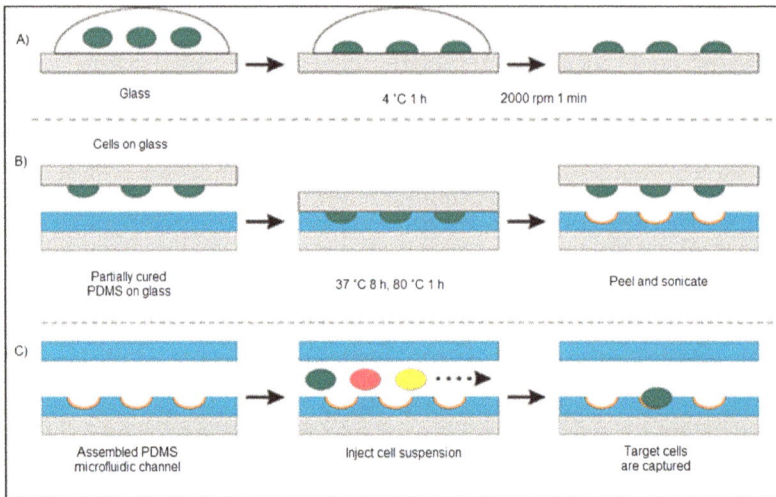

Figure 18. The schematic representation of the cell-imprinted polydimethylsiloxane process: (**A**) The template preparation; (**B**) polymer imprinting with the template and (**C**) cell sorting with the microfluidic chip [115].

Bers and co-workers synthesized a cell-specific surface-imprinted sensing system that was integrated with a heat transfer-based method to detect cells in cell mixtures [116]. They used a modified hamster ovarian cell line as a model. They also showed that their sensing system distinguished between different types of cells that only differ in the specific membrane protein expression. In addition, they disclosed that the detection sensitivity could be improved via exhibiting the sample surface and removing non-specific cells by purging between consecutive cell exposures.

Eersels et al. used an imprinting method with heat transfer resistance measurements to produce a cell-based sensor for macrophage and cancer cells detection [117]. Their approach was dependent on the difference between the heat transfer resistance at the port of the sensor influenced by the binding of a cell to the imprinted polymeric layer. They noted that the binding of cells caused in measurable extension of heat transfer resistance that indicated the cells acted as a thermally insulating layer. They calculated the LOD value as 10^4 cells/mL.

Mahmoudi et al. obtained cell-imprinted substrates based on mature and dedifferentiated chondrocytes [118]. They used rabbit adipose-derived mesenchymal stem cells seeded on these cell-imprinted substrates to adopt the specific shape and molecular characteristics of the cell types which had been used as a template for the cell-imprinting. Their data suggested that besides residual cellular fragments, which were presented on the template surface, the imprinted topography of the templates plays a role in the differentiation of the stem cells. All studies about cell detection were summarized in Table 6 according to the different parameters.

Table 6. Comparison of the sensor studies for cell detection.

Parameters	Reference			
	[115]	[116]	[117]	[118]
Target	*M. tuberculosis* bacilli	Chinese hamster ovarian cell	Macrophage and cancer cell	Mesenchymal stem cells
Linear dynamic range	NA	NA	10^4–10^6 cells/mL	30×10^3 cells
Buffers (Ads, Des)	pH 7.4, NA	pH 7.4, NA	pH 7.4, SDS (0.1%)	NA
Time	10 min	45 min	67 min	NA
Limit of detection	NA	NA	10^4 cells/mL	NA

NA: not available.

3.6. Bacteria Imprinted Sensors

There are a number of methods to identify bacteria in different media, however most of them require long incubation times and space for incubators. The need for an efficient, rapid, simple and cheap method for detection of microorganisms is fulfilled by bacteria-imprinted sensors [119]. İdil and collaborators prepared a label-free, selective and sensitive micro-contact-imprinted capacitive sensor for the detection of *Escherichia coli* (*E. coli*) [120]. After preparation of bacterial stamps, micro-contact *E. coli*-imprinted gold electrodes were prepared by using an amino acid-based recognition element, monomers and crosslinker under UV-polymerization (Figure 19). They performed real-time *E. coli* detection within the range of 1.0×10^2–1.0×10^7 CFU/mL and calculated the LOD value as 70 CFU/mL. They also detected *E. coli* with a recovery of 81–97% in samples, e.g. river water.

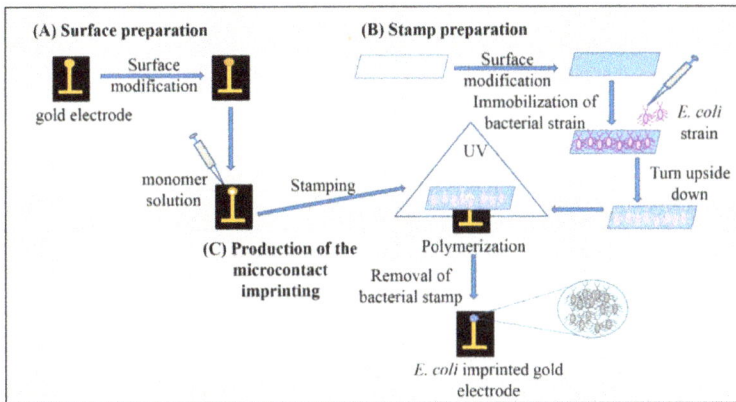

Figure 19. The scheme of the micro-contact imprinting of *E. coli* onto the polymer modified surfaces. The preparation of electrode surface (**A**), bacteria stamps (**B**), production of the micro-contact imprinting (**C**) [120].

As seen in Figure 20, Yılmaz et al. demonstrated a bacteria detection technique based on a micro-contact imprinting method on both SPR and QCM sensors [121]. *N*-methacryloyl-L-histidine methyl ester was employed in this study to obtain recognition comparable to that of natural antibodies. They also showed that the characterization of their SPR and QCM sensors by scanning electron microscopy. As seen in Figure 21, scanning electron microscopy images showed that SPR and QCM sensors had different sized imprinted cavities. They claimed that this might give rise to the random orientation of *E. coli* cells during fixation and local immersion distinctions during imprinting because of the rough glass slide surface.

Molecularly imprinted nanoparticles were visualized by Altintas and co-workers for endotoxin detection from *E. coli* with computational modeling [122]. Their process depended on the binding energy between endotoxin and each monomer. They showed that nano-MIPs were produced with functional groups to make the immobilization onto SPR sensor easy. The SPR surface could be regenerated more than 30 times without any significant loss in binding activity which made this method cost effective.

Hayden and Dickert developed artificial recognition sites for the monitoring of cells with moldable polymers [123]. They choose a mass-sensitive QCM sensor as transducer for the analysis due to its on-line acquisition capability and high sensitivity. The selectivity of the extremely hardy sensor permitted them to separate yeasts, Gram positive and negative bacteria. Their results encouraged them to enlarge the exploration of artificial interaction sites to the micrometer scale.

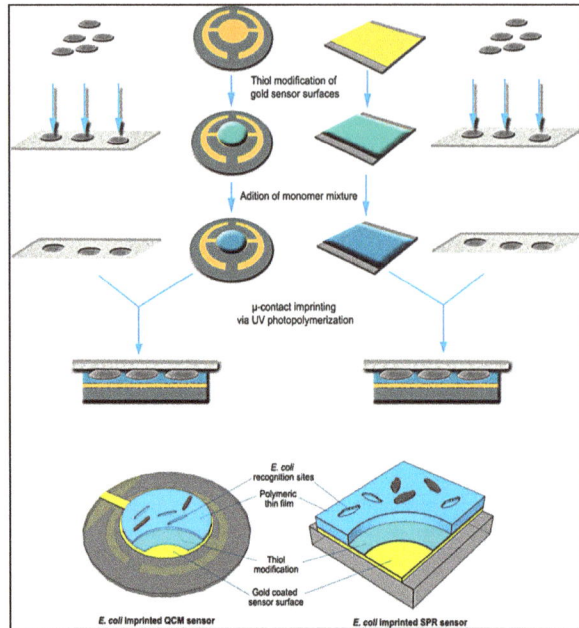

Figure 20. A schematic model of the imprinted SPR and QCM sensors [121].

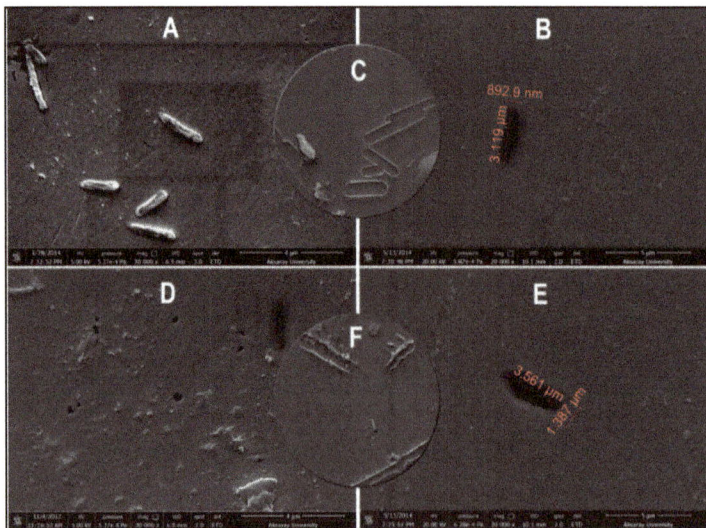

Figure 21. The surface morphologies of *E. coli*-imprinted (**A–C**) SPR and (**D–F**) QCM sensors [121].

Wan et al. designed a sensor that contains three quaternized magnetic nanoparticles fluorescent polymer systems to recognize and quantify bacteria [124]. The bacterial cell membranes disrupt the quaternized magnetic nanoparticle fluorescent polymer, generating a unique fluorescence response array. The response intensity of the array was dependent on the level of displacement determined by the relative quaternized magnetic nanoparticles fluorescent polymer binding strength and bacterial cells-magnetic nanoparticles interaction. Their approach has been used to measure bacteria within 20 min with an accuracy of 87.5% for 10^7 CFU/mL. Combined with UV-VIS measurements, the method can be successfully used to identify and detect eight different pathogen samples with an accuracy of 96.8%.

Qi et al. discussed the preparation of bacteria-based imprinted films on an impedimetric sensor [125]. They choose marine pathogen sulfate-reducing bacteria as a template. They deposited the chitosan doped with reduced graphene sheets, and helped the reduced graphene sheets-chitosan hybrid film as a platform for bacterial fixing. Furthermore, they also deposited a layer of chitosan film to embed the pathogen and then the bacterial template was washed out with acetone. All studies about bacteria detection were summarized in Table 7 according to the different parameters.

Table 7. Comparison of the sensor studies for bacteria detection.

Parameters	[120]	[121]	[122]	[123]	[124]	[125]
			Reference			
Target	*E. coli*	*E. coli*	Endotoxin from *E. coli*	*S. cerevisiae*	Bacteria	Marine pathogen sulfate-reducing bacteria
Linear dynamic range	1×10^2–1×10^7 CFU/mL	0.5–4.0 McFarland	4.4–5.3×10^{-10} M	2×10^7–5×10^9 cells/mL	10^7 cfu/mL	1.0×10^4–1.0×10^8 cfu/mL
Buffers (Ads, Des)	pH 7.4, NA	pH 7.4, Ethyl alcohol, 1 M lysozyme	HEPES (pH 7.0), NA	pH 6.0,	PBS, NA	0.2 M PBS buffer (pH 7.4), NA
Time	NA	7 min, 20 min	3.5 min	60 min	NA	6 min
Limit of detection	70 CFU/mL	3.72×10^5 CFU/mL, 1.54×10^6 CFU/mL	0.44 ng/mL	1.0×10^4 cells/mL	NA	0.7×10^4 cfu/mL

NA: not available.

3.7. Virus Imprinted Sensors

As viral contamination is a life-threatening issue, rapid and reliable detection of viruses as food borne pathogens, in the pharmaceutical industry and biological warfare conditions has crucial importance.

Altintas et al. presented a SPR-based microfluidics system to detect biological factors from water sources [126]. They used a new synthesis method (Figure 22) that was developed by the Cranfield University Biotechnology Group to obtain imprinted nanoparticles based on the existence of bacteriophage MS2 immobilized beads. They separated low and high affinity nanoparticles by using different temperatures and indicated the differences between high and low temperatures [127,128].

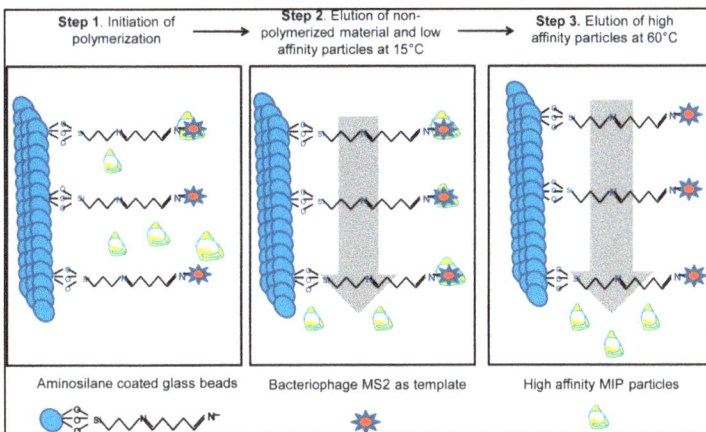

Figure 22. The fundamental of the synthesis by using a solid-phase method [126].

They initially coated the sensor with mercaptoundeconoic acid to acquire a self-assembled monolayer on the SPR surface. After that, they initiated the surface by using the mixture of N-(3-dimethylaminopropyl)-N′-ethyl carbodiimide and N-hydroxysuccinimide prior to the immobilization of imprinted nanoparticles (Figure 23). Their results showed that this method was really suitable to purify of water sources from biologically harmful factors such as viruses, bacteria, and fungi [126].

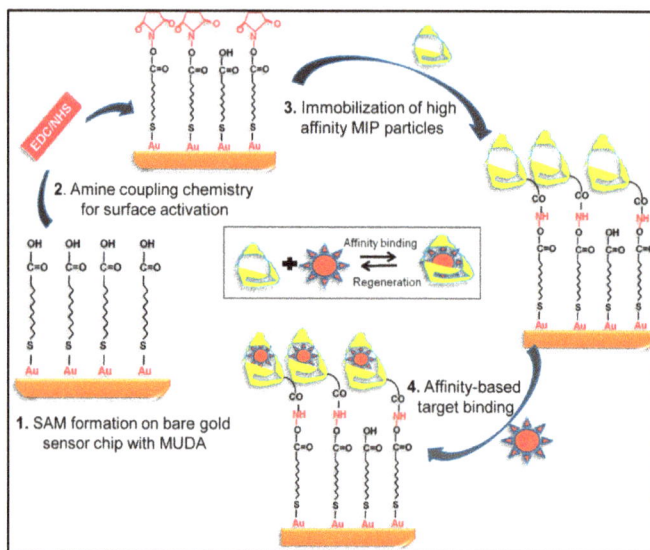

Figure 23. The SPR sensor for virus detection [126].

Jenik et al. adapted a sensor for detection of human rhinovirus and the foot-and-mouth disease virus [129]. Their procedures were guided by polyurethane layers that depict the geometrical properties of the template. Therefore, the imprinting process guided to an artificial antibody toward viruses, which did not only identify their receptor binding sites, but also discovered the whole virus as an entity.

Wangchareansak and collaborators applied molecular imprinting as a screening tool for diverse influenza subtypes [130]. MIPs for each of the subtypes led to a QCM sensor giving a LOD value as low as 10^5 particles/mL. Their analysis showed that the responses of the QCM sensor were correlated with the differences in hemagglutinin and neuraminidase patterns from databases.

The self-assembled monolayers were used to sense elements that were virions of poliovirus in a particular manner by Wang et al. [131]. They showed that these sensing elements are comprised of a gold-coated silicon chip with co-adsorbed hydroxyl-terminated alkanethiol molecules and template, where the thiol groups were bound to the substrate and self-assembled into extremely ordered monolayers. The molecules were extracted and the specific cavities were created in the monolayer matrix. All studies about virus detection were summarized in Table 8 according to the different parameters.

QCM and SPR sensors were developed to detect hemagglutinin which is a major protein of influenza A virus by Diltemiz et al. [132]. The authors modified QCM and SPR sensor surfaces with thiol groups and then a monomer mixture was immobilized onto sensor surfaces. They employed aqueous hemagglutinin solutions to determine the detection performance of the sensor for influenza A virus. They also calculated LOD values for QCM and SPR sensors as 4.7×10^{-2} µM, and 1.28×10^{-1} µM, in the 95% confidence interval.

Table 8. Comparison of the sensor studies for virus detection.

Parameters	Reference				
	[126]	[129]	[130]	[131]	[132]
Target	Waterborne virus	Picornavirus	Influenza A virus	Poliovirus	Influenza
Linear dynamic range	0.33–27 pmol	100–300 µg/mL	2.5×10^7–5×10^5 particles/mL	0–5000 $\times 10^8$ particles/mL	0.01–0.16 mM
Buffers (Ads, Des)	PBS, 0.1 M HCl and 20 mM NaOH	10 mM Tris (pH 7.2), NA	pH 7.2, NA	Dulbecco's PBS (pH 7.15), NA	HEPES (pH 7.4), Glycine/HCl (pH 2.0)
Time	8 min	75 min	80 min	60 min	25 min, 30 min
Limit of detection	5×10^6 pfu/mL	NA	10^5 particles/mL	NA	4.7×10^{-2} µM, 1.28×10^{-1} µM

NA: not available.

4. Conclusions

Molecular imprinting can be described as a method for creating synthetic polymers with bio-mimetic molecular recognition capability for diverse templates, where the template directs the positioning and orientation of the structural components via a cross-linking agent, followed by the elution of template to leave cavities which are complementary to the template molecules in shape, size and steric configuration. Macromolecules refer to molecules with molecular weights reaching tens of thousands of Dalton or more. Most macromolecules are made up of simple molecules. For example, the protein units are amino acids, and the nucleotide is the basic unit of nucleic acids. Detection of macromolecules has also been a great concern with the rapid development of biomedicine and proteomics. To establish a fast, simple, specific and high throughput detection approach for macromolecules has become one of the research focuses in analytical science. Compared to the small molecule imprinting, the macromolecular imprinting technique presents more difficulties and challenges. First of all, the elution and recognition of template molecules are the primary issues in the macromolecular imprinting technology because of low elution efficiency of the templates due to their large molecular size. Meanwhile, the highly cross-linked network structure usually limits the diffusion of the target molecule, leading to low binding capacity and long equilibration times. The sensors and recognition elements used in the sensor systems have an enormous impact on various biotechnology applications such as medicine, health-care, the pharmaceutical industry, environmental protection and monitoring, food analysis, defence and security. The combination of electronics and synthetic recognition elements prepared by molecularly imprinted polymers inspired by the natural mechanisms of antigen/antibody interactions has fuelled studies that encompass the development of novel, rapid, reliable, cheap, selective and sensitive diagnostic tools. In particular, the detection of macromolecules has required the development of several sensors that are based on molecularly imprinted polymers with sufficient sensitivity, selectivity and stability. The molecularly imprinted sensors have a "molecular key and lock" principle thus give the researchers the possibility to achieve fast, precise detection of target molecules. A huge mass of knowledge on molecularly imprinted sensors is now available addressing the practical needs for the detection of the broad spectrum of analytes, ranging from small molecules to large molecules like proteins, enzymes, nucleic acids and even whole cells, bacteria and viruses.

Hue et al. have addressed the perspectives of macromolecular imprinting sensors and their applications, including optical, electrochemical and mass-sensitive molecular imprinting sensors. In addition, the opportunities, challenges, and further research orientations of molecular imprinting sensors for macromolecules detection prospected. Difficulties and challenges arise from the drawbacks of unsatisfactory recognition efficiency due to relatively high mass transfer limitations, and the effective smaller size of the MIP was described by them. They presented a review giving a "visual" overview of the approaches to solving problems related to challenges of the bulk imprinting method and the emergence of surface imprinting and epitope imprinting techniques of the macromolecular imprinting methods to reduce the limitations of imprinting processes [133]. In addition, Iskierkoa and coworkers have focused on gathering, summarizing, and critically evaluating the results of the last decade on the separation and sensing of macromolecular compounds and microorganisms with the use of molecularly imprinted polymer synthetic receptors. Their article mainly considers chemical sensing of deoxyribonucleic acids, proteins and protein fragments as well as sugars and oligosaccharides. Moreover, it briefly discusses the fabrication of sensors for determination of bacteria and viruses that can ultimately be considered as extremely large macromolecules [134].

Molecularly imprinted polymers and their incorporation with various transducer platforms are among the most promising approaches for the detection of several macromolecules. The variety of molecular imprinting techniques used for the preparation of biomimetic sensors, including surface imprinting, micro-contact imprinting and epitope imprinting with various transducer platforms such as optical, electrochemical, impedance, current, potential and mass devices were overviewed for different macromolecule types, including enzymes, antibody/antigens, proteins, nucleic acids, cells, bacteria and viruses in detail in this review.

There is still a lack of good information on the molecular properties of molecularly imprinted polymers and also on means to improve the stability features of structures that are regarded as already very stable. A next step would be the further optimization of active surface and sensing platforms with uniform and easily reachable binding sites to decrease the difficulties in the accessibility of the binding site shape in the three dimensional polymer networks with increased mass transport resulting in adequate recognition properties and also better selectivity and sensitivity for a broader range of analytically significant molecules. The molecularly imprinted sensors are expected to create a revolution in researchers' understanding of usual sensor devices by providing a new generation of sensing materials.

Conflicts of Interest: The authors declare no conflict of interest.

References

1. Wulff, G.; Sarhan, A. Use of polymers with enzyme analogous structures for the resolution of racemates. *Angew. Chem. Int. Ed.* **1972**, *11*, 341–344.
2. Wulff, G.; Gross, T.; Schonfeld, R. Enzyme models based on molecularly imprinted polymers with strong esterase activity. *Angew. Chem. Int. Ed.* **1997**, *36*, 1962–1964. [CrossRef]
3. Algieri, C.; Drioli, E.; Guzzo, L.; Donato, L. Bio-mimetic sensors based on molecularly imprinted membranes. *Sensors* **2014**, *14*, 13863–13912. [CrossRef] [PubMed]
4. Verma, A.; Nakade, H.; Simard, J.M.; Rotello, V.M. Recognition and stabilization of peptide α-helices using templatable nanoparticle receptors. *J. Am. Chem. Soc.* **2004**, *126*, 10806–108074. [CrossRef] [PubMed]
5. Aubin-Tam, M.-E.; Hamad-Schifferli, K. Gold nanoparticle-cytochrome c complexes: The effect of nanoparticle ligand charge on protein structure. *Langmuir* **2005**, *21*, 12080–12084. [CrossRef] [PubMed]
6. Cabaleiro-Lago, C.; Quinlan-Pluck, F.; Lynch, I.; Lindman, S.; Minogue, A.M.; Thulin, E.; Walsh, D.M.; Dawson, K.A.; Linse, S. Inhibition of amyloid β protein fibrillation by polymeric nanoparticles. *J. Am. Chem. Soc.* **2008**, *130*, 15437–15443. [CrossRef] [PubMed]
7. Hoshino, Y.; Urakami, T.; Kodama, T.; Koide, H.; Oku, N.; Okahata, Y.; Shea, K.J. Design of synthetic polymer nanoparticles that capture and neutralize a toxic peptide. *Small* **2009**, *5*, 1562–1568. [CrossRef] [PubMed]
8. Yan, M.; Ramstrom, O. (Eds.) *Molecularly Imprinted Materials. Science and Technology*; CRC Press: Boca Raton, FL, USA, 2004.
9. Sellergren, B. (Ed.) *Molecularly Imprinted Polymers*; Elsevier: Amsterdam, The Netherlands, 2001.
10. Wulff, G. Molecular imprinting in cross-linked materials with the aid of molecular templates—A way towards artificial antibodies. *Angew. Chem. Int. Ed.* **1995**, *34*, 1812–1832. [CrossRef]
11. Haupt, K. Molecularly imprinted polymers: The next generation. *Anal. Chem.* **2003**, *75*, 376A–383A. [CrossRef] [PubMed]
12. Zimmerman, S.C.; Lemcoff, N.G. Synthetic hosts via molecular imprinting—Are universal synthetic antibodies realistically possible? *Chem. Commun.* **2004**, *10*, 5–14. [CrossRef] [PubMed]

13. Mosbach, K. The promise of molecular imprinting. *Sci. Am.* **2006**, *295*, 4, 86–91. [CrossRef]
14. Wackerlig, J.; Schirhagl, R. Applications of molecularly imprinted polymer nanoparticles and their advances toward industrial use: A review. *Anal. Chem.* **2016**, *88*, 250–261. [CrossRef] [PubMed]
15. Bole, A.L.; Manesiotis, P. Advanced materials for the recognition and capture of whole cells and microorganisms. *Adv. Mater.* **2016**, *28*, 5349–5366. [CrossRef] [PubMed]
16. Zeng, Z.; Hoshino, Y.; Rodriguez, A.; Yoo, H.; Shea, K.J. Synthetic polymer nanoparticles with antibody-like affinity for a hydrophilic peptide. *ACS Nano* **2010**, *4*, 199–204. [CrossRef] [PubMed]
17. Baggiani, C.; Giovannoli, C.; Anfossi, L.; Passini, C.; Baravalle, P.; Giraudi, G. A connection between the binding properties of imprinted and nonimprinted polymers: A change of perspective in molecular imprinting. *J. Am. Chem. Soc.* **2012**, *134*, 1513–1518. [CrossRef] [PubMed]
18. Asliyuce, S.; Uzun, L.; Rad, A.Y.; Unal, S.; Say, R.; Denizli, A. Molecular imprinting based composite cryogel membranes for purification of anti-hepatitis B surface antibody by fast protein liquid chromatography. *J. Chromatogr. B* **2012**, *889*, 95–102. [CrossRef] [PubMed]
19. Zhang, H.; Ye, L.; Mosbach, K. Non-covalent molecular imprinting with emphasis on its application in separation and drug development. *J. Mol. Recognit.* **2016**, *19*, 248–259. [CrossRef] [PubMed]
20. Kempe, M.; Mosbach, K. Molecular imprinting used for chiral separations. *J. Chromatogr. A.* **1995**, *694*, 3–13. [CrossRef]
21. Vidyasankar, S.; Arnold, F.H. Molecular imprinting: Selective materials for separations, sensors and catalysis. *Curr. Opin. Biotechnol.* **1995**, *6*, 218–224. [CrossRef]
22. Gupta, B.D.; Shrivastav, A.M.; Usha, S.P. Surface plasmon resonance-based fiber optic sensors utilizing molecular imprinting. *Sensors* **2016**, *16*, 1381. [CrossRef] [PubMed]
23. Meyer, E. Internal water molecules and H-bonding in biological macromolecules: A review of structural features with functional implications. *Protein Sci.* **1992**, *1*, 1543–1562. [CrossRef] [PubMed]
24. Karplus, M. Molecular dynamics of biological macromolecules: A brief history and perspective. *Biopolymers* **2003**, *68*, 350–358. [CrossRef] [PubMed]
25. Sali, A.; Glaeser, R.; Earnest, T.; Baumeister, W. From words to literatures in structural proteomics. *Nature* **2003**, *422*, 216–225. [CrossRef] [PubMed]
26. Guan, G.; Liu, B.; Wang, Z.; Zhang, Z. Imprinting of molecular recognition sites on nanostructures and its applications in chemosensors. *Sensors* **2008**, *8*, 8291–8320. [CrossRef] [PubMed]
27. Chen, L.; Xu, S.; Li, J. Recent advances in molecular imprinting technology: Current status challenges and highlighted applications. *Chem. Soc. Rev.* **2011**, *40*, 2922–2942. [CrossRef] [PubMed]
28. Xu, S.; Li, J.; Chen, L. Molecularly imprinted core–shell nanoparticles for determination of trace atrazine by reversible addition-fragmentation chain transfer surface imprinting. *J. Mater. Chem.* **2011**, *21*, 4346–4351. [CrossRef]
29. Saylan, Y.; Üzek, R.; Uzun, L.; Denizli, A. Surface imprinting approach for preparing specific adsorbent for IgG separation. *J. Biomat. Sci. Polym. Ed.* **2014**, *25*, 881–894. [CrossRef] [PubMed]
30. Tretjakov, A.; Syritski, V.; Reut, J.; Boroznjak, R.; Volobujeva, O.; Öpik, A. Surface molecularly imprinted polydopamine films for recognition of immunoglobulin G. *Microchim. Acta* **2013**, *180*, 1433. [CrossRef]
31. Songjun, L.; Shunsheng, C.; Whitcombe, M.J.; Piletsky, S.A. Size matters: Challenges in imprinting macromolecules. *Progress Polym. Sci.* **2014**, *39*, 145–163.
32. Menger, M.; Yarman, A.; Erdőssy, J.; Yildiz, H.B.; Gyurcsányi, R.E.; Scheller, F.W. MIPs and aptamers for recognition of proteins in biomimetic sensing. *Biosensors* **2016**, *6*, 35. [CrossRef] [PubMed]
33. Li, Y.; Yang, H.H.; You, Q.H.; Zhuang, Z.X.; Wang, X.R. Protein recognition via surface molecularly imprinted polymer nanowires. *Anal. Chem.* **2006**, *78*, 317–320. [CrossRef] [PubMed]
34. Ouyang, R.Z.; Lei, J.P.; Ju, H.X. Surface molecularly imprinted nanowire for protein specific recognition. *Chem. Commun.* **2008**, *44*, 5761–5763. [CrossRef] [PubMed]
35. Nematollahzadeh, A.; Sun, W.; Aureliano, C.S.A.; Lütkemeyer, D.; Stute, J.; Abdekhodaie, M.J.; Shojaei, A.; Sellergren, B. High-capacity hierarchically imprinted polymer beads for protein recognition and capture. *Angew. Chem. Int. Ed.* **2011**, *50*, 495–498. [CrossRef] [PubMed]
36. Bonini, F.; Piletsky, S.; Turner, A.P.F.; Speghini, A.; Bossi, A. Surface imprinted beads for the recognition of human serum albumin. *Biosens. Bioelectron.* **2007**, *22*, 2322–2328. [CrossRef] [PubMed]
37. Shiomi, T.; Matsui, M.; Mizukami, F.; Sakaguchi, K. A method for the molecular imprinting of hemoglobin on silica surfaces using silanes. *Biomaterials* **2005**, *26*, 5564–5571. [CrossRef] [PubMed]

38. Ugajin, H.; Oka, N.; Okamoto, T.; Kawaguchi, H. Polymer particles having molecule-imprinted skin layer. *Colloid Polym. Sci.* **2013**, *291*, 109–115. [CrossRef]

39. Inoue, J.; Ooya, T.; Takeuchi, T. Protein imprinted TiO$_2$-coated quantum dots for fluorescent protein sensing prepared by liquid phase deposition. *Soft Matter* **2011**, *7*, 9681–9684. [CrossRef]

40. Gao, R.X.; Kong, X.; Wang, X.; He, X.W.; Chen, L.X.; Zhang, Y.K. Preparation and characterization of uniformly sized molecularly imprinted polymers functionalized with core-shell magnetic nanoparticles for the recognition and enrichment of protein. *J. Mater. Chem.* **2011**, *21*, 17863–17871. [CrossRef]

41. Kan, X.W.; Zhao, Q.; Shao, D.L.; Geng, Z.R.; Wang, Z.L.; Zhu, J.J. Preparation and recognition properties of bovine hemoglobin magnetic molecularly imprinted polymers. *J. Phys. Chem. B* **2010**, *114*, 3999–4004. [CrossRef] [PubMed]

42. Moreira, F.T.C.; Dutra, R.A.F.; Noronha, J.P.C.; Cunha, A.L.; Sales, M.G. Artificial antibodies for troponin T by its imprinting on the surface of multi walled carbon nanotubes: Its use as sensory surfaces. *Biosens. Bioelectron.* **2011**, *28*, 243–250. [CrossRef] [PubMed]

43. Moreira, F.T.C.; Dutra, R.A.F.; Noronha, J.P.C.; Sales, M.G. Surface imprinting approach on screen printed electrodes coated with carboxylated PVC for myoglobin detection with electrochemical transduction. *Procedia Eng.* **2012**, *47*, 865–868. [CrossRef]

44. Moreira, F.T.C.; Sharma, S.; Dutra, R.A.F.; Noronha, J.P.C.; Cass, A.E.G.; Sales, M.G. Smart plastic antibody material (SPAM) tailored on disposable screen printed electrodes for protein recognition: Application to myoglobin detection. *Biosens. Bioelectron.* **2013**, *45*, 237–244. [CrossRef] [PubMed]

45. Fukazawa, K.; Ishihara, K. Fabrication of a cell-adhesive protein imprinting surface with an artificial cell membrane structure for cell capturing. *Biosens. Bioelectron.* **2009**, *25*, 609–614. [CrossRef] [PubMed]

46. Fukazawa, K.; Li, Q.; Seeger, S.; Ishihara, K. Direct observation of selective protein capturing on molecular imprinting substrates. *Biosens. Bioelectron.* **2013**, *40*, 96–101. [CrossRef] [PubMed]

47. Wang, S.; Ye, J.; Bie, Z.; Liu, Z. Affinity-tunable specific recognition of glycoproteins via boronate affinity-based controllable oriented surface imprinting. *Chem. Sci.* **2014**, *5*, 1135–1140. [CrossRef]

48. Sukumaran, P.; Vineesh, T.V.; Rajappa, S.; Lib, C.Z.; Alwarappan, S. Ionic liquid modified N-doped graphene as a potential platform for the electrochemical discrimination of DNA sequences. *Sens. Actuators B Chem.* **2017**, *247*, 556–563. [CrossRef]

49. Sener, G.; Ozgur, E.; Yılmaz, E.; Uzun, L.; Say, R.; Denizli, A. Quartz crystal microbalance based nanosensor for lysozyme detection with lysozyme imprinted nanoparticles. *Biosens. Bioelectron.* **2010**, *26*, 815–821. [CrossRef] [PubMed]

50. Li, F.; Li, J.; Zhang, S. Molecularly imprinted polymer grafted on polysaccharide microsphere surface by the sol-gel process for protein recognition. *Talanta* **2008**, *74*, 1247–1255. [CrossRef] [PubMed]

51. Bernard, A.; Delamarche, E.; Schmid, H.; Michel, B.; Bosshard, H.R.; Biebuyck, H. Printing patterns of proteins. *Langmuir* **1998**, *14*, 2225–2229. [CrossRef]

52. Kam, L.; Boxer, S.G. Cell adhesion to protein-micro patterned-supported lipid bilayer membranes. *J. Biomed. Mater. Res.* **2001**, *55*, 487–495. [CrossRef]

53. Csucs, G.; Michel, R.; Lussi, J.W.; Textor, M.; Danuser, G. Microcontact printing of novel co-polymers in combination with proteins for cell-biological applications. *Biomaterials* **2003**, *24*, 1713–1720. [CrossRef]

54. Shi, H.Q.; Ratner, B.D. Template recognition of protein-imprinted polymer surfaces. *J. Biomed. Mater. Res.* **2000**, *49*, 1–11. [CrossRef]

55. Shi, H.Q.; Tsai, W.B.; Garrison, M.D.; Ferrari, S.; Ratner, B.D. Template-imprinted nanostructured surfaces for protein recognition. *Nature* **1999**, *398*, 593–597. [PubMed]

56. Chou, C.; Rick, J.; Chou, T.C. C-reactive protein thin-film molecularly imprinted polymers formed using a micro-contact approach. *Anal. Chim. Acta* **2005**, *542*, 20–25. [CrossRef]

57. Lin, H.Y.; Hsu, C.Y.; Thomas, J.L.; Wang, S.E.; Chen, H.C.; Chou, T.C. The microcontact imprinting of proteins: The effect of cross-linking monomers for lysozyme ribonuclease A and myoglobin. *Biosens. Bioelectron.* **2006**, *22*, 534–543. [CrossRef] [PubMed]

58. Chen, Y.W.; Rick, J.; Chou, T.C. A systematic approach to forming micro-contact imprints of creatine kinase. *Org. Biomol. Chem.* **2009**, *7*, 488–494. [CrossRef] [PubMed]

59. Ertürk, G.; Hedström, M.; Tümer, M.A.; Denizli, A.; Mattiasson, B. Real time PSA detection with PSA imprinted (PSA-MIP) capacitive biosensors. *Anal. Chim. Acta* **2015**, *891*, 120–129. [CrossRef] [PubMed]

60. Osman, B.; Uzun, L.; Beşirli, N.; Denizli, A. Microcontact imprinted surface plasmon resonance sensor for myoglobin detection. *Mat. Sci. Eng. C* **2013**, *33*, 3609–3614. [CrossRef] [PubMed]

61. Rachkov, A.; Minoura, N. Recognition of oxytocin and oxytocin-related peptides in aqueous media using a molecularly imprinted polymer synthesized by the epitope approach. *J. Chromatogr. A* **2000**, *889*, 111–118. [CrossRef]

62. Rachkov, A.; Hu, M.; Bulgarevich, E.; Matsumoto, T.; Minoura, N. Molecularly imprinted polymers prepared in aqueous solution selective for [Sar^1,Ala^8]angiotensin II. *Analy. Chim. Acta* **2004**, *504*, 191–197. [CrossRef]

63. Rachkov, A.; Minoura, N.; Shimizu, T. Peptide separation using molecularly imprinted polymer prepared by epitope approach. *Anal. Sci.* **2001**, *17*, i609–i612.

64. Minoura, N.; Rachkov, A.; Higuchi, M.; Shimizu, T. Study of the factors influencing peak asymmetry on chromatography using a molecularly imprinted polymer prepared by the epitope approach. *Bioseparation* **2002**, *10*, 399–407. [CrossRef]

65. Ertürk, G.; Uzun, L.; Tümer, M.A.; Say, R.; Denizli, A. Fab fragments imprinted SPR biosensor for real-time human immunoglobulin G detection. *Biosens. Bioelectron.* **2011**, *28*, 97–104. [CrossRef] [PubMed]

66. Nihino, H.; Huang, C.S.; Shea, K.J. Selective proteins capture by epitope imprinting. *Ange. Chem.* **2006**, *45*, 2392–2396. [CrossRef] [PubMed]

67. Nicholls, I.A. Thermodynamic considerations for the design of and ligand recognition by molecularly imprinted polymers. *Chem. Lett.* **1995**, *24*, 1035–1036. [CrossRef]

68. Venton, D.L.; Gudipati, E. Influence of protein on polysiloxane polymer formation: Evidence for induction of complementary protein-polymer interactions. *Biochim. Biophys. Acta* **1995**, *1250*, 126–136. [CrossRef]

69. Burow, M.; Minoura, N. Molecular imprinting: Synthesis of polymer particles with antibody-like binding characteristics for glucose oxidase. *Biochem. Biophys. Res. Commun.* **1996**, *227*, 419–422. [CrossRef] [PubMed]

70. Hjerten, S.; Liao, J.L.; Nakazato, K.; Wang, Y.; Zamaratskaia, G.; Zhang, H.Y. Gels mimicking antibodies in their selective recognition of proteins. *Chromatographia* **1997**, *44*, 227–234. [CrossRef]

71. Hirayama, K.; Burow, M.; Morikawa, Y.; Minoura, N. Synthesis of polymer-coated silica particles with specific recognition sites for glucose oxidase by the molecular imprinting technique. *Chem. Lett.* **1998**, *27*, 731–732. [CrossRef]

72. Kempe, M.; Mosbach, K. Separation of amino acids, peptides and proteins on molecularly imprinted stationary phases. *J. Chromatogr.* **1995**, *691*, 317–323. [CrossRef]

73. Andersson, L.I.; Müller, R.; Vlatakis, G.; Mosbach, K. Mimics of the binding sites of opioid receptors obtained by molecular imprinting of enkephalin and morphine. *Proc. Natl. Acad. Sci. USA* **1995**, *92*, 4788–4792. [CrossRef] [PubMed]

74. Andersson, L.I.; Müller, R.; Mosbach, K. Molecular imprinting of the endogenous neuropeptide Leu5-enkephalin and some derivatives thereof. *Macromol. Rapid Commun.* **1996**, *17*, 65–71. [CrossRef]

75. Rachkov, A.; Minoura, N. Towards molecularly imprinted polymers selective to peptides and proteins. The epitope approach. *Biochim. Biophys. Acta* **2001**, *1544*, 255–266. [CrossRef]

76. Li, S.; Yang, K.; Deng, N.; Min, Y.; Liu, L.; Zhang, L.; Zhang, Y. Thermoresponsive epitope surface-imprinted nanoparticles for specific capture and release of target protein from human plasma. *ACS Appl. Mater. Interfaces* **2016**, *8*, 5747–5754. [CrossRef] [PubMed]

77. Guo, X. Surface plasmon resonance based biosensor technique: A review. *J. Biophoton.* **2012**, *5*, 483–501. [CrossRef] [PubMed]

78. Hernández-Ibáñez, N.; García-Cruz, L.; Montiel, V.; Foster, C.W.; Banks, C.E.; Iniest, J. Electrochemical lactate biosensor based upon chitosan/carbon nanotubes modified screen-printed graphite electrodes for the determination of lactate in embryonic cell cultures. *Biosens. Bioelectron.* **2016**, *77*, 1168–1174. [CrossRef] [PubMed]

79. Nascimento, N.M.; Juste-Dolz, A.; Grau-García, E.; Román-Ivorra, J.A.; Puchades, R.; Maquieira, A.; Moraisa, S.; Gimenez-Romero, D. Label-free piezoelectric biosensor for prognosis and diagnosis of Systemic Lupus Erythematosus. *Biosens. Bioelectron.* **2017**, *90*, 166–173. [CrossRef] [PubMed]

80. Tokel, O.; Yıldız, U.H.; Inci, F.; Durmus, N.G.; Ekiz, O.O.; Turker, B.; Cetin, C.; Rao, S.; Sridhar, K.; Natarajan, N.; et al. Portable microfluidic integrated plasmonic platform for pathogen detection. *Sci. Rep.* **2015**, *5*, 9152. [CrossRef] [PubMed]

81. Wang, X.; Ge, L.; Yu, Y.; Dong, S.; Li, F. Highly sensitive electrogenerated chemiluminescence biosensor based on hybridization chain reaction and amplification of gold nanoparticles for DNA detection. *Sens. Actuators B Chem.* **2015**, *220*, 942–948. [CrossRef]
82. Ji, J.; Gu, W.; Sun, C.; Sun, J.; Jiang, H.; Zhang, Y.; Sun, X. A novel recombinant cell fluorescence biosensor based on toxicity of pathway for rapid and simple evaluation of DON and ZEN. *Sci. Rep.* **2016**, *6*, 31270. [CrossRef] [PubMed]
83. Han, Q.; Wang, K.; Xu, L.; Yan, X.; Zhang, K.; Chen, X.; Wang, Q.; Zhang, L.; Pei, R. N-doped TiO$_2$ based visible light activated label-free photoelectrochemical biosensor for detection of Hg^{2+} through quenching of photogenerated electrons. *Analyst* **2015**, *140*, 4143–4147. [CrossRef] [PubMed]
84. Avci, O.; Lortlar Ünlü, N.; Yalçın Özkumur, A.; Ünlü, M.S. Interferometric reflectance imaging sensor (IRIS)—A platform technology for multiplexed diagnostics and digital detection. *Sensors* **2015**, *15*, 17649–17665. [CrossRef] [PubMed]
85. Nurul Najian, A.B.; Engku Nur Syafirah, E.A.R.; Ismail, N.; Mohamed, M.; Yean, C. Development of multiplex loop mediated isothermal amplification (m-LAMP) label-based gold nanoparticles lateral flow dipstick biosensor for detection of pathogenic *Leptospira*. *Anal. Chim. Acta* **2016**, *903*, 142–148. [CrossRef] [PubMed]
86. Yılmaz, F.; Saylan, Y.; Akgönüllü, S.; Çimen, D.; Derazshamshir, A.; Bereli, N.; Denizli, A. Surface plasmon resonance based nanosensors for detection of triazinic pesticides in agricultural foods. In *New Pesticides and Soil Sensors*; Elsevier: Cambridge, MA, USA, 2017; Volume 19, p. 679.
87. Justino, C.I.L.; Freitas, A.C.; Pereira, R.; Duarte, A.C.; Rocha Santos, T.A.P. Recent developments in recognition elements for chemical sensors and biosensors. *Trends Anal. Chem.* **2015**, *68*, 2–17. [CrossRef]
88. Rich, J.O.; Dordick, J.S. Imprinting enzymes for use in organic media. In *Methods in Biotechnology, Enzymes in Nonaqueous Solvents: Methods and Protocols*; Vulfson, E.N., Halling, P.J., Holland, H.L., Eds.; Humana Press Inc.: Totowa, NJ, USA, 2001; Volume 15.
89. Sener, G.; Uzun, L.; Say, R.; Denizli, A. Use of molecular imprinted nanoparticles as biorecognition element on surface plasmon resonance sensor. *Sens. Actuators B Chem.* **2011**, *160*, 791–799. [CrossRef]
90. Saylan, Y.; Yılmaz, F.; Derazshamshir, A.; Yılmaz, E.; Denizli, A. Synthesis of hydrophobic nanoparticles for real-time lysozyme detection using surface plasmon resonance sensor. *J. Mol. Recognit.* **2017**. [CrossRef] [PubMed]
91. Sunayama, H.; Takeuchi, T. Molecularly imprinted protein recognition cavities bearing exchangeable binding sites for post-imprinting site-directed introduction of reporter molecules for readout of binding events. *ACS Appl. Mater. Interfaces* **2014**, *6*, 20003–20009. [CrossRef] [PubMed]
92. Dibekkaya, H.; Saylan, Y.; Yılmaz, F.; Derazshamshir, A.; Denizli, A. Surface plasmon resonance sensors for real-time detection of cyclic citrullinated peptide antibodies. *J. Macromol. Sci. A Pure Appl. Chem.* **2016**, *53*, 585–594. [CrossRef]
93. Uzun, L.; Say, R.; Ünal, S.; Denizli, A. Production of surface plasmon resonance based assay kit for hepatitis diagnosis. *Biosens. Bioelectron.* **2009**, *24*, 2878–2884. [CrossRef] [PubMed]
94. Ertürk, G.; Özen, H.; Tümer, M.A.; Mattiasson, B.; Denizli, A. Microcontact imprinting based surface plasmon resonance (SPR) biosensor for real-time and ultrasensitive detection of prostate specific antigen (PSA) from clinical samples. *Sens. Actuators B Chem.* **2016**, *224*, 823–832. [CrossRef]
95. Uludag, Y.; Tothill, I.E. Cancer biomarker detection in serum samples using surface plasmon resonance and quartz crystal microbalance sensors with nanoparticle signal amplification. *Anal. Chem.* **2012**, *84*, 5898–5904. [CrossRef] [PubMed]
96. Chianella, I.; Guerreiro, A.; Moczko, E.; Caygill, J.S.; Piletska, E.V.; De Vargas Sansalvador, I.M.; Whitcombe, M.J.; Piletsky, S.A. Direct replacement of antibodies with molecularly imprinted polymer nanoparticles in ELISA—Development of a novel assay for vancomycin. *Anal. Chem.* **2013**, *85*, 8462–8468. [CrossRef] [PubMed]
97. Türkoğlu, E.A.; Yavuz, H.; Uzun, L.; Akgöl, S.; Denizli, A. The fabrication of nanosensor-based surface plasmon resonance for IgG detection. *Artif. Cells Nanomed. Biotechnol.* **2013**, *41*, 213–221. [CrossRef] [PubMed]
98. Gao, R.; Zhao, S.; Hao, Y.; Zhang, L.; Cui, X.; Liu, D.; Tang, Y. Facile and green synthesis of polysaccharide-based magnetic molecularly imprinted nanoparticles for protein recognition. *RSC Adv.* **2015**, *5*, 88436–88444. [CrossRef]

99. Moreira, F.T.C.; Sharma, S.; Dutra, R.A.F.; Noronha, J.P.C.; Cass, A.E.G.; Sales, M.G.F. Protein-responsive polymers for point-of-care detection of cardiac biomarker. *Sens. Actuators B Chem.* **2014**, *196*, 123–132. [CrossRef]

100. Chunta, S.; Suedee, R.; Lieberzeit, P.A. Low-density lipoprotein sensor based on molecularly imprinted polymer. *Anal. Chem.* **2016**, *88*, 1419–1425. [CrossRef] [PubMed]

101. Reddy, S.M.; Phan, Q.T.; El-Sharif, H.; Govada, L.; Stevenson, D.; Chayen, N.E. Protein crystallization and biosensor applications of hydrogel-based molecularly imprinted polymers. *Biomacromolecules* **2012**, *13*, 3959–3965. [CrossRef] [PubMed]

102. Bakhshpour, M.; Özgür, E.; Bereli, N.; Denizli, A. Microcontact imprinted quartz crystal microbalance nanosensor for protein C recognition. *Colloid Surfaces B* **2017**, *151*, 264–270. [CrossRef] [PubMed]

103. Wang, Z.; Li, F.; Xia, J.; Xia, L.; Zhang, F.; Bi, S.; Shi, G.; Xia, Y.; Liu, J.; Li, Y.; et al. An ionic liquid-modified graphene based molecular imprinting electrochemical sensor for sensitive detection of bovine hemoglobin. *Biosens. Bioelectron.* **2014**, *61*, 391–396. [CrossRef] [PubMed]

104. Wang, Y.T.; Zhou, Y.X.; Sokolov, J.; Rigas, B.; Levon, K.; Rafailovich, M. A potentiometric protein sensor built with surface molecular imprinting method. *Biosens. Bioelectron.* **2008**, *24*, 162–166. [CrossRef] [PubMed]

105. Moreira, F.T.C.; Dutra, R.A.F.; Noronha, J.P.C.; Sales, M.G.F. Myoglobin-biomimetic electroactive materials made by surface molecular imprinting on silica beads and their use as ionophores in polymeric membranes for potentiometric transduction. *Biosens. Bioelectron.* **2011**, *26*, 4760–4766. [CrossRef] [PubMed]

106. Çiçek, Ç.; Yılmaz, F.; Özgür, E.; Yavuz, H.; Denizli, A. Molecularly imprinted quartz crystal microbalance sensor (QCM) for bilirubin detection. *Chemosensors* **2016**, *4*, 21. [CrossRef]

107. Dechtrirat, D.; Jetzschmann, K.J.; Stöcklein, W.F.M.; Scheller, F.W.; Gajovic-Eichelmann, N. Protein rebinding to a surface-confined imprint. *Adv. Funct. Mater.* **2012**, *22*, 5231–5237. [CrossRef]

108. Nutiu, R.; Billen, L.P.; Li, Y. *Fluorescence-Signaling Nucleic Acid-Based Sensors, in Nucleic Acid Switches and Sensors*; Silverman, S.K., Ed.; Landes Bioscience and Springer Science & Business Media: Boston, MA, USA, 2006.

109. Diltemiz, S.E.; Denizli, A.; Ersöz, A.; Say, R. Molecularly imprinted ligand-exchange recognition assay of DNA by SPR system using guanosine and guanine recognition sites of DNA. *Sens. Actuators B Chem.* **2008**, *133*, 484–488. [CrossRef]

110. Ersöz, A.; Diltemiz, S.E.; Atılır Özcan, A.; Denizli, A.; Say, R. 8-OHdG sensing with MIP based solid phase extraction and QCM technique. *Sens. Actuators B Chem* **2009**, *137*, 7–11. [CrossRef]

111. Diltemiz, S.E.; Hür, D.; Ersöz, A.; Denizli, A.; Say, R. Designing of MIP based QCM sensor having thymine recognition sites based on biomimicking DNA approach. *Biosens. Bioelectron.* **2009**, *25*, 599–603. [CrossRef] [PubMed]

112. Taghdisi, S.M.; Danesh, N.M.; Ramezani, M.; Emrani, A.S.; Abnous, K. A novel electrochemical aptasensor based on Y-shape structure of dual-aptamer-complementary strand conjugate for ultrasensitive detection of myoglobin. *Biosens. Bioelectron.* **2016**, *80*, 532–537. [CrossRef] [PubMed]

113. Li, C.; Li, J.; Yang, X.; Gao, L.; Jing, L.; Ma, X. A label-free electrochemical aptasensor for sensitive myoglobin detection in meat. *Sens. Actuators B Chem.* **2017**, *242*, 1239–1245. [CrossRef]

114. Ren, K.; Zare, R.N. Chemical recognition in cell-imprinted polymers. *ACS Nano* **2012**, *6*, 4314–4318. [CrossRef] [PubMed]

115. Ren, K.; Banaei, N.; Zare, R.N. Sorting inactivated cells using cell-imprinted polymer thin films. *ACS Nano* **2013**, *7*, 6031–6036. [CrossRef] [PubMed]

116. Bers, K.; Eersels, K.; Grinsven, B.V.; Daemen, M.; Bogie, J.F.J.; Hendriks, J.J.A.; Bouwmans, E.E.; Püttmann, C.; Stein, C.; Barth, S.; et al. Heat-transfer resistance measurement method (HTM)-based cell detection at trace levels using a progressive enrichment approach with highly selective cell-binding surface imprints. *Langmuir* **2014**, *30*, 3631–3639. [CrossRef] [PubMed]

117. Eersels, K.; Grinsven, B.V.; Ethirajan, A.; Timmermans, S.; Monroy, K.L.J.; Bogie, J.F.J.; Punniyakoti, S.; Vandenryt, T.; Hendriks, J.J.A.; Cleij, T.J.; et al. Selective identification of macrophages and cancer cells based on thermal transport through surface-imprinted polymer layers. *ACS Appl. Mater. Interfaces* **2013**, *5*, 7258–7267. [CrossRef] [PubMed]

118. Mahmoudi, M.; Bonakdar, S.; Shokrgozar, M.A.; Aghaverdi, H.; Hartmann, R.; Pick, A.; Witte, G.; Parak, W.J. Cell-imprinted substrates direct the fate of stem cells. *ACS Nano* **2013**, *7*, 8379–8384. [CrossRef] [PubMed]

119. Starosvetsky, J.; Cohen, T.; Cheruti, U.; Bilanovic, D.; Armon, R. Effects of physical parameters on bacterial cell adsorption onto pre-imprinted sol-gel films. *J. Biomater. Nanobiotechnol.* **2012**, *3*, 499–507. [CrossRef]

120. Idil, N.; Hedström, M.; Denizli, A.; Mattiasson, B. Whole cell based microcontact imprinted capacitive biosensor for the detection of *Escherichia coli*. *Biosens. Bioelectron.* **2017**, *87*, 807–815. [CrossRef] [PubMed]
121. Yilmaz, E.; Majidi, D.; Ozgur, E.; Denizli, A. Whole cell imprinting based *Escherichia coli* sensors: A study for SPR and QCM. *Sens. Actuators B Chem.* **2015**, *209*, 714–721. [CrossRef]
122. Altintas, Z.; Abdin, M.J.; Tothill, A.M.; Karim, K.; Tothill, I.E. Ultrasensitive detection of endotoxins using computationally designed nanoMIPs. *Analy. Chim. Acta* **2016**, *935*, 239–248. [CrossRef] [PubMed]
123. Hayden, O.; Dickert, F.L. Selective microorganism detection with cell surface imprinted polymers. *Adv. Mater.* **2001**, *13*, 1480–1483. [CrossRef]
124. Wan, Y.; Sun, Y.; Qi, P.; Wang, P.; Zhang, D. Quaternized magnetic nanoparticles-fluorescent polymer system for detection and identification of bacteria. *Biosens. Bioelectron.* **2014**, *55*, 289–293. [CrossRef] [PubMed]
125. Qi, P.; Wan, Y.; Zhang, D. Impedimetric biosensor based on cell-mediated bioimprinted films for bacterial detection. *Biosens. Bioelectron.* **2013**, *39*, 282–288. [CrossRef] [PubMed]
126. Altintas, Z.; Gittens, M.; Guerreiro, A.; Thompson, K.A.; Walker, J.; Piletsky, S.; Tothill, I.E. Detection of waterborne viruses using high affinity molecularly imprinted polymers. *Anal. Chem.* **2015**, *87*, 6801–6807. [CrossRef] [PubMed]
127. Abdin, M.; Altintas, Z.; Tothill, I.E. In silico designed nanoMIP based optical sensor for endotoxins monitoring. *Biosens. Bioelectron.* **2015**, *67*, 177–183. [CrossRef] [PubMed]
128. Altintas, Z.; Guerreiro, A.; Piletsky, S.A.; Tothill, I.E. NanoMIP based optical sensor for pharmaceuticals monitoring. *Sens. Actuators B Chem.* **2015**, *213*, 305–313. [CrossRef]
129. Jenik, M.; Schirhagl, R.; Schirk, C.; Hayden, O.; Lieberzeit, B.; Blaas, D.; Paul, G.; Dickert, F.L. Sensing picornaviruses using molecular imprinting techniques on a quartz crystal microbalance. *Anal. Chem.* **2009**, *81*, 5320–5326. [CrossRef] [PubMed]
130. Wangchareansak, T.; Thititthanyanont, A.; Chuakheaw, D.; Gleeson, M.P.; Lieberzeit, P.A.; Sangma, C. Influenza A virus molecularly imprinted polymers and their application in virus sub-type classification. *J. Mater. Chem. B* **2013**, *1*, 2190–2197. [CrossRef]
131. Wang, Y.; Zhang, Z.; Jain, V.; Yi, J.; Mueller, S.; Sokolov, J.; Liuf, Z.; Levon, K.; Rigas, B.; Rafailovich, M.H. Potentiometric sensors based on surface molecular imprinting: Detection of cancer biomarkers and viruses. *Sens. Actuators B Chem.* **2010**, *146*, 381–387. [CrossRef]
132. Diltemiz, S.E.; Ersöz, A.; Hür, D.; Keçili, R.; Say, R. 4-Aminophenyl boronic acid modified gold platforms for influenza diagnosis. *Mater. Sci. Eng. C* **2013**, *33*, 824–830. [CrossRef] [PubMed]
133. Ma, X.-H.; LI, J.-P.; Wang, C.; Xu, G.-B. A review on bio-macromolecular imprinted sensors and their applications. *J. Anal. Chem.* **2016**, *44*, 152–159. [CrossRef]
134. Iskierko, Z.; Sharma, P.S.; Bartold, K.; Pietrzyk-Le, A.; Noworyta, K.; Kutner, W. Molecularly imprinted polymers for separating and sensing of macromolecular compounds and microorganisms. *Biotechnol. Adv.* **2016**, *34*, 30–46. [CrossRef] [PubMed]

MDPI

Article

Microcontact Imprinted Plasmonic Nanosensors: Powerful Tools in the Detection of *Salmonella paratyphi*

Işık Perçin [1], Neslihan Idil [1], Monireh Bakhshpour [2], Erkut Yılmaz [3], Bo Mattiasson [4,5] and Adil Denizli [2,*]

[1]   Department of Biology, Hacettepe University, 06800 Ankara, Turkey; ipercin@hacettepe.edu.tr (I.P.); nsurucu@hacettepe.edu.tr (N.I.)
[2]   Department of Chemistry, Hacettepe University, 06800 Ankara, Turkey; monir.b1985@gmail.com
[3]   Department of Biotechnology and Molecular Biology, Aksaray University, 68100 Aksaray, Turkey; yilmazerkut@aksaray.edu.tr
[4]   Department of Biotechnology, Lund University, 223 62 Lund, Sweden; bo.mattiasson@biotek.lu.se
[5]   CapSenze Biosystems AB, 223 63 Lund, Sweden
*   Correspondence: denizli@hacettepe.edu.tr; Tel.: +90-312-297-7963

Academic Editor: Giovanna Marrazza
Received: 12 April 2017; Accepted: 31 May 2017; Published: 13 June 2017

Abstract: Identification of pathogenic microorganisms by traditional methods is slow and cumbersome. Therefore, the focus today is on developing new and quicker analytical methods. In this study, a Surface Plasmon Resonance (SPR) sensor with a microcontact imprinted sensor chip was developed for detecting *Salmonella paratyphi*. For this purpose, the stamps of the target microorganism were prepared and then, microcontact *S. paratyphi*-imprinted SPR chips were prepared with the functional monomer N-methacryloyl-L-histidine methyl ester (MAH). Characterization studies of the SPR chips were carried out with ellipsometry and scanning electron microscopy (SEM). The real-time *Salmonella paratyphi* detection was performed within the range of 2.5×10^6–15×10^6 CFU/mL. Selectivity of the prepared sensors was examined by using competing bacterial strains such as *Escherichia coli*, *Staphylococcus aureus* and *Bacillus subtilis*. The imprinting efficiency of the prepared sensor system was determined by evaluating the responses of the SPR chips prepared with both molecularly imprinted polymers (MIPs) and non-imprinted polymers (NIPs). Real sample experiments were performed with apple juice. The recognition of *Salmonella paratyphi* was achieved using these SPR sensor with a detection limit of 1.4×10^6 CFU/mL. In conclusion, SPR sensor has the potential to serve as an excellent candidate for monitoring *Salmonella paratyphi* in food supplies or contaminated water and clearly makes it possible to develop rapid and appropriate control strategies.

Keywords: microcontact imprinting; SPR biosensor; N-methacryloyl-L-histidine methyl ester; *Salmonella paratyphi*

1. Introduction

Salmonella paratyphi (*S. paratyphi*) is known as one of the major globally distributed pathogenic bacteria and a leading cause of foodborne diseases [1]. These diseases represent a major threat due to their significantly increased incidence throughout the world [2]. Therefore, it is important to be able to detect pathogenic microorganisms in food to provide real-time quality results [3]. Rapid detection of these microorganisms is necessary in order to prevent the occurrence of foodborne diseases and when outburst of the disease appears, to reduce the spreading. Therefore it is important to develop strategies to ensure a safe food supply [4,5].

Conventional methods used to detect foodborne pathogenic bacteria are usually and laborious time consuming. Different experimental stages such as cultivation, biochemical identification and serological confirmation are all included in the routine procedures. Biochemical test kits, antibody-based methods, DNA/RNA-based assays and immunological methods have been developed and applied to detect foodborne pathogens [6]. There are a few developments presented nanosensor-based technologies, are exciting and promising developments since these detection tools have useful characteristics, making them suitable for quick assays combined with reliability and sensitivity [7–11].

Optical sensors are the most extensively used sensors for detecting foodborne pathogens. Use of such sensors offers several advantages, including high sensitivity and specificity, accuracy, relatively low cost, rapid response and portability [3]. In recent years, SPR sensors, the most frequently used optical sensors, have attracted particular and great attention for the detection of different target molecules such as prostate specific antigen, bisphenol-A and glucose [3,12–15].

These sensors make it possible to observe and quantify characteristics of biomolecular interactions on the surface of the sensor in real time with the advantage of label free detection [12]. Several assays to improve surface plasmon resonance (SPR)-based sensors for detecting *Salmonella sp.* have been presented in the literature [16–22]. Most assays were initially based on use of antibodies as recognition elements. In the last few years, molecularly imprinted polymers have been evaluated in connection to detection of microbial cells, and today one can state that they represent an applicable and efficient method in terms of creating specific and selective recognition sites for a target molecule in polymer matrices [23]. Functional and cross-linking monomers are used for co-polymerization together with the template molecule in order to form template-shaped three dimensional cavities [24]. Therefore, sensors relying on molecularly imprinted polymers (MIPs) offer several advantages such as high sensitivity, selectivity and portability, which make these polymers suitable to be applied in many fields [4,25,26]. To our knowledge there are no reports on assays based on whole cell imprinting related to the detection of *Salmonella paratyphi*.

In this study, a microcontact imprinted SPR biosensor for pathogenic bacteria, *S. paratyphi* was developed. Characterization studies were performed using ellipsometry and SEM. The selectivity of the resulting *S. paratyphi*-imprinted biosensors was examined by using other bacterial strains such as *Bacillus subtilis (B. subtilis)*, *Staphylococcus aureus (S. aureus)*, and *Escherichia coli (E. coli)*. The results obtained when using MIP and non-imprinted polymers (NIP) chips were compared in order to indicate the efficiency of the MIPs. In addition, real sample experiments with the microorganisms present in a real sample were performed with apple juice.

2. Materials and Methods

2.1. Materials

Strains obtained from the Culture Collection Laboratory of the Department of Biology Biotechnology Division, Hacettepe University (Turkey) were *S. paratyphi* ATCC 9150, *E. coli* ATCC 25922, *S. aureus* ATCC 25923, *B. subtilis* ATCC 23857. The following chemicals were obtained from Sigma Chemical Co. (St. Louis, MO, USA): allyl mercaptan, glutaraldehyde (50%, *w/v*), 3-amino-propyltriethoxysilane (APTES), 2-hydroxyethyl methacrylate (HEMA) and ethylene glycol dimethacrylate (EGDMA). Fluka (Buchs, Switzerland) provided α-α'-azoisobutyronitrile (AIBN). The aminoacid-modified acrylate N-methacryloyl L-histidine methyl ester (MAH) was supplied by Research Group Bioreg (Hacettepe University, Ankara, Turkey). All other chemicals were of analytical grade and purchased from Merck A.G. (Darmstadt, Germany).

2.2. Preparation of Bacteria

S. paratyphi, *E. coli*, *S. aureus* and *B. subtilis* strains were used in this study. The bacterial strains were inoculated into Luria-Bertani broth (100 mL in a 250 mL Erlenmeyer flask). After incubation at 37 °C for 18 h with constant shaking at 150 rpm, measurements of the viable counts were performed by serial 10-fold dilutions in sterile 10 mM phosphate buffered saline (PBS) (pH 7.4). Aliquots of each dilution (0.1 mL) were plated onto Tryptic Soy Agar plates in triplicate. The plates were incubated overnight at 37 °C and the colonies on the plate were counted. The concentration of bacteria was calculated in colony forming units per milliliter (CFU/mL). After incubation, one milliliter aliquot of each bacterial culture was centrifuged at 3300 g for 15 min at 4 °C and the culture supernatant was removed. Each bacterial pellet was washed with 1 mL sterile 10 mM PBS buffer (pH 7.4) by resuspending them in the buffer and then spin them down again for three times. The concentrated precipitate was resuspended in 1 mL sterile water.

2.3. Preparation of Microcontact S. paratyphi Imprinted SPR Chips

2.3.1. Preparation and Modification of the Glass Slides

Glass slides (26 × 75 mm) were cleaned for 5 min each in pure ethyl alcohol and then deionized water before treatment for 20 min in acidic Piranha solution (3:1, H_2SO_4/ H_2O_2, *v/v*). After washing with water the glass slides were dried with nitrogen gas. Chemical modification of the glass slides to introduce amino groups was performed with 3% APTES in toluene (*v/v*) for 2 h. Washing with toluene removed excess APTES before the glass plates were dried with nitrogen gas. The amino groups were derivatized by adding an excess of glutaraldehyde (3% *v/v*) in phosphate buffer (pH 7.4). After 2 h excess glutaraldehyde was removed by washing with phosphate buffer followed by wash with distilled water before being dried with nitrogen gas. Finally, the *S. paratyphi* cells (200 μL of a suspension of 0.5×10^8 CFU/mL) were added dropwise to the glass surface. The plates were left at room temperature over night for immobilization of the cells and subsequent drying. The slides were rinsed with deionized water and dried with nitrogen gas. They were kept at 4 °C in a closed Petri dish until use.

2.3.2. Modification of the SPR Chip Surfaces

SPR chips (GWC Technologies, S. Rosa Rd Madison, WI, USA) have a gold surface which was modified with allyl mercaptan (CH_2CHCH_2SH) according to Yılmaz et al. [27]. A solution of allyl mercaptan (3.0 M) was added dropwise to the SPR chips and incubated in a fume hood overnight. Excess allyl mercaptan was removed by washing with ethyl alcohol before the chips were dried in a vacuum oven (200 mmHg, 25 °C).

2.3.3. Microcontact Imprinting of *Salmonella paratyphi* onto the SPR Chips

The first step was to form a pre-polymerization complex between the monomers. MAH and $Cu(NO_3)_2 \cdot 2.5H_2O$ were mixed in the ratio of (1:1) for 1 h before a stock solution of HEMA (13 μL), EGDMA (40 μL), was mixed with the MAH-Cu(II) complex for 5 min. Then, the initiator AIBN was added into this stock solution. The SPR chip was placed horizontally and the monomer solution with the initiator was placed on the SPR chip surface. Theglass slide with immobilized *S. paratyphi* was brought into contact with the monomer solution on the chip. Polymerization was initiated by UV light (100 W and 365 nm) and lasted for 20 min under nitrogen atmosphere. Then the glass slide with the immobilized cells was removed and the sensor chip was cleaned with 10 mM phosphate buffer (pH 7.4). The chip was also treated with 10 mg/mL lysozyme solution (in PBS buffer, pH 7.4, 10 mM) for 30 min in order to remove any bacterial residues from surface of SPR chips.

2.4. Characterization of SPR Chips

Surface characterization of SPR chips were performed by observation using a JEM 1200 EX Scanning Electron Microscope (SEM, JEOL, Tokyo, Japan). The chip surface was cleaned with distilled water and then, dried with nitrogen gas and coated with Au/Pd. Ellipsometry measurements of the SPR chips' surfaces were performed using an auto-nulling imaging ellipsometer (Nanofilm EP3, Goettingen, Germany). A four-zone auto-nulling procedure integrating over a sample area of approximately 200 μm × 200 μm followed by a fitting algorithm has been carried out in order to analyse the SPR surface thickness. Phase models including air, polymeric film, gold, chromium and SF10 glass was assumed for SPR chips.

2.5. Real Time Salmonella paratyphi Detection

The reflectivity (ΔR) of light was influenced by binding of material to the sensor chip by SPR imager II system (GWC Technologies). *S. paratyphi* detection was carried out from aqueous bacterial suspension prepared in the concentration range of 2.5×10^6–15×10^6 CFU/mL. Ethyl alcohol (50%, *v/v*) and 10 mg/mL lysozyme solution were used in the regeneration step to disrupt the bacterial cell wall.

The first step involved equilibration of the sensor chip to 10 mM PBS (pH 7.4) for 200 s (flow rate: 150 µL/min). After reaching the stable resonance frequency, aqueous *S. paratyphi* suspensions (5 mL; 150 µL/min flow rate) were applied to the system and changes in reflectivity (ΔR%) were monitored online until the signal became stable (400 s). Ethyl alcohol (50%, *v/v*) and 10 mg/mL lysozyme solution (in pH 7.4, PBS, 100 mM) were applied to the SPR sensor system in order to remove bound bacterial cells from the polymer surface by breaking the interactions between the cell wall and the polymer (350 s). Lastly, reconditioning by washing away residues of the regeneration solution and products form the enzymatic degradation of the cell wall was performed with PBS (pH 7.4) for 200 s to make the system ready for new injections.

Control SPR chips with non-imprinted polymer (NIP) were also prepared according to the same protocol as that the glass slides used for the imprinting were the same chemistry as the glass plates used for production of MIPs, except that the NIP-related glass-plates did not carrying any immobilized *S. paratyphi*. The NIP chip was used to evaluate the non-specific effect of imprinting of the imprinting chemistry on the affinity of bacterial strains for the surface of the sensor chips without any selective cavities. The responses of these chips to the template and competitive bacterial strains were monitored continuously.

2.6. Selectivity of the Microcontact-Salmonella paratyphi Imprinted SPR Chip

Selectivity of SPR sensors was determined by their responses to *B. subtilis*, *S. aureus* and *E. coli* strains. *B. subtilis* and *S. aureus* were included due to the fact they have different cell wall structures. *E. coli* was included due preferred because of its similar cell wall structure and bacterial cellular morphology with *S. paratyphi*. The concentrations of each bacterial suspension were kept constant as 15×10^6 CFU/mL. The NIP-chips were applied in the same manner as mentioned above for determining the non-selective binding.

2.7. Real Sample Experiments and Reusability

Experiments were performed with real samples, in this case in apple juice. The juice was diluted 10 times with PBS (pH 7.4) and then, spiked with *S. paratyphi* in a range of concentrations (2.5; 5.0; 7.5; 10.0 $\times 10^6$ CFU/mL). *S. paratyphi* was detected repeatedly, using equilibration-injection-regeneration cycles for 5 times. The reusability of the system was examined by evaluating the change in reflectivity at repeated assays with the same concentration of *S. paratyphi* suspension (7.5×10^6 CFU/mL).

3. Results and Discussion

3.1. Characterization of Salmonella paratyphi-Imprinted SPR Chips

S. paratyphi imprinted SPR chips were prepared by microcontact imprinting method with the bacterial stamp and monomer mixture covering modified SPR chip surface (Figure 1). MAH (histidine containing specific monomer) was preferred as metal-complexing ligand in order to functionalize the polymer surface and generate specific recognition regions. By this way, selectivity was obtained towards some amino acids present on the cell wall. The interactions between MAH and the cell wall is via glutamic acid residues, diaminopimelic acid. Furthermore, if the cells carry some complex-bound divalent heavy metal ions, then interactions can be expected also there. The interactions between the imprinted cavities made for the target bacterial strain and the polymeric film resulted in high affinity of the bacterial cells for imprinted nano-cavities. It should be stressed that the concept of microcontact imprinting of particulate matter, e.g. bacterial cells, may lead to generation of recognition cavities with two different properties which both contribute to the efficiency in recognizing the target cells. The shape of the cavity is one clear contributing factor to the recognition of the cells and as a second factor, matching chemistry on the surface of the cavity which will selectively interact with the target structures. As is explained in the experimental part, there is a period for the monomers to bind to the surface of the cells before polymerization is initiated. During this period, a sterical arrangement takes place resulting in selective cavities. In connection to the polymerization these conditions are frozen and the MIP structure can after proper treatment start to bind complementary structures (cells of the same character as those used for imprinting). Upon removal of the print structure (the immobilized cells) one is left with a cavity with proper shape of the cells and with a chemistry of the surface of the cavity that matches structures on the cell wall of the target cells.

Molecularly imprinted polymers (MIPs) represent created artificial receptors and selectively recognition of the target molecule/cell with an efficiency similar to that of natural receptors. Natural receptors have the ability of recognizing the target molecules selectively, but less efficiently recognize by shape of particulate matter. The biological receptors are usually not stable under conditions outside the physiological range [28]. MIPs have great advantages over antibodies due to high stability, low cost and easy preparation [27]. The gold surfaces of SPR sensors were characterized by SEM and ellipsometry measurements. SEM analysis of the SPR sensor chips were performed by JEOL, JEM 1200 EX, (Tokyo, Japan). Bacteria-SEM images indicates the morphology of the *S. paratyphi* imprinted SPR chip surfaces. It is worth noticing that the shape of the cavities are very similar, even though the pictures are taken from different parts of a sensor chip (Figure 2). The thickness of imprinted and non-imprinted polymeric films on the SPR chip surface were determined with ellipsometry measurements as 88.7 ± 1.8 nm and 87.3 ± 0.7 nm, respectively (Figure 3A,B). That the polymer layers were formed on the sensor chips was established by SEM studies and ellipsometry contributed to determine the thickness of these layers.

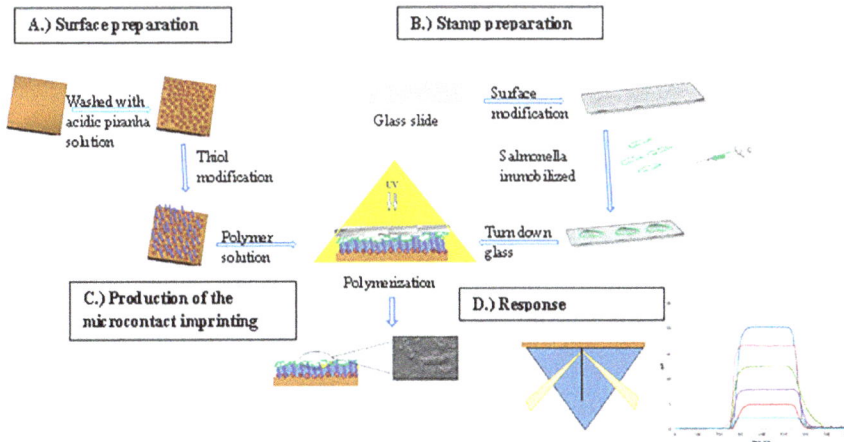

Figure 1. Schematic representation of microcontact imprinting of *S. paratyphi* onto the SPR chip. (**A**) preparation of SPR chip surface, (**B**) preparation of *S. paratyphi* stamps, (**C**) production of the microcontact imprinting and (**D**) response of the SPR sensor system.

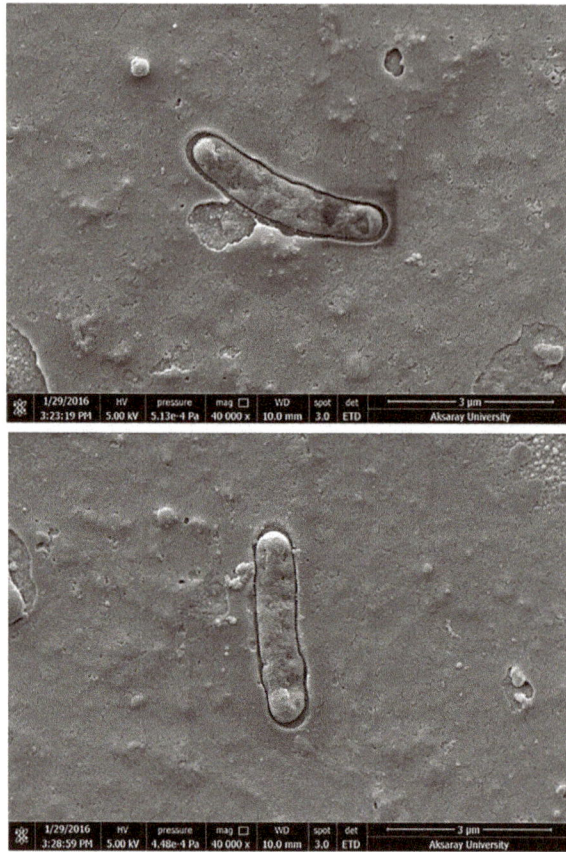

Figure 2. SEM images of microcontact *S. paratyphi* imprinted SPR chip surfaces indicating captured cells.

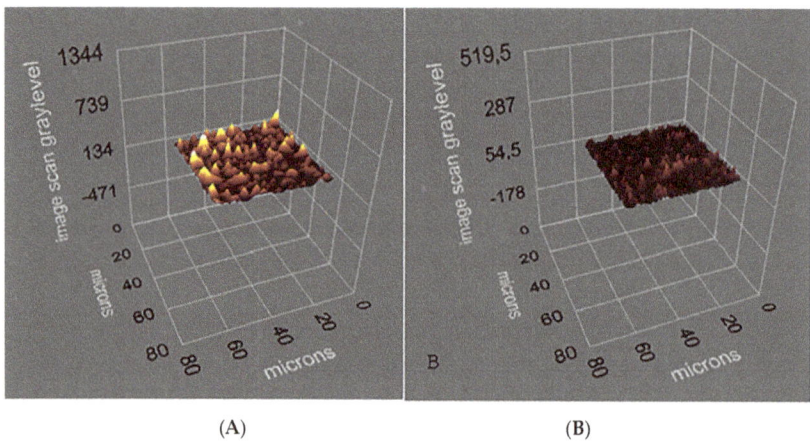

(A) (B)

Figure 3. Ellipsometry of (**A**) microcontact *S. paratyphi* imprinted and (**B**) non-imprinted SPR chip surfaces.

3.2. Real Time Detection of Salmonella paratyphi

In Figure 4A, the straight line indicates in the first 200 s indicated baseline. After *S. paratyphi* injection a change in (%ΔR) was monitored as a result of binding of *S. paratyphi* to the recognition cavities and the peak height indicated the amount of material binding of target microorganism. As can be seen from Figure 4B, an increase in response intensity (% resonance frequency shift (%ΔR)) in SPR was registered with the increasing concentrations of *S. paratyphi*.

S. paratyphi detection was performed with bacterial suspensions prepared in the concentration range of 2.5×10^6–15×10^6 CFU/mL and *S. paratyphi*-imprinted SPR sensor has a response with a linear relationship to the concentration of cells measured that fits the regression equation $y = 14.813x - 2.226$ ($R^2 = 0.9925$). The limit of detection (LOD) and the limit of quantification (LOQ) were determined to be 1.4×10^6 CFU/mL and 4.5×10^6 CFU/mL, based on IUPAC guidelines.

In the literature, there are many studies reporting the use of SPR sensors for the detection and quantification of microorganisms in buffer systems and simple matrices. In recent years, they have gained great attention in the fields of health science, drug discovery, diagnosis of infections, environmental and agricultural monitoring [26,29–31]. It has been proven that these sensors are also suitable for detecting target molecules/cells in complex media (e.g., blood, urine, stool, food, fruit juice) [32].

There are some challenges when using SPR sensors for quantifying cells, due to the size and morphology of the microorganisms. Their huge size leads to slow diffusion to the sensor surface and this could hamper the sensor response and limit detection capability [32].

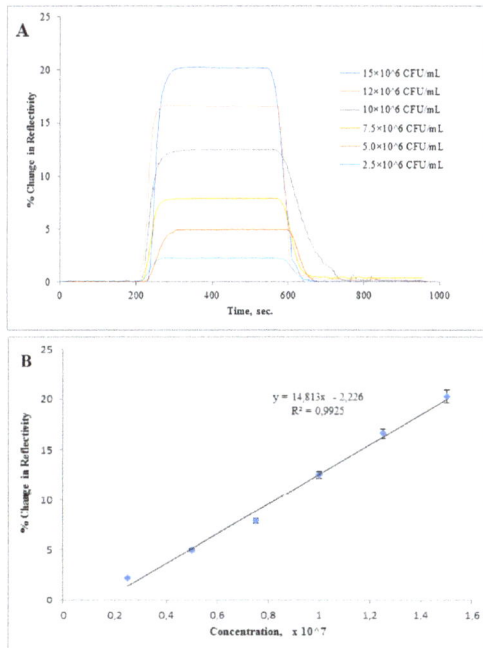

Figure 4. Real time responses of (**A**) microcontact *S. paratyphi* imprinted SPR sensor at different concentrations of *S. paratyphi*, (**B**) calibration curve of *S. paratyphi* obtained at a concentration of (0.25×10^7–1.5×10^7 CFU/mL) under experimental conditions, sample concentration: 1.5×10^7 CFU/mL, flow rate: 200 μL/min, running buffer: PBS buffer, regeneration buffer: 10 mg/mL lysozyme solution, 50% ethyl alcohol.

The pretreatment step during sample preparation has a significant impact on the detection of microorganisms with a SPR sensor. In a previous study, some treatment procedures using cell preparations such as living, heat killed, heat killed and soaked in 70% ethyl alcohol, or detergent- lysed cells were examined and LOD values were found to be respectively decreased. These differences in detection limits can be correlated to the effect of the treatments on size and morphology of the target cell [32,33].

A Biacore system was applied to indicate the detection of heat-killed *Salmonella* strains from groups A, B, D and E, according to Kauffmann-White typing. Antibodies were immobilized on a carboxymethylated Dextran chip by EDC/NHS coupling. A total of 53 different *Salmonella* serovars were detected at 10^7 CFU/mL. The detection limit in a suspension comprising half *S. enteritidis* and half *S. typhimurium* was demonstrated as 7×10^5 CFU/mL. In two different studies done by the same group, the detection of *S. typhimurium* [17] and *S. paratyphi* [18] was successfully performed. The technique used for *S. typhimurium* detection was similar to what was used when the group previously studied and application for the detection of *E. coli* O157:H7. In terms of *S. paratyphi* detection, the sensor surface was generated by capturing the antibody with self-assembled thiolated protein G. The detection range of these studies was shown as 10^2–10^7 CFU/mL. In another study, Koubova et al. used a custom-built SPR sensor for the detection of heat-killed and ethyl alcohol soaked *Salmonella enteritidis* [19]. The SPR chip surface was constructed by immobilizing antibodies raised against these bacterial strains. The LOD value was calculated as 1×10^6 CFU/mL.

Salmonella sp. are a widespread group of bacteria and food products, including pasteurised milk, can be contaminated by these strains. Therefore, it is necessary to detect these microorganisms rapidly since they are one of the main threats to food industry and of course to consumers of the food products. Mazumdar et al. used a cuvette based Plasmonic® SPR device in order to establish a fast and easy immunoassay for detecting *Salmonella typhimurium* in milk [34]. The method was set up as a sandwich model using a polyclonal antibody against the target microorganism. The proposed assay in this study was used to detect *S. typhimurium* in the concentration of 1.25×10^5 cells/mL both in milk and PBS buffer. It can be concluded that there were no differences in detection limits by changing the matrix. Among foodborne pathogens, *Salmonella* serotypes are the causative agents of salmonellosis. In addition to be a main threat to food industry as mentioned above, their detection in water is of great concern in public security. A look at other research reports, polyclonal anti-*Salmonella* antibody was used in order to recognize multiple *Salmonella* serovars simultaneously with the Plasmonic® SPR device. The detection limits were determined to be 2.5×10^5 cells/mL and 2.5×10^8 cells/mL for *S. typhimurium* and *S. enteritidis*, respectively. Milk spiked with both of these bacterial strains was used as a real sample to confirm the assay [20].

Our study pointed out that pathogenic microorganisms could be detected without using any antibody and microcontact imprinting has developed into a powerful tool to functionalize surfaces in order to quantify microorganisms. The potential use of microcontact imprinting in combination with SPR sensors provide unique properties. SPR sensors enable label-free, high sensitive and real time detection. These sensors have some additional advantages such as low volume sample requirement and quantitative analysis. When whole cell imprinting was taken into consideration along with the comparison of LOD values reported in literature, the LOD value obtained from SPR biosensor used in our study is among the lowest values in literature.

3.3. Selectivity of the Salmonella paratyphi Imprinted SPR Chip

The selectivity of generated SPR system was examined against *B. subtilis*, *S. aureus* and *E. coli* strains. Figure 5A shows the responses of sensor systems obtained from the application of all the tested bacterial strains to the microcontact-*S. paratyphi* imprinted SPR chips. As can be seen, ΔR values for competing bacterial strains were lower than that of *S. paratyphi*. Among these competing bacterial strains, the highest ΔR value was monitored after *E. coli* injection. *E. coli* belongs to the same family (Enterobacteriaceae) as *S. paratyphi* and has a similar size as *S. paratyphi*.

All tested bacterial strains other than *S. paratyphi* when introduced to the SPR chip caused low ΔR values in both MIP and NIP ones (Figure 5B). Higher responses against *S. paratyphi* indicated the unique characteristics of complementary cavities produced during the imprinting process. In comparison with non-imprinted ones, the responses against competitive bacterial strains were higher in imprinted sensors because of some similar features of the bacterial surfaces (Table 1).

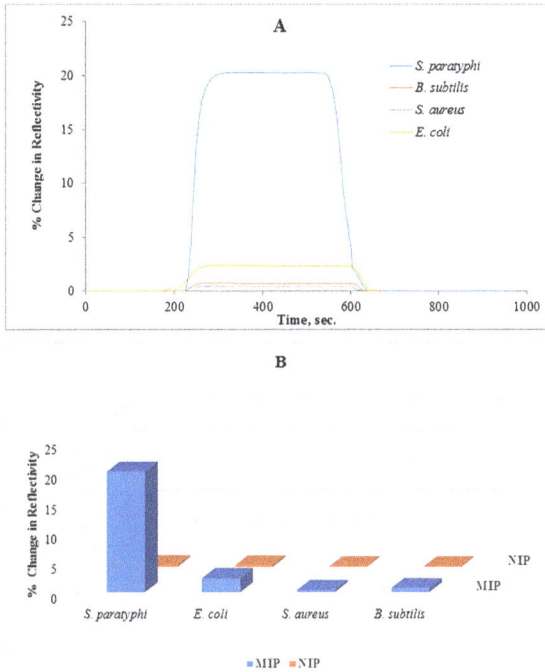

Figure 5. (**A**) Selectivity of microcontact-*S. paratyphi* imprinted SPR biosensor against competing bacterial strains; *Escherichia coli, Staphylococcus aureus, Bacillus subtilis*, (**B**) Imprinting efficiency of the microcontact-*S. paratyphi* imprinted SPR chips vs. non-imprinted SPR chips, experimental conditions; sample concentration: 1.5×10^7 CFU/mL, flow rate: 200 µL/min, running buffer: PBS buffer, regeneration buffer: 10 mg/mL lysozyme solution, 50% ethyl alcohol.

Table 1. Selectivity coefficients of *S. paratyphi*-MIP and NIP SPR chips (ΔR: SPR response for the *S. paratyphi*-MIP and NIP SPR chips, *k*: selectivity coefficient for *S. paratyphi* versus competing bacterial strains, *k'*: relative selectivity coefficient for *S. paratyphi*-MIP SPR chips versus NIP SPR chips).

Bacterial Strains	SPR Response, ΔR	SPR Response, ΔR	Selectivity Coefficient, *k*	Selectivity Coefficient, *k*	Relative Selectivity Coefficient, *k'*
	Imprinted	Non-imprinted	Imprinted	Non-imprinted	
S. paratyphi	20.25	0.25	-	-	-
S. aureus	0.4	0.28	50.63	0.89	56.70
E. coli	0.7	0.35	28.93	0.71	40.50
B. subtilis	2.3	0.35	8.80	0.71	12.33

3.4. Real Sample and Reusability Studies

Real sample experiments were carried out with apple juice which provides ambient media for bacterial growth. The apple juice samples were spiked with *S. paratyphi* at a range of concentrations

$(2.5 \times 10^6\text{--}10 \times 10^6$ CFU/mL). As seen in Figure 6A, the increase in concentration of *S. paratyphi* caused the increase in sensor response.

The reusability of the microcontact-*S. paratyphi* imprinted sensors was examined with apple juice spiked with *S. paratyphi*. Equilibration–adsorption–regeneration cycles were repeated for five times at the concentration of 7.5×10^6 CFU/mL (Figure 6B).

Idil et al. preferred both *E. coli* spiked $(1.0 \times 10^2\text{--}1.0 \times 10^4$ CFU/mL) apple juice and river water samples in real sample experiments to indicate the applicability of the microcontact-*E. coli* imprinted capacitive sensor [4]. In another study, apple juice was selected as real sample and apple juice samples spiked with *E. coli* at different concentrations in the range of 0.5–4.0 McFarland (approx. $1.5 \times 10^8\text{--}12 \times 10^8$ CFU/mL) were applied to the SPR system to confirm the developed sensor for *E. coli* detection [27]. Tokonami et al. used apple juice for real sample assays in order to verify if the generated system was able to detect target microorganisms at concentrations between 10^7 to 10^9 CFU/mL [35]. Son et al evaluated the feasibility of a miniature SPR sensor for detection of *Salmonella* enteritidis. Anti-*Salmonella* antibodies were attached on the SPR chip surface by using neutravidin. *Salmonella* was detected by SPR biosensor at concentrations down to 10^5 CFU/mL [36]. Generation of *E. coli* imprints was successful using ready-to use materials as well as ab initio synthesized polyurethanes [37]. Dilutions of *E. coli* suspensions, down to a limit of detection of 1.4×10^7 CFU/mL, were successfully measured using QCM.

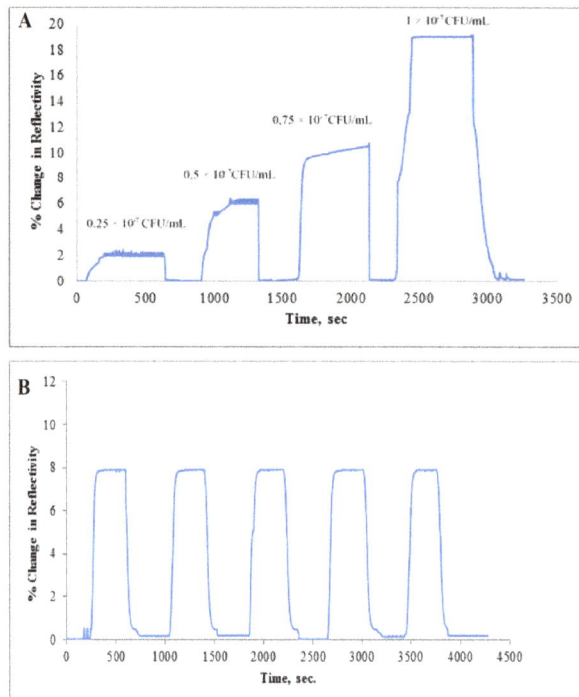

Figure 6. (**A**) Real-time responses of microcontact-*S. paratyphi* imprinted SPR chips against apple juice spiked with *S. paratyphi* at different concentrations in the range of $2.5 \times 10^6\text{--}10 \times 10^6$ CFU/mL, (**B**) Reusability of microcontact-*S. paratyphi* imprinted SPR chips at the concentration of 7.5×10^6 CFU/mL.

It is from the data presented in esp. Figure 6B quite obvious that one can use the same sensor chip repeatedly, provided proper regeneration is carried out between the assays. As discussed earlier in this paper, ethanol in combination with lysozyme turned out to be efficient so that the sensor chip after proper treatment could be reconditioned in PBS buffer before a new assay cycle started.

4. Conclusions

There is a growing need for selective recognition of microorganisms in complex samples due to the recognition of the importance of detection of microbial contaminants. Rapid, reliable, specific and cost-effective devices providing real-time screening are required. In this respect, SPR biosensors address these requirements with the advantage of biosensing in cell recognition. Physical properties, such as the shape, size and charge of the cell surface can influence cell recognition. These may gain advantage from chemical characteristics including complementary recognition molecules in order to increase selectivity. These two properties have to be taken into account for the development of successful devices. MIPs appear as promising tools employing both physical and chemical properties suitable for the development of successful (stable and sensitive) systems. In this study, the microcontact imprinting technique provided a simple surface patterning procedure combined with a SPR system. The created biosensing system had a good performance and gave sensitive and selective responses to the target bacteria at a concentration range from $(2.5 \times 10^6$–15×10^6 CFU/mL) with a CFU/mL detection limit and a linearity of $R^2 = 0.9925$. In conclusion, the proposed sensor technology has the potential to serve as an excellent candidate for monitoring *S. paratyphi* in contaminated water or food supplies and clearly indicates that it is possible to develop rapid and appropriate control strategies.

The present paper clearly illustrates the potential of combining microcontact MIPs since one can achieve both shape and chemical matching binding zones. In the future one can expect microcontact MIPs with even higher selectivity to be used. This can be done by reaching an even better matching between the chemical structure in the MIP cavities and the surface of the target cells.

Acknowledgments: We would like to thank Hacettepe University.

Author Contributions: I.P. conceived and designed the experiments, N.I., M.B. and E.Y. performed the experiments and analyzed the data, N.I. and M.B. wrote the paper, B.M commented on the work and updated the manuscript, A.D. supervised the overall work.

Conflicts of Interest: The authors declare no conflict of interest.

References

1. Zhao, Y.; Wang, H.; Zhang, P.; Sun, C.; Wang, X.; Wang, X.; Yang, R.; Wang, C.; Zhou, L. Rapid multiplex detection of 10 foodborne pathogens with an up-converting phosphor technology-based 10-channel lateral flow assay. *Sci. Rep.* **2016**, *6*, 21342. [CrossRef] [PubMed]
2. Ivnitski, D.; Abdel-Hamid, I.; Atanasov, P.; Wilkins, E. Biosensors for detection of pathogenic bacteria. *Biosens. Bioelectron.* **1999**, *14*, 599–624. [CrossRef]
3. Wang, Y.; Salazar, J.K. Culture-independent rapid detection methods for bacterial pathogens and toxins in food matrices. *Compr. Rev. Food Sci. Food Saf.* **2016**, *15*, 183–205. [CrossRef]
4. İdil, N.; Hedström, M.; Denizli, A.; Mattiasson, B. Whole cell based microcontact imprinted capacitive biosensor for the detection of *Escherichia coli. Biosens. Bioelectron.* **2017**, *87*, 807–815. [CrossRef] [PubMed]
5. Nguyen, P.D.; Tran, T.B.; Nguyen, D.T.X.; Min, J. Magnetic silica nanotube-assisted impedimetric immunosensor for the separation and label-free detection of *Salmonella typhimurium. Sens. Actuators B Chem.* **2014**, *197*, 314–320. [CrossRef]
6. Mandal, P.K.; Biswas, A.K.; Choi, K.; Pal, U.K. Methods for rapid detection of foodborne pathogens: An overview. *Am. J. Food Technol.* **2011**, *6*, 87–102. [CrossRef]
7. Zhao, X.; Lin, C.W.; Wang, J.; Oh, D.H. Advances in rapid detection methods for foodborne pathogens. *J. Microbiol. Biotechnol.* **2014**, *24*, 297–312. [CrossRef] [PubMed]
8. Zhang, H.; Zhang, Y.; Lin, Y.; Liang, T.; Chen, Z.; Li, J.; Yue, Z.; Lv, J.; Jiang, Q.; Yi, C. Ultrasensitive detection and rapid identification of multiple foodborne pathogens with naked eyes. *Biosens. Bioelectron.* **2015**, *71*, 186–193. [CrossRef] [PubMed]

9. Anderson, G.P.; Taitt, C.A.R. Biosensor-based detection of foodborne pathogens. In *Foodborne Pathogens: Microbiology and Molecular Biology*; Fratamico, P., Bhunia, A.K., Smith, J.L., Eds.; Caister Academic Press: Norfolk, UK, 2005.

10. Rasooly, A.; Herold, K.E. Biosensors for the analysis of food- and waterborne pathogens and their toxins. *J. AOAC Int.* **2006**, *89*, 873–883. [PubMed]

11. Geng, T.; Bhunia, A.K. Optical biosensors in foodborne pathogen detection. In *Smart Biosensor Technology*; Knopf, G.K., Bassi, A.S., Eds.; Taylor and Francis: Boca Raton, FL, USA, 2007.

12. Dey, D.; Goswami, T. Optical biosensors: A revolution towards quantum nanoscale electronics device fabrication. *Biomed. Res. Int.* **2011**. [CrossRef] [PubMed]

13. Yuan, Y.; Yang, X.; Gong, D.; Liu, F.; Hu, W.; Cai, W.; Huang, J.; Yang, M. Investigation for terminal reflection optical fiber SPR glucose sensor and glucose sensitive membrane with immobilized GODs. *Opt. Express* **2017**, *25*, 512–518. [CrossRef] [PubMed]

14. Shaikh, H.; Şener, G.; Memon, N.; Bhanger, M.I.; Nizamani, S.M.; Üzek, R.; Denizli, A. Molecularly imprinted surface resonance (SPR) based sensing of bisphenol A for its selective detection in aqueous systems. *Anal. Methods* **2015**, *7*, 4661–4670. [CrossRef]

15. Ertürk, G.; Özen, H.; Tümer, M.A.; Mattiasson, B.; Denizli, A. Microcontact imprinting based surface resonance (SPR) biosensor for real time and ultrasensitive detection of prostate specific antigen (PSA) from clinical samples. *Sens. Actuators B Chem.* **2016**, *224*, 823–832. [CrossRef]

16. Bokken, G.; Corbee, R.J.; van Knapen, F.; Bergwerff, A.A. Immunochemical detection of *Salmonella* group B, D and E using an optical surface plasmon resonance biosensor. *FEMS Microbiol. Lett.* **2003**, *222*, 75–82. [CrossRef]

17. Oh, B.K.; Kim, Y.K.; Park, K.W.; Lee, W.H.; Choi, J.W. Surface plasmon resonance immunosensor for the detection of *Salmonella typhimurium*. *Biosens. Bioelectron.* **2004**, *19*, 1497–1504. [CrossRef] [PubMed]

18. Oh, B.K.; Lee, W.; Kim, Y.K.; Lee, W.H.; Choi, J.W. Surface plasmon resonance immunosensor using self-assembled protein G for the detection of *Salmonella paratyphi*. *J. Biotechnol.* **2004**, *111*, 1–8. [CrossRef] [PubMed]

19. Koubova, V.; Brynda, E.; Karasova, L.; Skvor, J.; Homola, J.; Dostalek, J.; Tobiska, P.; Rosicky, J. Detection of foodborne pathogens using surface plasmon resonance biosensors. *Sens. Actuators B* **2001**, *74*, 100–105. [CrossRef]

20. Barlen, B.; Mazumdar, S.D.; Lezrich, O.; Kampfer, P.; Keusgen, M. Detection of *Salmonella* by surface plasmon resonance. *Sensors* **2007**, *7*, 1427–1446. [CrossRef]

21. Zhang, D.; Yan, Y.; Li, Q.; Yu, T.; Cheng, W.; Wang, L.; Ju, H.; Ding, S. Label-free and high-sensitive detection of *Salmonella* using a surface plasmon resonance DNA-based biosensor. *J. Biotechnol.* **2012**, *160*, 123–128. [CrossRef] [PubMed]

22. Nguyen, H.H.; Yi, S.Y.; Woubit, A.; Kim, M. A Portable surface plasmon resonance biosensor for rapid detection of *Salmonella typhimurium*. *Appl. Sci. Converg. Technol.* **2016**, *25*, 61–65. [CrossRef]

23. Bereli, N.; Andaç, M.; Baydemir, G.; Say, R.; Galaev, I.Y.; Denizli, A. Protein recognition via ion-coordinated molecularly imprinted supermacroporous cryogels. *J. Chromatogr. A* **2008**, *1190*, 18–26. [CrossRef] [PubMed]

24. Osman, B.; Uzun, L.; Beşirli, N.; Denizli, A. Microcontact imprinted surface resonance sensor for myoglobin detection. *Mater. Sci. Eng. C* **2013**, *33*, 3609–3614. [CrossRef] [PubMed]

25. Ertürk, G.; Hedström, M.; Tümer, A.; Denizli, A.; Mattiasson, B. Real time PSA detection with PSA imprinted (PSA-MIP) capacitive biosensors. *Anal. Chim. Acta* **2015**, *891*, 120–129. [CrossRef] [PubMed]

26. Bakhshpour, M.; Özgür, E.; Bereli, N.; Denizli, A. Microcontact imprinted quartz crystal microbalance nanosensor for protein C recognition. *Colloids Surf. B* **2017**, *151*, 264–270. [CrossRef] [PubMed]

27. Yılmaz, E.; Majidi, D.; Özgür, E.; Denizli, A. Whole cell imprinting based *Escherichia coli* sensors: A study for SPR and QCM. *Sens. Actuators B Chem.* **2015**, *209*, 714–721. [CrossRef]

28. Bole, A.L.; Manesiotis, P. Advanced materials for the recognition and capture of whole cells and microorganisms. *Adv. Mater.* **2016**, *28*, 5349–5366. [CrossRef] [PubMed]

29. Yılmaz, E.; Özgür, E.; Bereli, N.; Türkmen, D.; Denizli, A. Plastic antibody based surface plasmon resonance nanosensors for selective atrazine detection. *Mater. Sci. Eng. C* **2017**, *73*, 603–610. [CrossRef] [PubMed]

30. Saylan, Y.; Akgönüllü, S.; Çimen, D.; Derazshamshir, A.; Bereli, N.; Yılmaz, F.; Denizli, A. Surface plasmon resonance based nanosensors for detection of triazine pesticides. *Sens. Actuators B Chem.* **2017**, *241*, 446–454. [CrossRef]

31. Atay, S.; Pişkin, K.; Yılmaz, F.; Çakır, C.; Yavuz, H.; Denizli, A. Quartz crystal microbalance based biosensors for detecting highly metastatic breast cancer cells. *Anal. Methods* **2016**, *8*, 153–161. [CrossRef]

32. Taylor, A.D.; Yu, Q.; Chen, S.; Homola, J.; Jiang, S. Comparison of *E. coli* O157:H7 preparation methods used for detection with surface plasmon resonance sensor. *Sens. Actuators B Chem.* **2005**, *107*, 202–208. [CrossRef]

33. Taylor, A.D.; Ladd, J.; Chen, S.; Homola, J.; Jiang, S. Quantitative and simultaneous detection of four foodborne bacterial pathogens with a multi-channel SPR sensor. *Biosens. Bioelectron.* **2006**, *22*, 752–758. [CrossRef] [PubMed]

34. Mazumdar, S.D.; Hartmann, M.; Kampfer, P.; Keusgen, M. Rapid method for detection of *Salmonella* in milk by surface plasmon resonance (SPR). *Biosens. Bioelectron.* **2007**, *22*, 2040–2046. [CrossRef] [PubMed]

35. Tokonami, S.; Nakado, Y.; Takahashi, M.; İkemizu, M.; Kadoma, T.; Saimatsu, K.; Dung, L.Q.; Shilgi, H.; Nagaoka, T. Label-free and selective bacteria detection using a film with transferred bacterial configuration. *Anal. Chem.* **2013**, *85*, 4925–4929. [CrossRef] [PubMed]

36. Son, J.R.; Kim, G.; Kothapalli, A.; Morgan, M.T.; Ess, D. Detection of *Salmonella enteridis* using a miniature optical surface plasmon resonance biosensor. *J. Phys.* **2007**, *61*, 1086–1090.

37. Poller, A.M.; Spieker, E.; Lieberzeit, P.A.; Preininger, C. Surface imprints: Advantageous application of ready2use materials for bacterial quartz-crystal microbalance sensors. *Appl. Mater. Interf.* **2017**, *9*, 1129–1135. [CrossRef] [PubMed]

sensors

MDPI

Review

Gas Sensors Based on Molecular Imprinting Technology

Yumin Zhang [1,2], Jin Zhang [2,*] and Qingju Liu [2,*]

[1] School of Physics and Astronomy, Yunnan University, 650091 Kunming, China; Zhang_Yumin88@163.com
[2] School of Materials Science and Engineering, Yunnan Key Laboratory for Micro/Nano Materials & Technology, Yunnan University, 650091 Kunming, China
* Correspondence: zhj@ynu.edu.cn (J.Z.); qjliu@ynu.edu.cn (Q.L.); Tel.: +86-871-6503-2713 (Q.L.)

Received: 21 March 2017; Accepted: 29 June 2017; Published: 4 July 2017

Abstract: Molecular imprinting technology (MIT); often described as a method of designing a material to remember a target molecular structure (template); is a technique for the creation of molecularly imprinted polymers (MIPs) with custom-made binding sites complementary to the target molecules in shape; size and functional groups. MIT has been successfully applied to analyze; separate and detect macromolecular organic compounds. Furthermore; it has been increasingly applied in assays of biological macromolecules. Owing to its unique features of structure specificity; predictability; recognition and universal application; there has been exploration of the possible application of MIPs in the field of highly selective gas sensors. In this present study; we outline the recent advances in gas sensors based on MIT; classify and introduce the existing molecularly imprinted gas sensors; summarize their advantages and disadvantages; and analyze further research directions.

Keywords: gas sensor; molecular imprinting technology; quasi-molecular imprinting technology

1. Introduction

1.1. Significance of Gas Sensors

In the past few decades, the air quality of the globe has been declining, and many people die each year from indoor air pollution. The second United Nations Environment Assembly pointed that there are many people dying each year from indoor air pollution. Usually, the indoor air pollutants are composed of a series of volatile organic compounds (VOCs), which are generally harmful to human body. Those with low molecular weights (less than 100 Da) are particularly toxic, such as acetone [1], benzene [2], methanol [3,4], formaldehyde [5,6] and so on. Therefore, an accurate way to detect the toxic air pollutants is needed urgently.

The traditional detection methods include spectrophotometry [7], chromatography [8], electrochemical methods [9], catalytic luminescence methods [10] and gas sensors [11]. Among them, the spectrophotometric method has the advantages of fast detection and low cost. However, this method requires a professional spectrophotometer, cannot be commercialized and needs extensive application. At the same time, there are also disadvantages, such as low accuracy, strict preparation of reagents and being easily affected by other factors (such as temperature and time of detection). The chromatography method is precise and fast, but there are also problems. This method needs special equipment that is expensive, in addition to having a large volume and high cost. It is also hard to separate isomers of the target reagent using chromatography. The electrochemical methods possess good stability and sensitivity, but the electrochemical sensors are more expensive. Furthermore, it has a short lifespan, and the detection process is susceptible to interference. The catalytic luminescence method, developed in recent years, is a new method for detection of toxic gases. Although the detection is of high sensitivity and good selectivity, the wide application of this method is restricted due to the complex equipment required and its high cost. Gas sensor methods are used to detect toxic gases with a high sensitivity and simple operations. The gas sensors are small devices available at a low cost, being suitable for real time monitoring and useful for detection of indoor air pollutants. Among many types of gas sensors, oxide semiconductor gas sensors are the mainstream products. The semiconductor gas sensor has been widely favored for the past twenty years due to its high sensitivity, stable performance, low price, small size, ease of use and so on. However, despite all these advantages, the semiconductor gas sensor still has some drawbacks, such as non-ideal selectivity for certain gases. When molecular imprinting technology (MIT) is used in gas sensors, there is an improvement in selectivity.

1.2. Molecular Imprinting Technology in Gas Sensors

How can MIT improve selectivity? We should first learn about the features of MIT, which first appeared in 1930s. MIT provides a straightforward route for creating binding sites for a desired template, comparable to those of natural antibodies [12,13]. During the fabrication process, the template and functional monomers first form a self-assembled complex by covalent or noncovalent bonds where the functional monomers surround the template. Polymerization is then processed to support this self-organized configuration in place, followed by removal of the imprinted template from the polymeric networks, which thereby leaves behind binding sites complementary to the template. In this way, the match between the template and binding sites constitutes an induced molecular memory, which makes the prepared imprinted polymers capable of recognizing the templates as illustrated in Figure 1. In comparison with antibodies, molecularly imprinted polymers (MIPs) have the merits of easy preparation, reusability, and robustness for chemical and physical stresses. The molecularly imprinted gas sensors (MIGS) can be used under harsh conditions, including elevated temperature and pressure, presence of metal ions and organic solvent. Furthermore, MIGS can be fabricated using standard microchip fabrication protocols, such as photolithography, and have an extended lifetime. MIGS could benefit from the new development of nanosized-imprinted materials ("plastic antibodies"). The nanosized-imprinted materials can significantly improve binding kinetics, which represents a challenge in the development of sensors for real-time analysis. Furthermore, it would be a straightforward application in microfluidic devices. The essentials of MIT include functional monomers, interactions, initiator and elution [14,15]. As discussed above, molecular imprinting has the features of conformation reservation, recognition specificity, environmental tolerance and reusability, which have laid a solid foundation for the application of MIT in the fields of gas sensors [16].

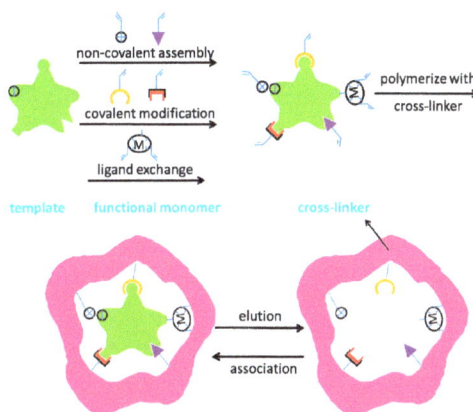

Figure 1. Schematic diagram of molecular imprinting process.

1.3. Synthesis Strategies of Molecularly Imprinted Gas Sensors

The synthesis of MIGS can be divided into two categories: the molecular imprinting method, and the quasi-molecular imprinting method.

1.3.1. Molecular Imprinting Method

The methods in this section contain all the reagents needed to be used in preparing the MIPs: template molecule, functional monomer, cross-linker, initiator and solvent. Functional monomers and a mass of cross-linkers are typically polymerized in the presence of the template molecules. Subsequently, the templates were removed from the polymeric matrix, which leaves behind some cavities. There are binding sites inside the cavities that are complementary to the template molecules in size, shape and functionality [17–20]. During these processes, the specific groups of functional monomers are oriented toward the desired binding sites within the cavities according to the structure, size and shape of the template [21]. In this manner, the preparation of MIPs is based on three steps: (1) the pre-interaction between the template molecule and the functional monomer(s); (2) the formation of a rigid polymeric matrix around the template molecule and the functional monomer(s) with a considerable number of cross-linkers; and (3) the removal of the template molecule [22,23].

Gas sensors in this category possess good stability and good selectivity. However, they have a long response and recovery time, due to the binding/elution process between the polymer matrix and the target gas molecules (the template molecules).

1.3.2. Quasi-Molecular Imprinting Method

This sort of method is based on the MIT concept, although it does not strictly follow the process. Essentially, there is no exact functional monomer, initiator or cross-linker. The samples are prepared following normal ways, such as the sol–gel method or hydrothermal method. The samples are just synthesized or dried in a specific atmosphere formed with the target gases to make the sample recognize the target gases to enhance the selectivity.

2. Molecularly Imprinted Gas Sensors

2.1. Molecularly Imprinted Organic Gas Sensors

MIT has usually been used in the recognition or separation of biomacromolecules. Studies for recognition of toxic gases started in the early 21st century. MIT combined with quartz crystal microbalance (QCM) could potentially improve both the selectivity and sensitivity of gas sensors, especially the sensitivity. The limit of detection of QCM-based sensors can be less than other sensors fabricated with the same materials. Hu et al. [24] reported that a sensor fabricated by the piezoelectric method combined with MIT could selectively detect formaldehyde molecules. The idea was based on combining the tiny mass detection of QCM with the MIPs. Noncovalent MIP production was utilized in this experiment. Formaldehyde was used as a template, while methacrylic acid (MAA), ethylene glycol dimethacrylate (EGDMA), 2,2'-azobis(2,4-dimethyl)valeronitrile (AMVN) and toluene were used as the functional monomer, cross-linker, initiator and solvent, respectively. The schematic diagram of the sensor setup is shown in Figure 2. The sensor performance (sensitivity or selectivity) was characterized by resonance frequency. In terms of the interaction between molecularly imprinted binding sites and template, the selectivity of molecularly imprinted sample to formaldehyde was partially enhanced compared to the non-imprinted ones, as shown in Figure 3.

Matsuguchi et al. [25] used toluene and *p*-xylene as both template and solvent. This study used MAA, divinyl-benzene and benzoyl peroxide as the functional monomer, cross-linker and initiator, respectively, to fabricate a molecularly imprinted QCM-based toluene/*p*-xylene sensor. The gas measure setup is shown in Figure 4. The response to toluene or *p*-xylene and selectivity of the sensor was determined by the amount of absorbed toluene/*p*-xylene ($\triangle W(T)$, Figure 5). The results show that the device exhibits a good selectivity and high response to target gases, indicating that MIT could improve the selectivity of sensors to certain gases.

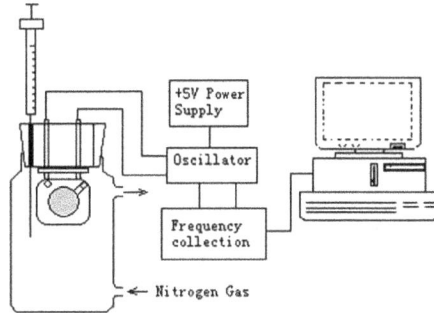

Figure 2. The schematic diagram of the sensor setup [24].

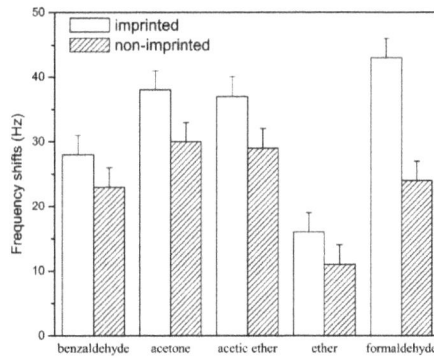

Figure 3. The resonance frequency response of the sensors to several analytes [24].

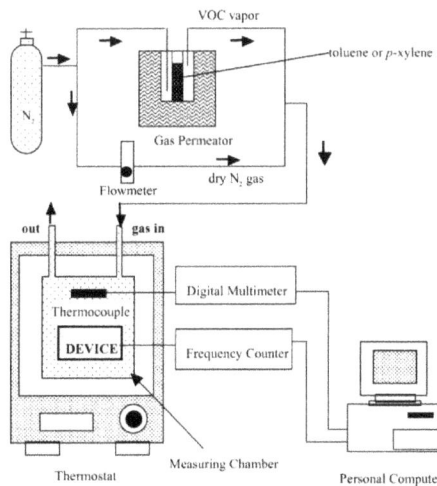

Figure 4. Experimental setup used to measure the volatile organic compounds sensor response [25].

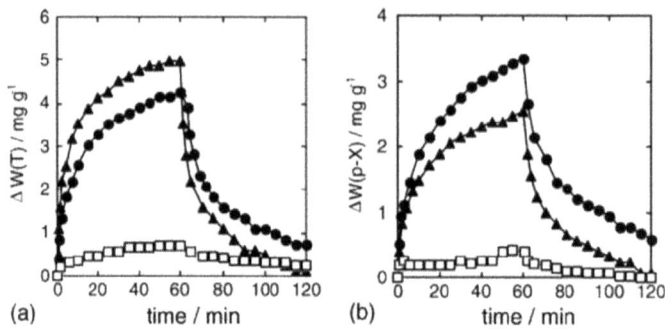

Figure 5. Sensor responses of MIP–PMMA and un-MIP–PMMA blend films measured using the QCM method at 30 °C: (**a**) toluene vapor sorption and (**b**) *p*-xylene vapor absorption (●: *p*-xylene-MIP-PMMA; ▲: toluene- MIP-PMMA; and □: un-MIP-PMMA) [25].

Hawari et al. [26] demonstrated an e-nose sensor that could detect Limonene volatiles using MIP as the sensing material. In this research, Limonene, MAA, EGDMA, azodiisobutyronitrile (AIBN) and tetrahydrofuran (THF) were used as the template, functional monomer, cross-linker, initiator and solvent, respectively. The MIPs were prepared by spin-coating on an interdigitated electrode (IDE), with the properties of the sensor estimated by capacitance. The results showed that the sensor responds to Limonene volatile gases. Hawari et al. [27] also use the Interdigitated Electrode (IDE) structure to fabricate a sensor for detecting Alpha Pinene volatile gases by using MIP. Alpha Pinene, MAA, AIBN and EGDMA were respectively used as the template, functional monomer, initiator and cross-linker. The emission of Alpha Pinene was monitored by placing the IDE–MIP sensor into an actual ripe Harumanis mango. The results showed that the IDE–MIP sensor exhibited high sensitivity and selectivity responses towards alpha pinene when compared to non-imprinted polymers.

Ji et al. [28] fabricated a 2-methylisoborneol (MIB) QCM-based gas sensor using MIT. MIB was used as the template, MAA as functional monomer, EGDMA as cross-linker and AMVN as initiator and hexane as the solvent. The results showed that the QCM coated with a MIB imprinted polymer exhibit responses that were 1.1 ± 1.3 times higher than those of sensors coated with a non-imprinted polymer. The responses of the imprinted sensors to MIB were always the highest, and the detection limit was lowered to approximately 200 μg L^{-1}. Bunte et al. [29] prepared a MIP-coated QCM sensor to detect 2,4,6-trinitrotoluene (2,4,6-TNT) vapor. A PAA-MIP synthesized with chloroform showed best adsorption properties for TNT vapor. Direct measurements of the mass attachment with respect to the frequency decrease in the coated QCMs during vapor treatment showed a TNT-uptake of about 150 pg per μg MIP per hour. Lieberzeit et al. [30] examined a formaldehyde gas sensor using MIPs. A co-polymer thin film was prepared with styrene, MAA and EGDMA, before the thin film was coated with QCM to form a sensor. The sensor exhibits a detection limit of 500 ppb formaldehyde in dry air. These MIPs showed specific behaviors when tested against a range of VOCs, such as acetaldehyde, formic acid, dichloromethane and methanol, which can be seen in Figure 6. The device possesses great selectivity to formaldehyde.

Alizadeh et al. [31] combined the imprinted polymeric particles with graphene to produce a nanocomposite chemo-resistor gas sensor. MAA was used as the first functional monomer, vinyl benzene as the second functional monomer, nitrobenzene as the template, divinylbenzene as cross-linker, and 2,20-azobisisobutyronitrile as radical initiator. The results showed that the sensor could recognize acetonitrile specifically, as shown in Figure 7. Wen et al. [32] presents a 300 MHz surface acoustic wave (SAW) gas sensor coated with MIPs, which was combined with a SAW oscillator. The MIP was prepared using o-phenylenediamine (o-PD) as a functional monomer and sarin acid as a molecular template, before being co-polymerized using cyclic voltammetry (CV). The sensitivity for the detection of dimethyl methyl phosphonate (DMMP) concentrations in a range of 1–100 mg/m^3 was evaluated as approximately 96 Hz/mg/m^3, with the threshold detection limit being up to 0.5 mg/m^3. Walke et al. [33] reported a MIP to monitor explosive 2,4,6-TNT. Organically modified sol–gel polymer films with a thickness of submicrons were deposited on a waveguide surface as the sensing layer. Molecularly imprinted sol–gels were created for TNT using covalently bound template molecules linked to the matrix through 1 or 2 carbamate linkages. González-Vila et al. [34] focused on an optical fiber for formaldehyde gas detection. The optical fiber was coated with a layer of MIP based on polypyrrole, which is a conductive polymer. During the test, light is scattered when the target molecule attaches to the cavities present in the polymer. The sensor possesses good selectivity for formaldehyde. Jakoby et al. [35] applied a MIP thin film to a Love wave gas sensor. The sensor showed a certain response to 2-methoxy 3-methyl pyrazine (MMP).

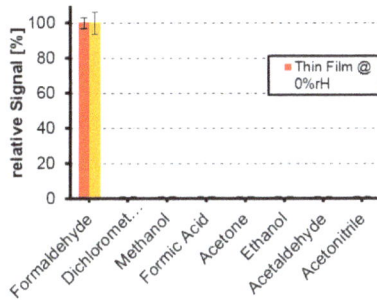

Figure 6. Selectivity patterns of MIP thin films and MIP nano-particles layers showing relative signals [30].

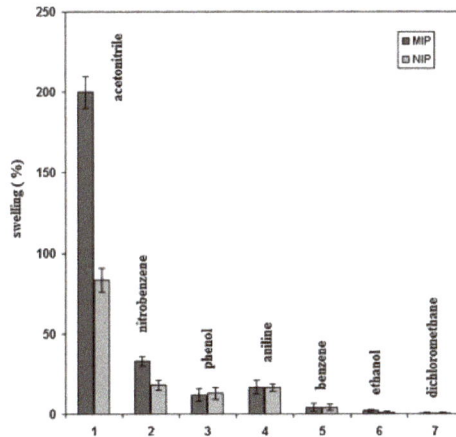

Figure 7. Comparison of the swelling intensities of the MIP and NIP in acetonitrile (1 mL) and other organic compounds (1 mM of them in acetonitrile, 10 mL) [31].

González-Vila et al. [36] fabricated a QCM gas sensor combined with sol–gel films for detection of parathion. Authors used silanes as functional monomers and parathion as the template to prepare MIPs. Following this, the MIPs were mixed with sol–gel films and then placed on a QCM. The molecularly imprinted QCM gas sensor possesses good selectivity for parathion, as the signal of the sensor to parathion is 3 times larger than that to other analytes. Jha et al. [37] designed a sensor array with volatile acids imprinted with 3-element QCM for the recognition of the odorous organic acids: propanoic acid, hexanoic acid and octanoic acid. Each element in QCM is in charge of detecting the odor of one acid and they all show good selectivity. Jha et al. [38] also developed a QCM sensor array to identify primary aldehydes in human body odor. Imahashi et al. [39] prepared a molecular imprinting filtering layer using benzaldehyde as template molecules. The molecular imprinting filtering layer was coated on a substrate to form a sensor. The sensor shows good selectivity to benzaldehyde, as shown in Figure 8.

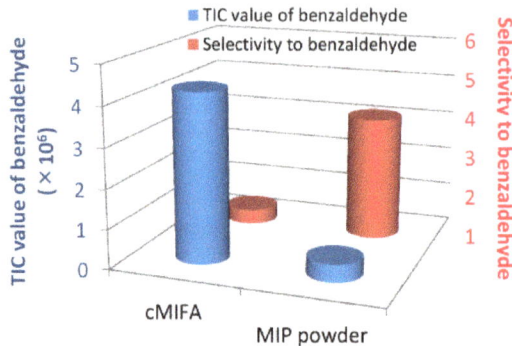

Figure 8. The comparison of the adsorption capability and selectivity of molecularly imprinted filtering adsorbents and MIP powder [39].

Although some progress has been made in the early reports of MIGS, MIPs prepared in these reports usually use organic raw materials during the whole process and suffer from some drawbacks, such as poor electrical conductivity, moderate sensitivity and selectivity, and incomplete template removal. Therefore, the test process needs the help of QCM. Recently, new polymerization strategies have been proposed to deal with obtaining imprinted materials in order to improve the gas molecule detection of MIPs. Among these strategies, the sol–gel technique seems to be one of the simplest and most well-suited methods to obtain well-conducted MIPs.

2.2. Molecularly Imprinted Organic/Inorganic Hybrid Gas Sensors

In order to improve the conductivity of the MIPs, organic/inorganic hybrid MIPs have been examined lately using the sol–gel technique. Molecular imprinting-based sol–gel technology combines molecular imprinting and sol–gel techniques. The sol–gel technique is usually used to prepare solid-containing oxides or other compounds of inorganic or metal alkoxides. This process involves creating a solution, applying the sol–gel cure and heat treatment. The molecularly imprinted sol–gel technology is the method used to prepare a rigid material with an inorganic network structure, into which the template molecules are introduced via the sol–gel process. After the template molecules are removed, they show good affinity with the template molecules.

Liu et al. [40,41] achieved a series of VOC sensors (formaldehyde, benzene, acetone, methanol and so on) based on Ag-doped $LaFeO_3$ (ALFO) using the methods of the MIT combined sol–gel technique. In contrast to some single-metal oxide semiconductors, $LaFeO_3$ (LFO) is a common perovskite-type oxide that exhibits p-type semiconducting behavior [42], and is a promising material with an abundance of functionalities, especially in the field of gas sensing. LFO possesses great potential for detecting pollutant gases because of its specific chemical and physical characteristics, including its large surface area, rich active oxygen lattice, good thermostability [43], controllable structure [44] and strong reducibility [45–49]. Thus, LFO is a more attractive gas-sensing material than other metal oxides. In addition, after LFO is combined with Ag, some Ag remains in the form of single atoms to act as catalyst mixtures in the matrix. Indeed, some of the Ag fills areas between the grains of the matrix, working to decrease the contact potential barrier and enhance the interfacial effects. This leads to a lower resistance and, thus, a lower operating temperature [50]. Therefore, ALFO is chosen to be used as the cross-linker (matrix) in Liu's work. Most importantly, the selectivity of ALFO towards acetone, benzene, methanol and formaldehyde is successfully modulated by MIT. Using acetone as the template, N'N-methylene-bis-acrylamide (MBA) as the functional monomer and AIBN as the initiator, ALFO was prepared as an acetone gas-sensing material based on MIT. Similarly, using benzene, methanol and formaldehyde as the template, as well as choosing a proper functional monomer (formaldehyde for benzene, MAA for methanol and acrylamide for formaldehyde) and AIBN as the initiator, ALFO was then prepared as benzene, methanol and formaldehyde gas-sensing materials. Following this, these sensing materials were fabricated into heater-type gas sensors, based on which gas sensing properties were tested. The selectivity of each sensor is very good, which can be seen from Figure 9a. The relationship between response and concentration was also reported (Figure 9b). The increasing responses for the four sensors are closely linear to the concentration of each analyte, which indicates that the sensors can be used for continuous real-time monitoring of low concentration of VOCs. The response and recovery times of the molecular imprinted sensors to different concentrations of acetone, benzene, methanol and formaldehyde are shown in Figure 10. The response and recovery times were 44 s and 78 s (acetone); 63 s and 48 s (benzene); 40 s and 50 s (methanol); and 40 s and 54 s (formaldehyde). Nowadays, the MIT field has been dominated by the use of recognition and separation for organic macromolecules, such as proteins (molecular weight: 40,000–220,000 Da), enzymes (130,000–140,000 Da) and so on. However, for the small organic molecules, including VOCs (molecular weight is less than 100 Da), there are very few relevant reports. This is a great breakthrough in small organic molecule imprinting with the sol–gel technique based on a metal oxide semiconductor (ALFO). The simplicity and efficacy of this method has profound applications for the construction of applied gas sensors and has enormous potential.

Tang et al. [51] used electro-polymerization to synthesize the MIP layer on the TiO$_2$ nano tube array/Ti sheet. The MIGS selectively detects formaldehyde in the ppm range at room temperature. Its sensitivity to formaldehyde is higher than that of acetone, acetaldehyde, acetic acid and ethanol.

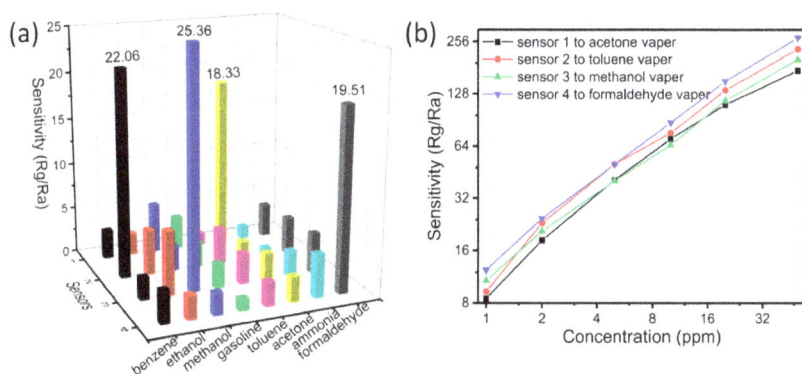

Figure 9. (**a**) The response of each sensor to eight types of analytes (2.5 ppm). (**b**) Response-concentration relationships to the target gases, Sensor 1 is the acetone sensor, sensor 2 is toluene sensor, sensor 3 is methanol sensor and sensor 4 is formaldehyde sensor.

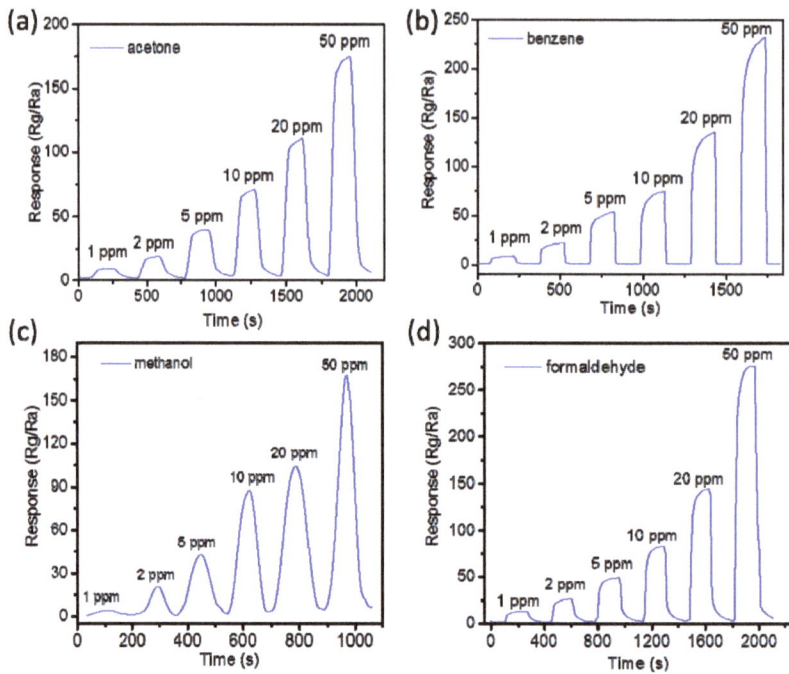

Figure 10. Rand recovery time of the: (**a**) molecular imprinted acetone sensor, (**b**) benzene sensor, (**c**) methanol sensor and (**d**) formaldehyde sensor.

3. Quasi-Molecularly Imprinted Gas Sensors

Another category of MIGS is prepared with quasi-MIT. According to the features of MIT, Huang et al. [52] suggested that a similar mechanism, the quasi-molecular-imprinting method, can be used in fabrication of gas sensors. Huang et al. put the target molecules in the preparation process in order to make the prepared materials recognize the target molecules to enhance the selectivity. They successively produced a CO sensor [52], ethanol sensor [53] and acetone sensor [54] using quasi-MIT.

To prepare the CO sensor, the authors first prepared SnO_2 mesoporous nanomaterial using the hydrothermal method [52]. Following this, the prepared SnO_2 film sensors were divided into two groups. One of the groups was dried in a carbon monoxide (which is the target gas) atmosphere to imprint CO on the surface of the material, while another was dried in the air for comparison. During the gas response test, the two groups of SnO_2 sensors were exposed to CO gas with various concentrations ranging from 50 to 3000 ppm. The results shown in Figure 11 demonstrate that the imprinted sensor shows a faster response and recovery as well as a higher response compared to non-imprinted ones in all CO concentrations. It illustrates that quasi-MIT can enhance gas sensor performance.

For the ethanol sensor [53], the authors designed ethanol gas sensors based on SnO_2 using the quasi-molecular-cluster imprinting mechanism. The target molecules (ethanol) were introduced during the SnO_2 synthesis and device fabrication to acquire the specific structure that was more suitable for the adsorption and desorption of target molecules (ethanol gas). Under the same experimental conditions, nonimprinted sensors were prepared with deionized water. The synthesized SnO_2 sensors coated by the imprinted and nonimprinted films were both exposed to ethanol gas, with concentrations ranging from 4 ppm to 100 ppm. The imprinted sensor exhibited the best and fastest response and recovery (Figure 12).

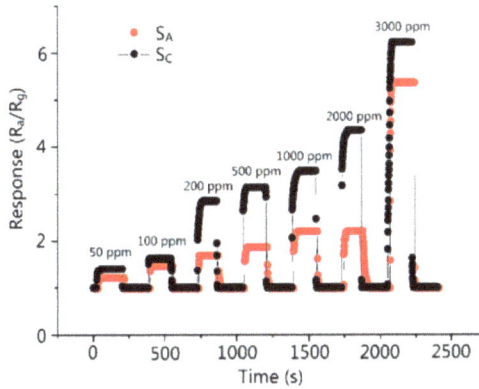

Figure 11. Responses to 50–3000 ppm of CO gas of imprinted (S_C) and nonimprinted (S_A) sensors at 300 °C [52].

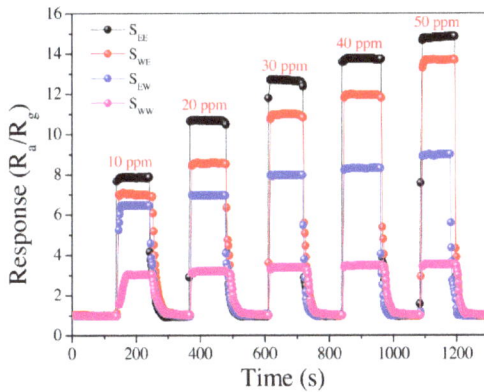

Figure 12. Responses to 10, 20, 30, 40 and 50 ppm of ethanol gas of imprinted and nonimprinted sensors at 300 °C [53].

For the acetone sensor [54], the method was very similar to that of the ethanol sensor mentioned above. Acetone solutions were introduced during the synthesis of SnO_2 nanomaterials or/and device fabrication to produce appropriate structures more suitable for the adsorption and desorption of acetone gas. Similarly, there were imprinted and nonimprinted sensors. The synthesized sensors were exposed to acetone gas with various concentrations from 50 ppb to 100 ppm. The sensor produced by incorporating acetone both during the nanomaterial synthesis and device fabrication exhibits the best performance in terms of sensitivity, response recovery and reproducibility (Figure 13).

The quasi-MIT concept is novel, and it is easy to implement. In this way, even inorganic gas sensors could be achieved. Although the mechanism is not very clear, and the author's speculations have not been completely confirmed, quasi-MIT provides a new idea in the molecular imprinted gas sensors, especially those for inorganic gases.

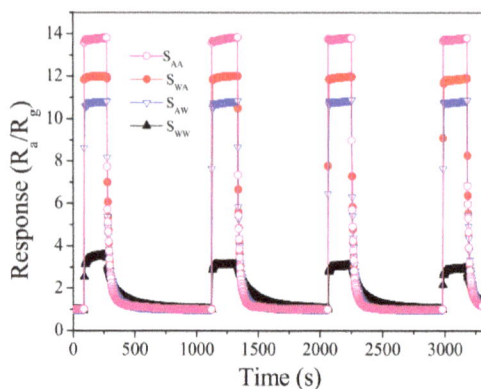

Figure 13. Responses to 50 ppm acetone gas for S_{AA}, S_{WA}, S_{AW} and S_{WW} at 250 °C [54].

4. Conclusions

The molecular imprinting technique has good application prospects in the field of gas sensors, as molecular imprinted nanomaterials are suitable in sensing devices. For better performance, there are several problems that need to be resolved, such as the lack of a general synthesizing protocol, the non-uniformity of binding sites, the inorganic gas molecule recognition and improper adherence with the sensor surface. Exploring composites and hybrids of organic/inorganic polymers for developing imprinted materials in gas sensing would be of great value. An estimated trend in the gas sensors is toward miniaturization and sensor arrays. For example, certain sensors based on detection of VOCs integrated in one sensor array with different sensing units could be applied to simultaneous detection of multiple VOCs. In addition, materials, such as $LaFeO_3$ or SnO_2, can be fabricated as gas sensors to respond to different gas molecules with MIT and the (quasi-)molecular imprinting technique. In summary, any target molecules could be recognized by molecularly imprinted gas sensors as needed. In addition, by integrating different molecular imprinting gas sensors in an array, multiple kinds of gases could be detected simultaneously. This effectively expands the applications of molecular imprinting techniques in the field of gas sensors.

Acknowledgments: This work was supported by National Natural Science Foundation of China (No. 51562038 and 51402257).

Conflicts of Interest: The authors declare no conflict of interest.

References

1. Shah, R.R.; Abbott, N.L. Principles for measurement of chemical exposure based on recognition–driven anchoring transitions in liquid crystals. *Science* **2001**, *293*, 1296–1299. [CrossRef] [PubMed]
2. Talin, A.A.; Centrone, A.; Ford, A.C.; Foster, M.E.; Stavila, V.; Haney, P.; Léonard, F.; Allendor, M.D. Tunable Electrical Conductivity in Metal-Organic Framework Thin-Film Devices. *Science* **2014**, *343*, 66–69. [CrossRef] [PubMed]
3. Potyrailo, R.A.; Ghiradella, H.; Vertiatchikh, A.; Dovidenko, K.; Cournoyer, J.; Olson, R.E. Morpho butterfly wing scales demonstrate highly selective vapour response. *Nat. Photonics* **2007**, *1*, 123–128. [CrossRef]
4. Prieto, G.; Zečević, J.; Friedrich, H.; Jong, K.P.; Jongh, P.E. Towards stable catalysts by controlling collective properties of supported metal nanoparticles. *Nat. Mater.* **2013**, *12*, 34–39. [CrossRef] [PubMed]
5. Moghaddam, A.; Olszewska, W.; Wang, B.; Tregoning, J.S.; Helson, R.; Sattentau1, Q.J.; Openshaw, P.J.M. A potential molecular mechanism for hypersensitivity caused by formalin-inactivated vaccines. *Nat. Med.* **2006**, *12*, 905–907. [CrossRef] [PubMed]
6. Liu, S.Q.; Saijo, K.; Todoroki, T.; Ohno, T. Induction of human autologous cytotoxic T lymphocytes on formalin-fixed and paraffin-embedded tumour sections. *Nat. Med.* **1995**, *1*, 267–271. [CrossRef] [PubMed]

7. Bricker, C.E.; Johnson, H.R. Spectrophotometric method for determining formaldehyde. *Ind. Eng. Chem. Anal. Ed.* **1945**, *17*, 400–402. [CrossRef]
8. Beasley, R.K.; Hoffmann, C.E.; Rueppel, M.L.; Worley, J.W. Sampling of formaldehyde in air with coated solid sorbent and determination by high performance liquid chromatography. *Anal. Chem.* **1980**, *52*, 1110–1114. [CrossRef]
9. Herschkovitz, Y.; Eshkenazi, I.; Campbell, C.E.; Rishpon, J. An electrochemical biosensor for formaldehyde. *J. Electroanal. Chem.* **2000**, *491*, 182–187. [CrossRef]
10. Zhou, K.W.; Ji, X.L.; Zhang, N.; Zhang, X.R. On-line monitoring of formaldehyde in air by cataluminescence-based gas sensor. *Sens. Actuators B Chem.* **2006**, *119*, 392–397. [CrossRef]
11. Zhang, Y.M.; Zhang, J.; Chen, J.L.; Zhu, Z.Q.; Liu, Q.J. Improvement of response to formaldehyde at Ag–LaFeO$_3$ based gas sensors through incorporation of SWCNTs. *Sens. Actuators B Chem.* **2014**, *195*, 509–514. [CrossRef]
12. Ju, H.X.; Qiu, Z.M.; Ding, S.J. *Bio-Analytical Chemistry*; Science Press: Beijing, China, 2007; pp. 272–290.
13. Ye, L.; Mosbach, K. Molecular Imprinting: Synthetic Materials as Substitutes for Biological Antibodies and Receptors. *Chem. Mater.* **2008**, *20*, 859–868. [CrossRef]
14. Chen, L.X.; Wang, X.Y.; Lu, W.H.; Wu, X.Q.; Li, J.H. Molecular imprinting: Perspectives and applications. *Chem. Soc. Rev.* **2016**, *45*, 2137–2211. [CrossRef] [PubMed]
15. Haupt, K.; Mosbach, K. Molecularly imprinted polymers and their use in biomimetic sensors. *Chem. Rev.* **2000**, *100*, 2495–2504. [CrossRef] [PubMed]
16. Chen, W.; Ma, Y.; Pan, J.; Meng, Z.; Pan, G.; Sellergren, B. Molecularly Imprinted Polymers with Stimuli-Responsive Affinity: Progress and Perspectives. *Polymers* **2015**, *7*, 1689–1715. [CrossRef]
17. Liu, R.; Guan, G.; Wang, S.; Zhang, Z. Core-shell nanostructured molecular imprinting fluorescent chemosensor for selective detection of atrazine herbicide. *Analyst* **2011**, *136*, 184–190. [CrossRef] [PubMed]
18. Tan, X.; Li, B.; Liew, K.Y.; Li, C. Electrochemical fabrication of molecularly imprinted porous silicate film electrode for fast and selective response of methyl parathion. *Biosens. Bioelectron.* **2010**, *26*, 868–871. [CrossRef] [PubMed]
19. Casey, C.N.; Campbell, S.E.; Gibson, U.J. Phenylalanine detection using matrix assisted pulsed laser evaporation of molecularly imprinted amphiphilic block copolymer films. *Biosens. Bioelectron.* **2010**, *26*, 703–709. [CrossRef] [PubMed]
20. Tamayo, F.G.; Casillas, J.L.; Martin-Esteban, A. Evaluation of new selective molecularly imprinted polymers prepared by precipitation polymerisation for the extraction of phenylurea herbicides. *J. Chromatogr. A* **2005**, *1069*, 173–181. [CrossRef] [PubMed]
21. Hong, C.C.; Chang, P.H.; Lin, C.C.; Hong, C.L. A disposable microfluidic biochip with on-chip molecularly imprinted biosensors for optical detection of anesthetic propofol. *Biosens. Bioelectron.* **2010**, *25*, 2058–2064. [CrossRef] [PubMed]
22. Stephenson, C.J.; Shimizu, K.D. Colorimetric and fluorometric molecularly imprinted polymer sensors and binding assays. *Polym. Int.* **2007**, *56*, 482–488. [CrossRef]
23. Morelli, I.; Chiono, V.; Vozzi, G.; Ciardelli, G.; Silvestri, D.; Giusti, P. Molecularly imprinted submicronspheres for applications in a novel model biosensor-film. *Sens. Actuators B Chem.* **2010**, *150*, 394–401. [CrossRef]
24. Feng, L.; Liu, Y.J.; Zhou, X.D.; Hu, J.M. The fabrication and characterization of a formaldehyde odor sensor using molecularly imprinted polymers. *J. Colloid Interface Sci.* **2005**, *284*, 378–382. [CrossRef] [PubMed]
25. Matsuguchi, M.; Uno, T. Molecular imprinting strategy for solvent molecules and its application for QCM-based VOC vapor sensing. *Sens. Actuators B Chem.* **2006**, *113*, 94–99. [CrossRef]
26. Hawari, H.F.; Samsudin, N.M.; Ahmad, M.N.; Shakaff, A.Y.M.; Ghani, S.A.; Wahab, Y. Recognition of Limonene Volatile Using Interdigitated Electrode Molecular Imprinted Polymer Sensor. In Proceedings of the Third International Conference on Intelligent Systems Modelling and Simulation, Kota Kinabalu, Malaysia, 8–10 February 2012; pp. 723–726.
27. Hawari, H.F.; Samsudin, N.M.; Md Shakaff, A.Y.; Ghani, S.A.; Ahmad, M.N.; Wahab, Y.; Hashim, U. Development of Interdigitated Electrode Molecular Imprinted Polymer Sensor for Monitoring Alpha Pinene Emissions from Mango Fruit. *Procedia Eng.* **2013**, *53*, 197–202. [CrossRef]
28. Ji, H.-S.; McNiven, S.; Ikebukuro, K.; Karube, I. Selective piezoelectric odor sensors using molecularly imprinted polymers. *Anal. Chim. Acta* **1999**, *390*, 93–100. [CrossRef]

29. Bunte, G.; Hurttlen, J.; Pontius, H.; Hartlieb, K.; Krause, H. Gas phase detection of explosives such as 2,4,6-trinitrotoluene by molecularly imprinted polymers. *Anal. Chim. Acta* **2007**, *591*, 49–56. [CrossRef] [PubMed]

30. Hussain, M.; Kotova, K.; Lieberzeit, P.A. Molecularly Imprinted Polymer Nanoparticles for Formaldehyde Sensing with QCM. *Sensors* **2016**, *16*, 1011. [CrossRef] [PubMed]

31. Alizadeh, T.; Hamedsoltani, L. Graphene/graphite/molecularly imprinted polymer nanocomposite as the highly selective gas sensor for nitrobenzene vapor recognition. *J. Environ. Chem. Eng.* **2014**, *2*, 1514–1526. [CrossRef]

32. Wen, W.; Shitang, H.; Shunzhou, L.; Minghua, L.; Yong, P. Enhanced sensitivity of SAW gas sensor coated molecularly imprinted polymer incorporating high frequency stability oscillator. *Sens. Actuators B Chem.* **2007**, *125*, 422–427. [CrossRef]

33. Walker, N.R.; Linman, M.J.; Timmers, M.M.; Dean, S.L.; Burkett, C.M.; Lloyd, J.A.; Keelor, J.D.; Baughman, B.M.; Edmiston, P.L. Selective detection of gas-phase TNT by integrated optical waveguide spectrometry using molecularly imprinted sol-gel sensing films. *Anal. Chim. Acta* **2007**, *593*, 82–91. [CrossRef] [PubMed]

34. González-Vila, Á.; Debliquy, M.; Lahem, D.; Zhang, C.; Mégret, P.; Caucheteur, C. Molecularly imprinted electropolymerization on a metal-coated optical fiber for gas sensing applications. *Sens. Actuators B Chem.* **2017**, *244*, 1145–1151. [CrossRef]

35. Jakoby, B.; Ismail, G.; Byfield, M.; Vellekoop, M. A novel molecularly imprinted thin film applied to a Love wave gas sensor. *Sens. Actuators A Phys.* **1999**, *76*, 93–97. [CrossRef]

36. Sharon, M.; Amalya, Z.; Iva Turyan, A.; Mandler, D. Parathion Sensor Based on Molecularly Imprinted Sol–Gel Films. *Anal. Chem.* **2005**, *76*, 120–126.

37. Jha, S.K.; Liu, C.; Hayashi, K. Molecular imprinted polyacrylic acids based QCM sensor array for recognition of organic acids in body odor. *Sens. Actuators B Chem.* **2014**, *204*, 74–87. [CrossRef]

38. Jha, S.K.; Hayashi, K. A quick responding quartz crystal microbalance sensor array based on molecular imprinted polyacrylic acids coating for selective identification of aldehydes in body odor. *Talanta* **2015**, *134*, 105. [CrossRef] [PubMed]

39. Imahashi, M.; Chiyomaru, Y.; Hayashi, K. Ultrathin reconfigurable molecular filter for gas-selective sensing. In Proceedings of the 2013 IEEE Sensors, Baltimore, MD, USA, 3–6 November 2013.

40. Zhang, Y.M.; Liu, Q.J.; Zhang, J.; Zhu, Q.; Zhu, Z.Q. A highly sensitive and selective formaldehyde gas sensor using a molecular imprinting technique based on Ag–LaFeO$_3$. *J. Mater. Chem. C* **2014**, *2*, 10067–10072. [CrossRef]

41. Zhu, Q.; Zhang, Y.M.; Zhang, J.; Zhu, Z.Q.; Liu, Q.J. A new and high response gas sensor for methanol using molecularly imprinted technique. *Sens. Actuators B Chem.* **2015**, *207*, 398–403. [CrossRef]

42. Traversa, E.; Matsushuma, S.; Okada, G.; Sadaoka, Y.; Sakai, Y.; Watanabe, K. NO$_2$ sensitive LaFeO$_3$ thin films prepared by r.f. sputtering. *Sens. Actuator B Chem.* **1995**, *24*, 661–664. [CrossRef]

43. Wetchakun, K.; Samerjai, T.; Tamaekong, N.; Liewhiran, C.; Siriwong, C.; Kruefu, V.; Wisitsoraat, A.; Tuantranont, A.; Phanichphant, S. Semiconducting metal oxides as sensors for environmentally hazardous gases. *Sens. Actuators B Chem.* **2011**, *160*, 580–591. [CrossRef]

44. Natile, M.M.; Ponzoni, A.; Concina, I.; Glisenti, A. Chemical Tuning versus Microstructure Features in Solid-State Gas Sensors: LaFe$_{1-x}$Ga$_x$O$_3$, a Case Study. *Chem. Mater.* **2014**, *26*, 1505–1513. [CrossRef]

45. Toan, N.N.; Saukko, S.; Lantto, V. Gas sensing with semiconducting perovskite oxide LaFeO$_3$. *Phys. B Condens. Matter* **2003**, *327*, 279–282. [CrossRef]

46. Wang, X.F.; Qin, H.W.; Sun, L.H.; Hu, J.F. CO$_2$ sensing properties and mechanism of nanocrystalline LaFeO$_3$ sensor. *Sens. Actuators B Chem.* **2013**, *188*, 965–971. [CrossRef]

47. Wiglusz, R.J.; Kordek, K.; Malecka, M.; Ciupa, A.; Ptak, M.; Pazik, R.; Pohl, P.; Kaczorowski, D. A new approach in the synthesis of La$_{1-x}$Gd$_x$FeO$_3$ perovskite nanoparticles-structural and magnetic characterization. *Dalton Trans.* **2015**, *44*, 20067–20074. [CrossRef] [PubMed]

48. Murade, P.A.; Sangawar, V.S.; Chaudhari, G.N.; Kapse, V.D.; Bajpeyee, A.U. Acetone gas-sensing performance of Sr-doped nanostructured LaFeO$_3$ semiconductor prepared by citrate sol–gel route. *Curr. Appl. Phys.* **2011**, *11*, 451–456. [CrossRef]

49. Doroftei, C.; Popa, P.D.; Iacomi, F. Synthesis of nanocrystalline La-Pb-Fe-O perovskite and methanol-sensing characteristics. *Sens. Actuators B Chem.* **2012**, *161*, 977–981. [CrossRef]

50. Zhang, Y.M.; Lin, Y.T.; Chen, J.L.; Zhang, J.; Zhu, Z.Q.; Liu, Q.J. A high sensitivity gas sensor for formaldehyde based on silver doped lanthanum ferrite. *Sen. Actuators B Chem.* **2014**, *190*, 171–176. [CrossRef]

51. Tang, X.; Raskin, J.P.; Lahem, D.; Krumpmann, A.; Decroly, A.; Debliquy, M. A Formaldehyde Sensor Based on Molecularly-Imprinted Polymer on a TiO$_2$ Nanotube Array. *Sensors* **2017**, *17*, 675. [CrossRef] [PubMed]

52. Li, C.J.; Lv, M.; Zuo, J.L.; Huang, X.T. SnO$_2$ Highly Sensitive CO Gas Sensor Based on Quasi-Molecular-Imprinting Mechanism Design. *Sensors* **2015**, *15*, 3789–3800. [CrossRef] [PubMed]

53. Tan, W.H.; Yu, Q.X.; Ruan, X.F.; Huang, X.T. Design of SnO$_2$-based highly sensitive ethanol gas sensor based on quasi molecular-cluster imprinting mechanism. *Sens. Actuators B Chem.* **2015**, *212*, 47–54. [CrossRef]

54. Tan, W.H.; Ruan, X.F.; Yu, Q.X.; Yu, Z.T.; Huang, X.T. Fabrication of a SnO$_2$-based acetone gas sensor enhanced by molecular imprinting. *Sensors* **2015**, *15*, 352–364. [CrossRef] [PubMed]

MDPI AG

St. Alban-Anlage 66

4052 Basel, Switzerland

Tel. +41 61 683 77 34

Fax +41 61 302 89 18

http://www.mdpi.com

Sensors Editorial Office

E-mail: sensors@mdpi.com

http://www.mdpi.com/journal/sensors

www.ingramcontent.com/pod-product-compliance
Lightning Source LLC
Chambersburg PA
CBHW051848210326

41597CB00033B/5820